LATE STAGES OF STELLAR EVOLUTION

INTERNATIONAL ASTRONOMICAL UNION

UNION ASTRONOMIQUE INTERNATIONALE

SYMPOSIUM No. 66

(COPERNICUS SYMPOSIUM V)

HELD IN WARSAW, POLAND, 10–12 SEPTEMBER 1973

LATE STAGES OF STELLAR EVOLUTION

EDITED BY

R. J. TAYLER

*School of Mathematical and Physical Sciences, The University of Sussex,
Falmer, Brighton BN1 9QH, England*

CO-EDITED BY

J. E. HESSER

Observatorio Interamericano de Cerro Tololo, Casilla 63-D, La Serena, Chile

D. REIDEL PUBLISHING COMPANY

DORDRECHT-HOLLAND / BOSTON-U.S.A.

1974

Published on behalf of
the International Astronomical Union
by
D. Reidel Publishing Company, P.O. Box 17, Dordrecht, Holland

Sold and distributed in the U.S.A., Canada, and Mexico
by D. Reidel Publishing Company, Inc.
306 Dartmouth Street, Boston,
Mass. 02116, U.S.A.

Library of Congress Catalog Card Number 74–77970

ISBN-13:978-90-277-0471-9   e-ISBN-13:978-94-010-2237-8
DOI: 10.1007/978-94-010-2237-8

TABLE OF CONTENTS

PREFACE

IAU Symposium No. 66 was held in Warsaw from September 10th to September 12th 1973, in connection with the Extraordinary General Assembly of the IAU. It was arranged by IAU Symposium No. 35 and the Scientific Organising Committee consisted of A. G. Massevitch (Chairman), A. V. Tutukov (Secretary), H. M. van Horn, N. Dallaporta, J. P. Ostriker, B. Paczyński, G. Ruben, E. Schatzman, R. J. Tayler and A. Weigert.

This volume contains the full texts of all of the invited papers presented at the Symposium, apart from that delivered by R. P. Kraft, which is published in abstract because it is appearing in full elsewhere. In addition the short communications given at the Symposium are published in abstract. I attempted to take down all of the discussion as it occurred and all contributors to the discussion were asked to provide copies of their remarks. From these sources an edited version of the discussion has been produced. As the final version has not been seen by the contributors, I should be held responsible for all errors. At Warsaw, some of the short communications did not immediately follow the invited paper to which they referred. In the printed version they and any discussion relating to them are placed in the most logical position. A small number of short communications, which were circulated in abstract at Warsaw but which were not delivered orally, are also included in the published version. The volume closes with a personal view of the present state of the subject by M. Schwarzschild.

The detailed scientific planning of the conference was mainly in the hands of Massevitch and Tutukov and the local organisation was handled by J. Smak and the Local Organising Committee for the Extraordinary General Assembly together with Paczyński, the local member of the Scientific Organising Committee. We are grateful to all of these persons for the smooth running of the Symposium. The chairmen of the six sessions of the Symposium were S. Piotrowski, R. J. Tayler, N. Dallaporta, V. Weidemann, G. Ruben and A. G. Massevitch. I am grateful to most of the contributors of the Symposium for sending me the written versions of their papers very promptly.

R. J. TAYLER

NUCLEAR REACTIONS AND NEUTRINOS
IN STELLAR EVOLUTION

W. DAVID ARNETT

Dept. of Astronomy, University of Texas, Austin, Tex., U.S.A.

Much of the material discussed in this review represents a summary of some chapters in a monograph I am writing (Arnett, 1974); a more detailed discussion will be found there.

1. Neutrinos

A fundamental question for astrophysics is whether or not there exists a direct $e - \nu$ coupling in the weak interaction. As first emphasized by Pontecorvo (1959) this would imply efficient cooling processes for late stages of stellar evolution. Such an interaction is predicted by the conserved vector current (CVC) theory of weak interactions proposed by Feynman and Gell-Mann (1958). Dicus (1972) has shown that the theory of leptons of Stephen Weinberg (1971) gives cooling processes similar to those predicted by CVC, but with an uncertainty related to the precise value of the mass of the charged vector (W) meson which mediates the interaction.

Beaudet *et al.* (1967) have used CVC theory and numerically evaluated the composition-independent neutrino emission rates; they give analytic fits for the rate of radiation of energy by these processs. Also using CVC theory, Festa and Ruderman (1969) have examined neutrino-pair bremsstrahlung in a plasma consisting of degenerate electrons and nondegenerate ions; this process is probably important at high density and low temperature.

Figure 1 displays the rate of energy loss due to neutrino emission per gram, ε_ν, multiplied by the number of nucleons per electron, μ_e. This quantity $\mu_e \varepsilon_\nu$ is convenient because it is independent of μ_e. In Figure 1, $\mu_e \varepsilon_\nu$ is plotted vs the logarithm to base 10 of ϱ/μ_e where ϱ is the usual nucleon mass density. The curves are parametrized by temperature T, and labeled by $\log T(\mathrm{K})$. The solid lines refer to the Beaudet *et al.* rates; the dashed curves represent the Festa-Ruderman rate for a gas of pure $^{12}\mathrm{C}$.

At the high densities encountered in the late stages of stellar evolution, electron capture processes can become important. In the *Urca process* a nucleus alternatively captures an electron and undergoes a beta decay, meanwhile emitting a neutrino and an antineutrino. Thus a cyclic (but nonreversible) process occurs. In *neutronization* increasing density induces electron capture and causes a diminution in the number of electrons per nucleon present in the plasma. In general this is a noncylic process. Both types of process are currently under active study. In both cases the nature of the process depends upon previous evolution. For the Urca process, the abundance of Urca-active nuclei is of vital importance, and depends upon previous thermonuclear processing which can destroy or produce such nuclei. Neutronization might be less sensitive to previous evolution than the Urca process, but it too depends upon the

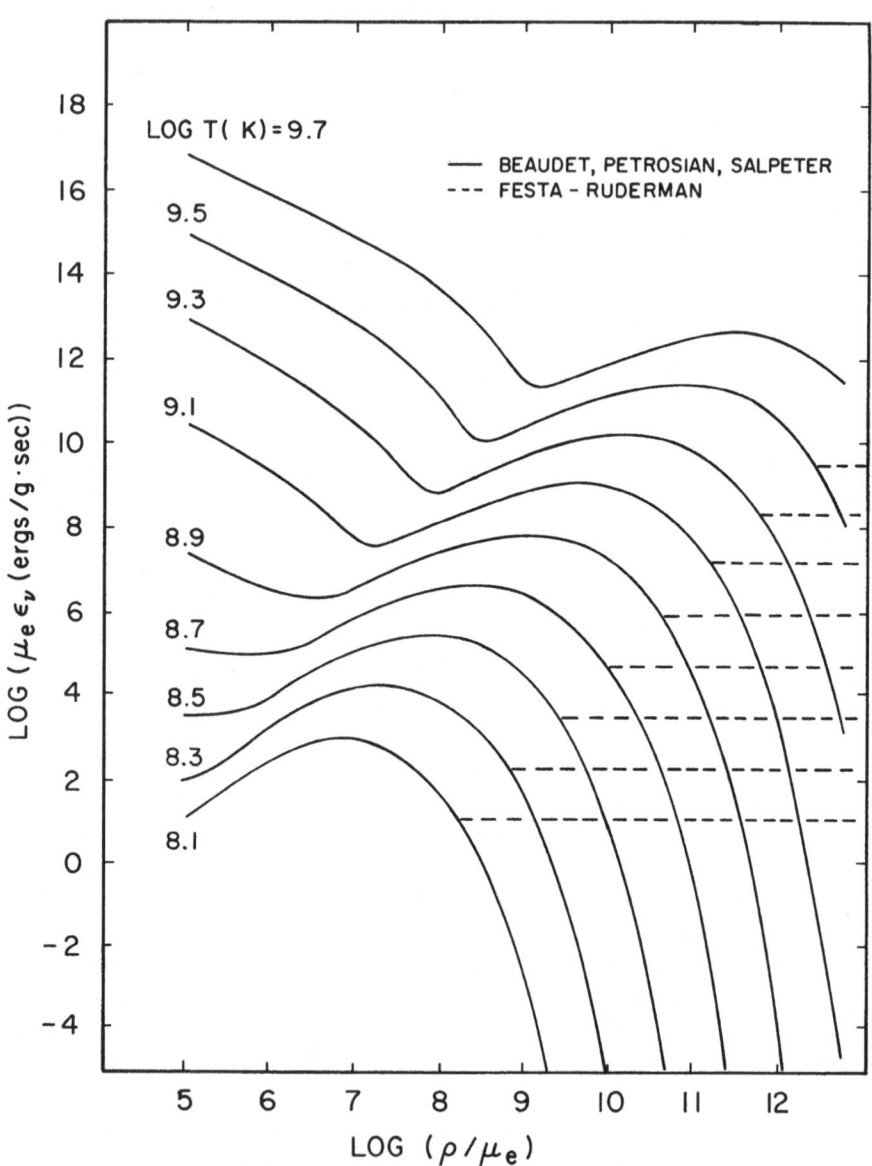

Fig. 1. Energy losses due to neutrino emission, as predicted by conserved vector current (CVC) theory.

composition, size and nature of the stellar core. Since these topics will be discussed in some detail in other papers (see especially those by Paczyński and by Imshennik and Nadyozhin), I will end regretfully my discussion of this fascinating and important topic here.

2. An Overview of the Stages of Thermonuclear Burning

The thermonuclear evolution of stellar matter may be thought of as consisting of a sequence of stages, in which the ashes of one stage become the fuel of the next. To describe such a sequence the stellar evolutionist must know at least two things: (1) the *rate of energy generation* by the thermonuclear consumption of a given fuel, and (2) the *composition* of the ashes which will become the fuel for the next stage. The evolutionary change in stellar composition is an initial-value problem; in general errors can amplify with successive evolution. Consequently it is vital to accurately represent the earlier burning stages if we wish to explore the later ones.

By far the vast bulk of stellar evolutionary work done to date involves no stage beyond hydrogen and helium burning. There is only one primary product of hydrogen burning, ^{4}He. Clearly the next major burning stage after hydrogen burning must involve the consumption of ^{4}He. Since the energy generation rate for helium burning is fairly insensitive to the nature of the ashes formed, one can get through both hydrogen and helium burning with a rather crude treatment of nucleosynthesis, missing only some fairly subtle but conceptually important effects. However, if for example no ^{12}C is formed, then there is no carbon burning stage at all! This is a *qualitative* as well as a quantitative difference. For the later stages of stellar evolution it appears that the question of the composition produced as well as that of energy generation rate must be carefully considered.

Table I summarizes the primary thermonuclear burning stages in stars. Note that if a direct $e - \nu$ coupling in the weak interaction does exist, then stages after helium burning are dominated by neutrino cooling rather than photon diffusion.

There is one more point that should be stressed. Massive stars $(M \gtrsim 3M_\odot)$ spend so little time in late burning stages (carbon burning and beyond) that the HR diagram is no longer such a useful test of evolutionary theory of these objects; this is due to poor statistics. However these stars all have a pronounced characteristic: they burn nuclear fuel at a prodigous rate. Their thermonuclear ashes may reveal their history.

TABLE I

Thermonuclear burning stages

Fuel	$T/10^9$ (K)	Ashes	q (ergs/g-fuel)	Cooling
^{1}H	0.02	^{4}He, ^{14}N	$(5$ to $8) \times 10^{18}$	Photons
^{4}He	0.2	^{12}C, ^{16}O, ^{22}Ne	7×10^{17}	Photons
^{12}C	0.8	^{20}Ne, ^{24}Mg, ^{16}O;	5×10^{17}	Neutrinos
		^{23}Na, ^{25,26}Mg		Neutrinos
	0.4	^{20}Ne, ^{23}Na	–	Neutrinos
^{20}Ne	1.5	^{16}O, ^{24}Mg, ^{28}Si;	1.1×10^{17}	Neutrinos
^{16}O	2	^{28}Si, ^{32}S;	5×10^{17}	Neutrinos
^{28}Si	3.5	^{56}Ni, $A \sim 56$ Nuclei	$(0$ to $3) \times 10^{17}$	Neutrinos
^{56}Ni	6–10	n, ^{4}He, ^{1}H;	-8×10^{18}	Neutrinos
$A \sim 56$ Nuclei	(depends on ϱ)	photodisintegration and neutronization		

In order to even attempt to read this history, and in a sense replace the HR diagram with an abundance table as our observational constraint, we must calculate abundances correctly.

3. Minimum Reaction Networks

The set of coupled nonlinear differential equations which govern abundances of nuclei undergoing thermonuclear reactions is referred to as a *reaction network*. As a star evolves to higher temperature and density an increase in the number of possible reactions results in more complex reaction networks. In principle all nuclei should be included in the reaction network. In practice the size of the network can be determined by an accuracy criterion (such as, all nuclei having abundances greater than ε are to be calculated to an accuracy of δ, where ε and δ are some chosen numbers). Clearly the accuracy needed depends upon the use to be made of the results. Con siderable computational economy can be obtained by judicious choice of the reaction network to be used. Consider all networks giving an error of size δ or less for any species having an abundance ε or greater. Any member of this set will be called an 'equivalent' network to any other member of the set. For efficiency we wish to find the minimum equivalent network, that is, the one with the fewest reactions and nuclear species. It should be noted that the question of accuracy will imply in practice the calibration of a smaller network by a larger, more general one.

Guided by these ideas I have developed what I consider to be the simplest acceptable network for helium and subsequent burning stages. The term 'acceptable' is a time dependent quantity; as we learn more we will require better treatments of the physics. The energy generation rate ε is well represented through oxygen burning; the composition for nuclei of abundance by mass of $\gtrsim 10\%$ is reasonably good up to the onset of oxygen burning. Beyond this point single nuclei are used to represent the Si to Ca and the Ti to Zn quasi equilibrium groups (denoted 'Si' and 'Ni' respectively). See Woosley *et al.* (1974) for details of how the nucleosynthesis actually occurs.

Table II lists the reactions which I now consider necessary to represent the helium, carbon, neon, oxygen and silicon burning stages. With the new evolutionary models now becoming available it should be possible to improve the silicon burning algorithm (although there is no evidence at present to indicate that the algorithm listed is inadequate). To better explain the approximations, Figure 2 presents the recommended groups of nuclei in a graphical format.

4. Helium Burning

As the first nuclear burning stage with more than one principal product, helium burning is a stage for which correct treatment of nucleosynthesis is vital. Renewed work on the triple-alpha reaction and especially new experimental work by Dyer (1973) at Caltech on $^{12}C(\alpha, \gamma)\,^{16}O$ has substantially improved the empirical basis of our helium burning calculations. Table III gives new rates suggested for these reactions. These expressions were derived from those given by Fowler *et al.* (1967 and private

TABLE II

Minimum Reaction networks
for energy generation

Fuel	Reactions
He	$3 \rightarrow {}^{12}C$
	${}^{12}C(\alpha, \gamma) {}^{16}O$
	${}^{16}O(\alpha, \gamma) {}^{20}Ne$ (small)
C	${}^{12}C({}^{12}C, \alpha) {}^{20}Ne$
	${}^{12}C({}^{12}C, p) {}^{23}Na (p, \alpha) {}^{20}Ne$
	${}^{16}O (\alpha, \gamma) {}^{20}Ne$
	${}^{20}Ne (\alpha, \gamma) {}^{24}Mg$
Ne	${}^{20}Ne (\gamma, \alpha) {}^{16}O$
	${}^{16}O (\gamma, \alpha) Ne^{20}$
	${}^{20}Ne (\alpha, \gamma) {}^{24}Mg$
	${}^{24}Mg (\alpha, \gamma) {}^{28}Si$ (small)
O	${}^{16}O ({}^{16}O, \alpha) {}^{28}Si$
	${}^{16}O ({}^{16}O, p) {}^{31}P (p, \alpha) {}^{28}Si$ etc.
	${}^{24}Mg (\alpha, \gamma) {}^{28}Si$
	${}^{28}Si + \alpha \rightarrow$ 'Si' = Si to Ca group
	ε is o.k. if we use ${}^{16}O$, ${}^{24}Mg$ and 'Si'
'Si'	'Si' $\rightarrow$ 'Ni' $\equiv A \sim 56$ nuclei
	as discussed by Bodansky et al. (1968)
	and Clayton (1968)

communication in August, 1973); the simplified expressions given here are valid *only* for hydrostatic helium burning temperatures. The ${}^{12}C(\alpha, \gamma) {}^{16}O$ rate reflects analysis of Dyer's data (by P. Dyer and by T. Tombrello) which attempts to abstract the contribution of the 7.115 MeV level in the ${}^{16}O$ compound nucleus from the laboratory measurement. The changes in the 3α rate and the ${}^{12}C(\alpha, \gamma) {}^{16}O$ rate are in different directions insofar as the nucleosynthesis of ${}^{12}C$ and ${}^{16}O$ is concerned. The change in the ${}^{12}C/{}^{16}O$ ratio in ashes of helium burning is therefore less than would be expected from either change taken alone. The reader should be warned that experiments currently underway and as yet unpublished may be reflected in further changes in these rates in the near future. Helium burning is one of the most difficult problems in experimental nuclear astrophysics; the current rate of progress in this area is exciting.

5. Carbon Burning

Recent experimental data on the ${}^{12}C + {}^{12}C$ reaction has been obtained at energies relevant to astrophysics by Patterson, Winkler, and Zaidins (1969, PWZ) and Mazarakis and Stephens (1972, 1973, MS). The Patterson et al. data went down to a center of mass energy of 3.25 MeV while that of Mazarakis and Stephens extended

W. DAVID ARNETT

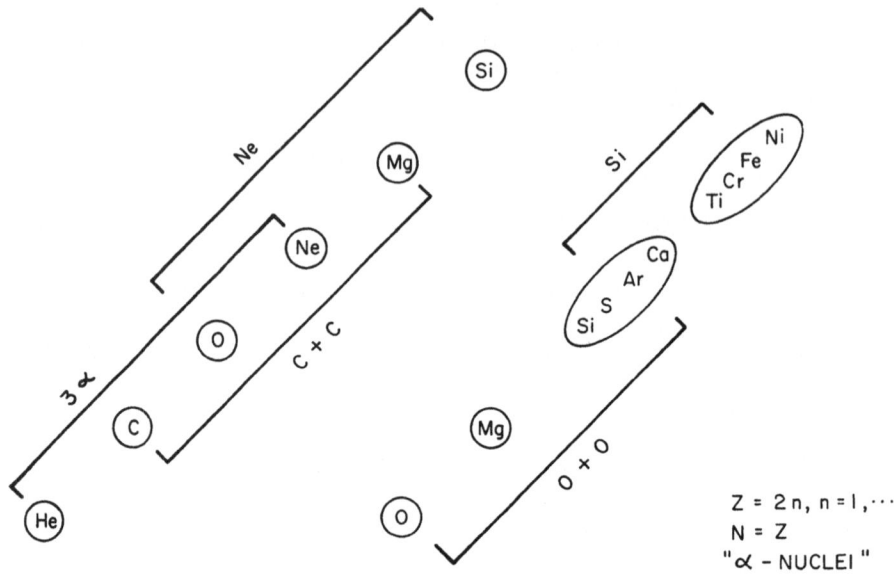

Fig. 2. Schematic diagram of groups of nuclei to be used for various burning stages.

TABLE III

New reaction rates for helium burning

$3\alpha \rightarrow {}^{12}C$. Fowler *et al.* (1967) and private communication (1973).
For $T_9 \sim 0.2$

$$a^2 \langle \sigma v \rangle = \frac{1}{T_9} 3 \,[2.48 \times 10^{-8} \exp(-4.4113/T_9) + 1.81 \times 10^{-8} \exp(-27.425/T_9)]$$

$$(\dot{Y}_c)_{3\alpha} = \varrho^2 Y_\alpha{}^3 a^2 \langle \sigma v \rangle /6; \quad Y_i \equiv X_i/A_i \equiv N_i/\varrho a$$

${}^{12}C\,(\alpha, \gamma)\,{}^{16}O$. Same source. For $T_9 \sim 0.2$.
$$a \langle \sigma v \rangle = 6.87 \times 10^7 \exp(-32.12/T_9{}^{1/3})/T_9{}^2$$

down to 2.45 MeV. Since the optimum bombarding energy is

$$E_0 = 2.41\, T_9^{12/3} \text{ MeV}$$

this corresponds to a temperature $T_9 \simeq 1$. The energy spread is

$$\varDelta = 1.005\, T_9^{5/6} \text{ MeV}.$$

These data are shown in Figure 3 in terms of the cross section factor

$$S = \sigma E\, e^{2\pi\eta},$$

where σ is the cross section, E the center of mass energy, and $\eta = Z_1 Z_2 e^2 / v$ is the Gamow factor for nuclei of proton number Z_1 and Z_2 with relative velocity v. The range of energy which determines the reaction rate is shown for $T_9 = 0.3$, 0.8 and 2.0. While the reaction rate is well determined for $T_9 = 2.0$ (i.e., explosive carbon burning), the rate is uncertain at $T_9 = 0.8$ and very uncertain at $T_9 = 0.3$. The early expression of Reeves (1966) is a fair average of the experimental data to date; it is shown as a solid curve. The fit of Patterson *et al.* (1969), shown as a dashed curve, represents their low energy data but not that of Mazarakis and Stephens (1972, 1973). If the rise in S at $E \simeq 2.5$ MeV is due to a resonant-like structure in the cross section, then the PWZ fit might be appropriate for lower energies. Michaud (1972) has discussed the experimental data in terms of an optical model; he suggests an increasing cross section factor with lower energy as shown by the dot-dashed curve.

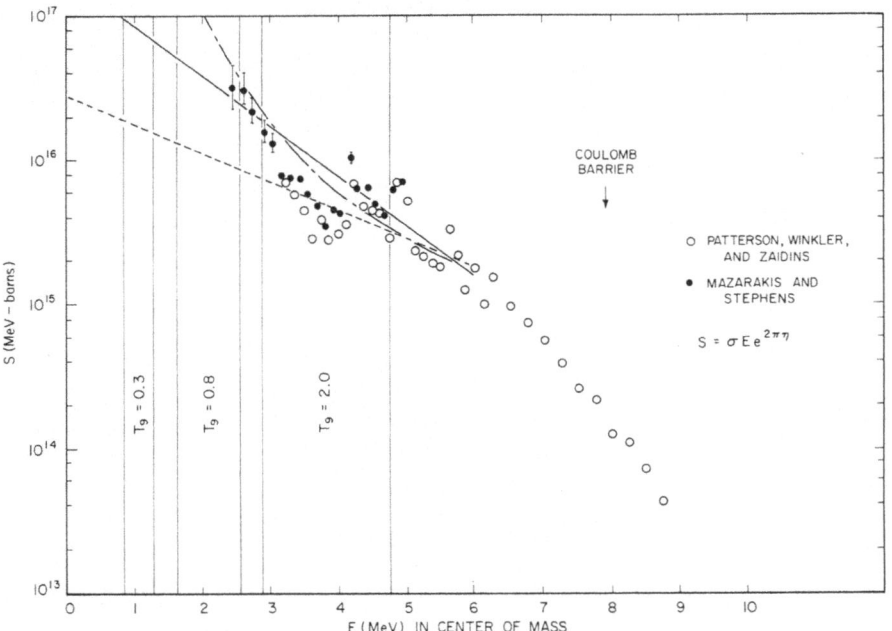

Fig. 3.   Cross section factor for $^{12}C + ^{12}C$ as a function of center of mass energy.

Because the low energy measurements of $^{12}C + ^{12}C$ cross section have revealed unexpected phenomena, there is a fundamental uncertainty in the nuclear physics theory. This implies an uncertainty in which extrapolation method to use, and therefore an uncertainty in the $^{12}C + ^{12}C$ reaction below $E \simeq 2.5$ MeV. In view of this uncertainty three rates for $^{12}C + ^{12}C$ are listed in Table IV. Fortunately for $T_9 \gtrsim 0.8$ which is typical of hydrostatic carbon burning in most stars all three rates are fairly close in magnitude.

TABLE IV

Reaction rates for ^{12}C $+$ ^{12}C

'Low':	Patterson *et al.* (1969)
	$a\langle\sigma v\rangle = T_9^{-2/3}\exp(61.053 - \tau)$
	$\tau = 84.173\,(1 + 0.0372\,T_9)^{1/3}/T_9^{1/3}$
'Middle':	Reeves (1966) (*Not* Reeves 1965!)
	$a\langle\sigma v\rangle = T_9^{-2/3}\exp(63.216 - \tau)$
	$\tau = 84.173\,(1 + 0.070\,T_9)^{1/3}/T_9^{1/3}$
'High':	Michaud (1972)
	$a\langle\sigma v\rangle = T_9^{-2/3}\exp(57.248 - 79.469/T_9^{1/3}),\ 0.3 \leqslant T_9 \leqslant 0.9$
	$= T_9^{-2/3}\exp(52.586 - 74.968/T_9^{1/3}),\ 0.9 \leqslant T_9 \leqslant 0.9$
	(The lower T expression has been slightly modified to fit
	more smoothly at $T_9 = 0.9$)

6. Neon and Oxygen Burning

The driving reaction for neon burning is ^{20}Ne(γ, α) ^{16}O; at $T_9 \simeq 1.5$ its reaction rate has been reasonably well known for some time. Toevs *et al.* (1971) have recently reconfirmed the rate given earlier by Fowler *et al.* (1967). The reaction rates for the (α, γ) reactions on ^{16}O, ^{20}Ne and ^{24}Mg can be found in the latter reference and unpublished work by these authors (an updated review is planned). Table V gives the rate for ^{20}Ne(γ, α) ^{16}O for $T_9 \simeq 1.5$ and a fit to the ^{16}O $+$ ^{16}O rate.

TABLE V

Some reaction rates for neon and oxygen burning

^{20}Ne(γ, α) ^{16}O.	Fowler *et al.* (1967) and Toevs *et al.* (1971)
	$\lambda_{\gamma\alpha}(^{20}\text{Ne}) = 2.31 \times 10^{12}\exp(-65.247/T_9) +$
	$+ 2.21 \times 10^{13}\exp(-67.131/T_9)$
	For $T_9 \simeq 1.5$
^{16}O $+$ ^{16}O.	A fit to the data of Spinka and Winkler (1972)
	$a\langle\sigma v\rangle = T_9^{-2/3}\exp(86.338 - \tau)$
	$\tau = 135.958\,(1 + 0.053\,T_9)^{1/3}/T_9^{1/3}$

Before discussing the ^{16}O $+$ ^{16}O reaction it is enlightening to consider ^{12}C $+$ ^{16}O (Figure 4) and then compare these as well as ^{12}C $+$ ^{12}C. The behavior of the cross section factor for ^{12}C $+$ ^{16}O as taken from the data of Patterson *et al.* (1971) and of Kuehner and Almqvist (1964). The solid curve is the fit of Woosley *et al.* (1971), which is very similar to that of Michaud (1972). Although the fine scale resonant structure is not seen, a large scale resonant 'hump' at $E \simeq 6.5$ MeV appears which is not unlike that seen in ^{12}C $+$ ^{12}C at $E \simeq 4.7$ MeV or so (see Figure 3). In Figure 5 is shown the data of Spinka and Winkler (1972) for the ^{16}O $+$ ^{16}O reaction. This is a particularly difficult reaction from the experimental point of view because of the many possible exit channels. In analogy with ^{12}C $+$ ^{12}C and ^{12}C $+$ ^{16}O one might suspect

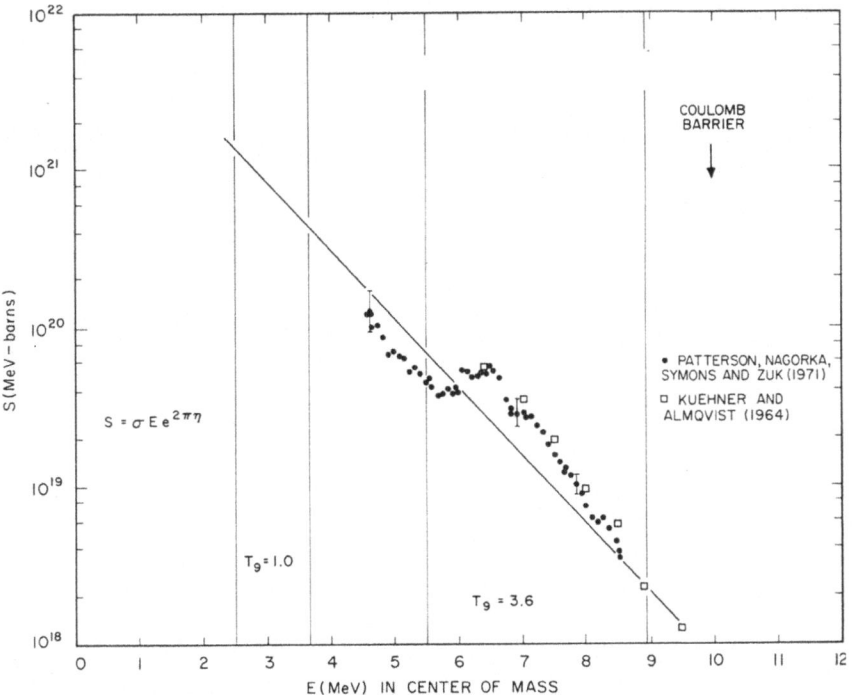

Fig. 4. Cross section factor for $^{12}C + ^{16}O$ as a function of center of mass energy.

that the experimental results extend down to the region of the 'hump', and that S will increase again from 6 to 4 MeV. This is merely a speculation.

Reeves (1965) used the higher energy data then available (unfortunately it was meager) to obtain the rate illustrated by the dotted line. This is most probably a severe overestimate at hydrostatic oxygen burning temperatures; use of this rate is *not* recommended. The dashed line presents the expression of Fowler and Hoyle (1964), which has also been widely used, for comparison. The recommended rate is the dot-dashed line (see Table V). This has the same slope as the expression of Truran and Arnett (1970) but is lower by about a factor of two. Michaud (1972) gets an expression which is virtually identical to this new rate except that it also should be reduced by a factor of two to account for a recalibration of the Spinka-Winkler data just prior to publication.

7. Some Structural Effects

The particular nature of the neutrino emission (as predicted for example by the conserved vector current theory) and of the nuclear energy generation causes some important structural changes in the late stages of stellar evolution. A detailed analysis of some of these effects has been published (Arnett, 1972).

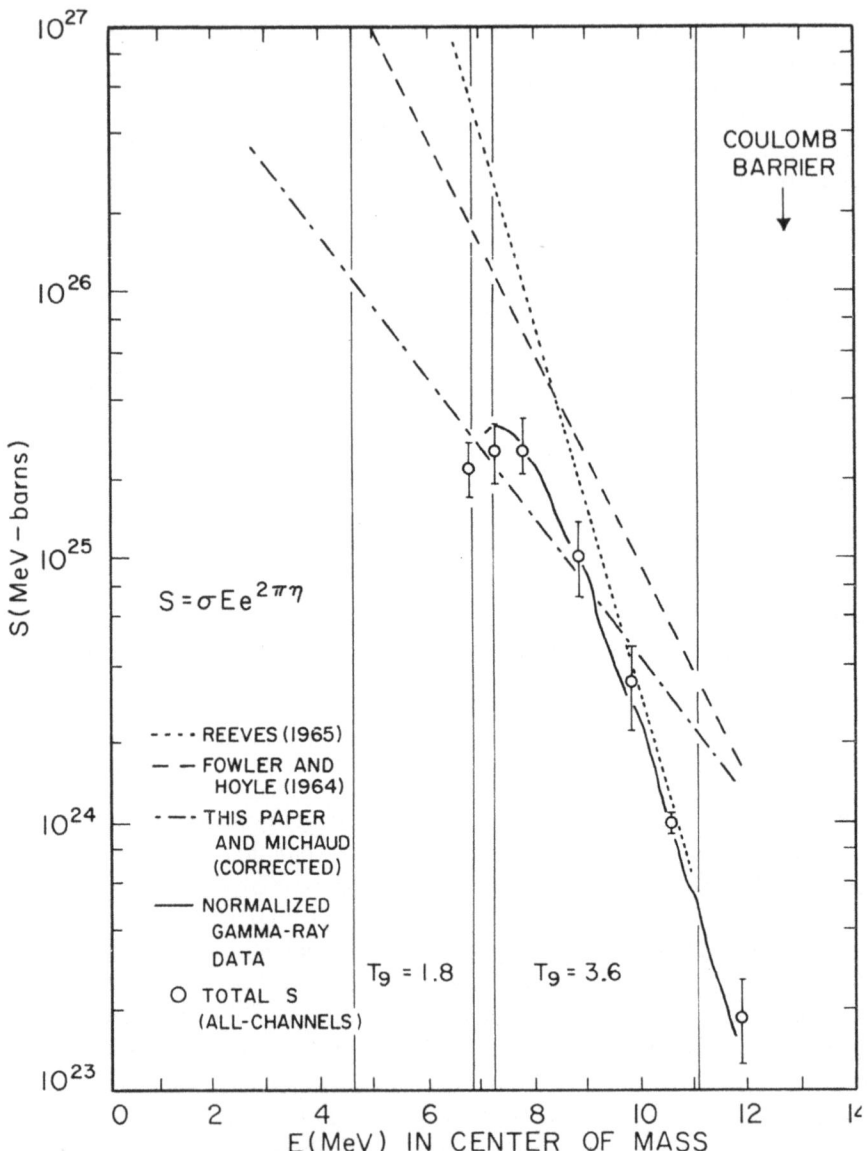

Fig. 5.   Cross section factor for $^{16}O + ^{16}O$ as a function of center mass energy
(Spinka and Winkler, 1972).

(A) Energy loss by neutrinos due to a direct $e - \nu$ coupling is a local process, determined by the local temperature and electron number density. If we approximate this energy loss (per gram) by

$$\varepsilon_\nu \propto T^k$$

then $k \gg 1$. Since the internal energy of matter in the relevant range of temperature T and density ϱ is roughly $E \propto T$, the hot inner regions of a star encounter the most severe cooling. Before a nuclear fuel ignites the only energy source is gravitational contraction, but this rate of energy supply is proportional to the local compression. For homologous contraction the fractional rate of energy supply is uniform. The central regions show a marked tendency to establish an isothermal temperature structure because the outlying layers show a larger fraction temperature increase than the inner layers. A strong, positive entropy gradient develops between the inner and outer regions.

(B) If we approximate the nuclear energy generation rate by

$$\varepsilon \propto T^m,$$

then $m > k \gg 1$. There is a critical temperature $T = T_{crit}$ at which $\varepsilon - \varepsilon_v = 0$. For $T > T_{crit}$ the nuclear rate dominates and heating occurs (in the absence of rapid expansion); similarly for $T < T_{crit}$ the neutrino rate dominates and cooling occurs (in the absence of rapid contraction). Suppose $T > T_{crit}$, so that nuclear energy is rapidly released. If photon diffusion and electron conduction of heat are ineffective (roughly $T \gtrsim 0.5 \times \times 10^9$ K) then the nuclear energy goes into heating and expanding the matter. Convective instability develops, and the hot buoyant blob floats upward, transporting energy, expanding and cooling.

(C) Consider the thermal balance for the convective core. Note that since $m > k$ the nuclear energy release is more concentrated toward the region of highest temperature (the centre of the star if the matter is not highly degenerate). If the mass contained in the convective core M_{cc} is small, then

$$\int_0^{M_{cc}} \varepsilon \, dm > \int_0^{M_{cc}} \varepsilon_v \, dm$$

and the entropy of the core increases. This allows the core to grow, enclosing more mass, until

$$\int_0^{M_{cc}} \varepsilon \, dm = \int_0^{M_{cc}} \varepsilon_v \, dm.$$

Note that this does *not* imply thermal equilibrium for the star as a whole, i.e.,

$$\int_0^M \varepsilon \, dm \neq \int_0^M \varepsilon_v \, dm$$

in general. Neglect of this latter point is probably the cause of some controversy with regard to analysis of the pulsational stability of these objects.

(D) The conditions above give rise to a phenomenon of 'core-size reduction'. For each successive stage the mass converted to 'ashes' is a small fraction of that potentially available as 'fuel'. The core gets smaller with each stage, and is surrounded by a growing mantle of unburned fuel. How small can the core – that is the central region which has undergone the most advanced burning – actually get? If its mass is less than about 1.4 $M_\odot$ (the Chandrasekhar limit, roughly speaking) it can support itself by its electron degeneracy pressure. Such a core does not contract and heat up to ignite the next burning stage. We have a 'waiting point'. The core is separated from the fuel of the last burning stage by a shell burning that fuel. If this burning increases the core mass then the core mass eventually rises to $M \gtrsim 1.4 \, M_\odot$, contraction, heating and ignition of the next burning stage can occur.

These processes are vital to our understanding of nucleosynthesis, presupernovae and gravitational collapse. It appears that the physical processes occurring in advanced evolution of stars set the stage in a very particular way for the stellar death scene.

8. Results of Some Recent Evolutionary Calculations Using These Rates

How do these topics relate to actual evolutionary calculations of stars? Table VI summarizes some aspects of a set of evolutionary calculations of helium stars of mass $M_\alpha = 4, 8, 16, 32, 64$, and $100 \, M_\odot$. These may be thought of as helium cores enclosed in a hydrogen rich envelope (corresponding masses are given in Table VI as $M/M_\odot$), or as stars which have lost their hydrogen rich envelope as probably would occur for example in a close binary system. The most massive objects ($M_\alpha = 64$ and $100 \, M_\odot$) became unstable to the electron-pair instability ($\gamma < \frac{4}{3}$ because of copious $e^\pm$ production). The lower masses (which are probably the astronomically relevant ones)

TABLE VI

Summary of the calculations for various helium core masses M_α (Abundances are by mass)

$M_\alpha/M_\odot$	4	8	16	32	64[b]	100[b]	Solar system	Galactic cosmic rays
$M/M_\odot$	15	22	36	70	120	170	–	–
He/O	11	1.9	.66	.41	.12	.16	25.0	7.3
C/O	1.1	.35	.21	.17	.070	.039	0.41	0.71
Ne/O	.23	.46	.33	.091	.014	.0071	0.20	0.24
Mg/O	.41	.15	.10	.047	.056	.044	0.075	0.33
Si/O[c]	7.2[a]	.28[a]	.055[a]	.073[a]	.095	.72	0.082	0.33
Fe/O[d]	2.1[a]	.80[a]	.24[a]	.14[a]	0.	.25	0.140	0.76
Stage	Core Si flash	Core Si flash	Core collapse	Core collapse	Post explosion	Post explosion	–	–
Central core mass/$M_\odot$	~ 1.4	~ 1.4	~ 1.4	~ 1.4	2.2 $M_\odot$ (Si) remnant	No remnant–		–

[a] Will be significantly modified by remnant formation and explosive processing.
[b] Abundances in ejected matter only.
[c] Si refers to the quasi equilibrium cluster from Si through Ca.
[d] Fe refers to the quasi equilibrium cluster around $A = 56$ (the 'iron' peak).

all evolved to core collapse with nearly identical core structure and mass ($M_{core} \simeq \simeq 1.4\ M_\odot$ of iron group elements).

The abundances of ^{4}He, ^{12}C, ^{20}Ne, ^{24}Mg as well as the 'Si' and the 'Fe' (or 'Ni') quasi equilibrium groups are tabulated, relative to ^{16}O, for all M_α. If each core becomes a neutron star (at least in a statistical sense), and each neutron star thus formed accelerates an equal number of the surrounding nuclei to cosmic ray energies, the resulting abundance distribution agrees well with the observed abundance distribution for galactic cosmic rays. (This number average abundance distribution is close to that shown for $M_\alpha = 4\ M_\odot$; compare with the column in Table VI called 'Galactic Cosmic Rays'). Even the curious C/O ratio in the cosmic rays appears naturally. See Arnett and Schramm (1973) for more details.

Similarly, if we do a mass average of $M_\alpha = 4, 8, 16, 32\ M_\odot$ ($M = 15, 22, 36$ and $70\ M_\odot$) over a realistic initial mass function then the resulting abundances agree with the 'cosmic' (solar system) abundances for C, O, Ne and Mg to within a factor of two. If we assume the remnant mass (in a statistical sense) is $\sim 1.4\ M_\odot$, then the abundance of Si + Fe is about right also. If we use even modestly realistic models of galactic evolution we find that the absolute as well as the relative abundance of these elements are correctly predicted.

These results are startling. The calculations are more than sufficiently general to allow widely divergent abundance distributions to appear. This may be an important clue as to the nature of the late stages of stellar evolution. By such abundance arguments we may learn something about gravitational collapse as well as nucleosynthesis and the late stages of stellar evolution.

Acknowledgement

The assistance of the National Science Foundation and the National Academy of Sciences is gratefully acknowledged.

References

Arnett, W. D.: 1972, *Astrophys. J.* **173**, 393.
Arnett, W. D.: 1974, *Thermonuclear Evolution of Stars and Galaxies*, Univ. of Chicago Press, Chicago.
Arnett, W. D. and Schramm, D. N.: 1973, *Astrophys. J.* **185**, L47.
Beaudet, G., Petrosian, V., and Salpeter, E. E.: 1967, *Astrophys. J.* **150**, 979.
Bodansky, D., Clayton, D. D., and Fowler, W. A.: 1968, *Astrophys. J. Suppl.* **16**, 299.
Clayton, D. D.: 1968, *Principles of Stellar Evolution and Nucleosynthesis* (McGraw-Hill: New York).
Dicus, D,: 1972, *Phys. Rev.* **D6**, 941.
Dyer, P.: 1973, in D. N. Schramm and W. D. Arnett (eds.), *Explosive Nucleosynthesis*, Univ. of Texas Press, Austin, p. 195.
Festa, G. C. and Ruderman, M. A.: 1969, *Phys. Rev.* **180**, 1227.
Feynman, R. P. and Gell-Mann, M.: 1958, *Phys. Rev.* **109**, 193.
Fowler, W. A., Caughlan, G., and Zimmerman, B. A.: 1967, *Ann. Rev. Astron. Astrophys.* **5**, 525.
Fowler, W. A. and Hoyle, F.: 1964, *Astrophys. J. Suppl.* **9**, 201.
Kuehner, J. A. and Almqvist, E.: 1964, *Phys. Rev.* **134**, B1229.
Mazarakis, M. and Stephens, W.: 1972, *Astrophys. J. Letters* **171**, L97.
Mazarakis, M. and Stephens, W.: 1973, *Phys. Rev.* **C7**, 1280.

Michaud, G.: 1972, *Astrophys. J.* **175**, 751.
Patterson, J. R., Nagorka, B. N., Symons, G., and Zuk, W. M.: 1971, *Nucl. Phys.* **A164**, 545.
Patterson, J. R., Winkler, H., and Zaidins, C.: 1969, *Astrophys. J.* **157**, 367.
Pontecorvo, B.: 1959, *Zh. Eksper. Teor. Fiz.* **36**, 615; 1960 *Soviet Phys. JETP* **9**, 1148.
Reeves, H.: 1965, *Stars and Stellar Systems* **8**, 113.
Reeves, H.: 1966, *Astrophys. J.* **146**, 447.
Spinka, H. and Winkler, H.: 1972, *Astrophys. J.* **174**, 455.
Toevs, J. W., Fowler, W. A., Barnes, C. A., and Lyons, P.: 1971, *Astrophys. J.* **169**, 421.
Truran, J. W. and Arnett, W. D.: 1970, *Astrophys. J.* **160**, 181.
Weinberg, S.: 1971, *Phys. Rev. Letters* **27**, 1688.
Woosley, S. E., Arnett, W. D., and Clayton, D. D.: 1971, *Phys. Rev. Letters* **27**, 213.
Woosley, S. E., Arnett, W. D., and Clayton, D. D.: 1974, *Astrophys. J. Suppl.* **26**, 231.

SCREENING FACTORS IN LATE STELLAR EVOLUTION*

(Abstract)

HAROLD C. GRABOSKE, Jr. and HUGH E. DeWITT

Lawrence Livermore Laboratory, University of California, Calif., U.S.A.

Recent studies (DeWitt *et al.*, 1973; Graboske *et al.*, 1973) have modified the screening factors for nuclear reactions. The theory develops a general statistical mechanical framework within which the various analytical models of Salpeter are combined with Monte Carlo and perturbation calculations for specific mixtures. The result is a general effective screening function in terms of the reacting charges Z_1, Z_2 and the plasma properties ϱ, T and $\bar{z}$. This screening function greatly enhances the screening effects for presupernova models (Couch and Arnett, 1973; Bruenn, 1973) reducing carbon ignition densities in the 4–8 $M_\odot$ range by a factor of two, and even more for degenerate O and Si burning.

Current studies of screening have verified the accuracy of the general screening function relation for a wider range of physical conditions. Extensive Monte Carlo calculations by Hansen (1973) and by DeWitt and Graboske at Livermore have considerably extended and improved the accuracy of the numerical data used to determine the screening function. These new numerical results produce very slight changes in the screening function, of the order of one percent in the range $0.8 \leqslant \xi_{12} \leqslant \infty$, where $\xi_{12} = Z_1 Z_2 / \bar{z}^2$. Only for $\xi_{12} \ll 1$ are changes as large as 5% in the screening function (corresponding to a factor of 50 in the screening factor at carbon ignition densities). For late evolution reactions, however, ξ_{12} is of order unity for all cases of interest (e.g. C^{12}, C^{12} in a C–O mixture), and the modifications required for $\xi_{12} \to 0$ are not relevant. The original screening function should continue to be sufficient. A second area investigated is the effect on the screening on nonuniform electron distributions. All previous strong screening studies assume that a uniform electron distribution always exists, an assumption strictly valid only at very high density. Using Hubbard's Monte Carlo method incorporating the Linhard plasma dielectric function, calculations were made to determine the effect on screening of electron clumping around the reacting nuclei. The results show significant enhancement of the screening function over the uniform distribution result; in certain lower density regions, as much as twenty percent increase in the screening function. But, again, for reactions in late evolutionary stages in the region of degenerate ignition, the density is sufficiently high that electron uniformity is a reasonable approximation. The region where electron nonuniformity is significant is more closely related to the case of intermediate screening, and it is quite likely that this area will be relevant to weakly degenerate carbon and oxygen ignition, for example, in the 8–15 $M_\odot$ range

* Work performed under the auspices of the U.S. Atomic Energy Commission. This paper was presented by H. C. Graboske.

where ignition occurs at lower densities and higher temperatures, before neutrino effects can radically cool and condense the interiors.

References

Bruenn, S. W.: 1973, in D. N. Schramm and W. D. Arnett (eds.), *Explosive Nucleosynthesis*, University of Texas Press, Austin.

Couch, R. G. and Arnett, W. D.: 1973, in D. N. Schramm and W. D. Arnett (eds.), *Explosive Nucleosynthesis*, University of Texas Press, Austin.

DeWitt, H. E., Graboske, H. C. Jr., and Cooper, M. S.: 1973, *Astrophys. J.* **181**, 439.

Graboske, H. C., DeWitt, H. E., Grossman, A. S., and Cooper, M. S.: 1973, *Astrophys. J.* **181**, 457.

Hansen, J. P.: 1973, *Phys. Rev.*, in press.

THE EXPLOSIVE r-PROCESS

(Abstract)

DAVID N. SCHRAMM

University of Texas, Austin, Tex., U.S.A.

Previous r-process calculations were done using constant temperature and neutron density. However, the r-process presumably occurs under dynamic conditions in supernovae. In this work, more realistic, dynamic r-process calculations with time varying temperature and density are presented. It is shown that these dynamical calculations eliminate the previous need for arbitrary smoothing of the calculated abundances in order to fit the observed r-process abundances. It is also shown that for certain conditions it is possible to make all three observed r-process abundance peaks in the same dynamical event. Previous non-dynamical calculations were always forced to have more than one event in order to fit the observed peaks. It is felt that these dynamic calculations are able to remove many of the major uncertainties regarding the r-process. In determining the r-process conditions, one may also be determining the conditions at the mass cut separating the ejected supernova material from the neutron star remnant. In order to determine these conditions, a detailed examination of the nuclear reaction network which builds 'seed' nuclei from neutrons, protons and alphas is being carried out, as is a careful examination of the β-rates appropriate to the r-process region.

Tayler (ed.), Late Stages of Stellar Evolution, 17. All Rights Reserved.
Copyright © 1974 by the IAU.

HOT CNO-Ne CYCLES

(Abstract)

JEAN AUDOUZE

Laboratoire René Bernas, Orsay, Radio Astronomie, Observatoire de Meudon,
Meudon and SEP, CEN, Saclay, France

Recent studies of the CNO–Ne cycles at temperatures higher than 10^8 K are summarised. At these temperatures nuclear reactions (induced by protons or alpha particles for instance) are more rapid than the beta decay reactions. As a consequence the behaviour and the results of the nuclear processing become very different from those of the CNO cycle at lower temperature which transforms ^{12}C and ^{16}O into ^{14}N.

Audouze *et al.* (1973) have shown that in hydrogen rich zones the relative slowness of the beta decay reactions (favouring the beta unstable nuclei) induces large enhancements of ^{13}C, ^{15}N, ^{17}O and ^{21}Ne at temperatures $10^8 < T < 5 \times 10^8$ K in time scales $1 < t < 10^4$ s, while temperatures as large as $5 \times 10^8 < T < 10^9$ are necessary to induce overabundances of ^{18}O and ^{19}F.

The hot CNO–Ne process is likely to occur in at least two different astronomical objects: (1) Starrfield *et al.* (1972) have shown that explosions of novae can be triggered by hot CNO processes occurring at the surface of the white dwarf precursor of the nova. In these conditions, the matter ejected from the nova is largely enriched in ^{13}C, ^{15}N and ^{17}O. (2) Audouze and Fricke (1973) have studied the nucleosynthetic effects during the implosion-explosion of supermassive stars (of mass $\sim 10^5 \, M_\odot$) and have shown that these objects, which may be present in the condensed nuclei of many galaxies, can be responsible for the formation of ^{13}C, ^{15}N and ^{17}O provided that one per cent of the galactic matter has been processed in such massive objects.

References

Audouze, J. and Fricke, K. J.: 1973, *Astrophys. J.* **186**, 239.
Audouze, J., Truran, J. W., and Zimmerman, B. A.: 1973, *Astrophys. J.* **184**, 493.
Starrfield, S., Truran, J. W., Sparks, W. N., and Kutter, G. S.: 1972, *Astrophys. J.* **176**, 169.

DISCUSSION FOLLOWING PAPER BY ARNETT

Fowler to Schramm: What is the atomic mass of the fourth peak which your last graph showed? Does it reach into the region of the superheavy elements?

Schramm: It is near atomic mass 281 and the neutron magic number 184 and is therefore not composed of truly superheavy nuclei.

Fowler: I wish to emphasize the point raised by Dr Arnett in his excellent presentation of nuclear processes in the late stages of stellar evolution – the extrapolation of the S-factor to low energies in the heavy ion reactions $^{12}C + ^{12}C$, $^{12}C + ^{16}O$ and $^{16}O + ^{16}O$ is very uncertain! Michaud claims that S continues to rise at low energy due to reactions under the coulomb barrier. On the other hand the rise at the lowest observable energies may be a 'shape-resonance' and may be followed at still lower energies by a marked decrease. Work at Caltech on the reaction $^{16}O + ^{28}Si$ is attempting to prove or disprove Michaud's effect. It will be some time before the experimentalists can resolve the issue.

Hesser to Arnett: Perhaps the study of HR diagrams will yet be able to provide some useful constraints for the theories which you, and others, have been discussing, but to push the observations to such a point will require much hard work. For instance, with the new large reflectors in the southern hemisphere, it may be possible to study the luminosity function of stars on their way to becoming white dwarfs in the nearest globular clusters.

As an example of the type of work that may be done when we leave aside our classical notions that all stars in clusters must lie on 'sequences' we have the beautiful work of Zinn and Norris at Yale. They are studying those stars that lie to the left of the giant branch and above the horizontal branch.

The difficulty of finding candidate stars (many clusters must be searched) and of obtaining the subsequent data should not discourage us in our attempts to use the observed HR diagram, with its many unexplained nuances, as valuable guide for theory in the future.

Tayler (ed.), Late Stages of Stellar Evolution, 19. *All Rights Reserved.*
Copyright © 1974 by the IAU.

CIRCULATION AND MIXING

R. KIPPENHAHN

Universitäts-Sternwarte, Göttingen, F.R.G.

In the preceding lecture Dr Arnett has shown us how exciting the inner life of a star can be when during late stages of evolution higher nuclear reactions take place. But in order to enjoy the variety of different nuclear species formed there stellar material from the deep interior has to be brought up to the surface, where the observers can see it.

When we produce evolutionary sequences of stellar models we simplify in many respects. In particular, we assume that nuclear material stays where it is being formed unless convection brings it to other places.

But there are indications that material from the very interior is occasionally brought to the surface – the technetium stars are probably the most striking examples (Aller, 1961, p. 159). How can material from a stellar core reach the surface? Maybe during more advanced evolutionary stages stars blow off their envelopes into space and the core material then becomes visible. Maybe convection brings the matter from the deep interior up to the surface of the stars. But it is also possible that other mixing processes become important. The most well known of these is meridional motion in radiative regions of rotating stars. This is a special case of a more common situation where, as a consequence of deviation from spherical symmetry in a star, hydrostatic equilibrium and thermal radiative equilibrium are incompatible, in which case meridional motions always occur. In the first two chapters of this paper we will deal with meridional motions of this type.

Since thermal conduction can be included in the stellar structure equation of radiative transport our considerations also hold in degenerate regions of stars, where conduction is the leading transport mechanism.

But also in the pure spherically symmetric case, where hydrostatic equilibrium and thermal radiative equilibrium can be fulfilled simultaneously meridional motion can occur. If in such a case a nonspherical perturbation is applied which is so slow that hydrostatic equilibrium is still a good approximation but which perturbs the thermal part of the stellar structure equations, then again circulation will occur which either enhances the perturbation or diminishes it. In the first case one has a secular instability which can give rise to meridional circulation. In Section 3 we give examples for this case of circulation.

1. Meridional Circulation

1.1. EDDINGTON-VOGT CIRCULATION

In the following we deal first with slowly rotating stars. The ratio χ_r of centrifugal to gravitational acceleration in the equatorial plane at distance r from the center is therefore assumed to be small. Then meridional motions occur in radiative regions with velocities of the order of

$$v_{EV} \approx \frac{\nabla_{ad}}{\delta \left(\nabla_{ad} - \nabla \right)} \frac{L_r \chi_r}{M_r} \frac{3}{4\pi r G \varrho} \qquad (1)$$

(see A1) with

$$\nabla = \frac{d \ln T}{d \ln P}, \quad \nabla_{ad} = \left(\frac{d \ln T}{d \ln P} \right)_{ad}, \quad \delta = -\left(\frac{\partial \ln \varrho}{\partial \ln T} \right)_P \qquad (2)$$

while M_r, L_r, r, G, ϱ have the usual meaning. δ has to be determined from the equation of state. For the perfect gas one has $\delta = 1$. With increasing degeneracy $\delta \rightarrow 0$ which is of importance if Equation (1) is applied to degenerate regions in a star.

The existence of meridional circulation in radiative regions has been postulated independently by Eddington (1925) and by Vogt (1925), by discussing von Zeipel's theorem. The first numerical estimate was given by Sweet (1950); therefore this type of motion is sometimes called Eddington-Sweet circulation.

If we take the mean over the whole star Equation (1) yields

$$\overline{v_{EV}} \approx \frac{1}{\delta \left(\nabla_{ad} - \nabla \right)} \frac{L}{M} \frac{\overline{\chi_r}}{g} \qquad (3)$$

where bars indicate mean values over the star, while $g = GM/R^2$ is the surface gravity at the equator. In Equation (3) we have neglected dimensionless factors of the order one. If we introduce the Kelvin-Helmholtz time scale $\tau_{KH} \approx E_G/L$ of the star where $E_G \approx GM^2/R$ is the absolute value of the gravitational energy of the star we find

$$\overline{v_{EV}} \approx \frac{1}{\delta \left(\nabla_{ad} - \nabla \right)} \frac{R}{\tau_{KH}} \overline{\chi_r}. \qquad (4)$$

Connected with this mean velocity there is a characteristic time scale given by

$$\tau_{EV} \approx \frac{R}{\overline{v_{EV}}} \approx \frac{\overline{\delta \left(\nabla_{ad} - \nabla \right)}}{\overline{\chi_r}} \tau_{KH}. \qquad (5)$$

The Eddington-Vogt circulation follows from the fact that hydrostatic equilibrium in a rotating star and thermal radiative equilibrium cannot be fulfilled simultaneously. Formally this can be seen if one writes down the equations for rotational perturbations. One then finds that the system is overdetermined until a new variable, the velocity of meridional motion is introduced (Kippenhahn, 1963).

The estimates for v_{EV}, given here, are derived from a first order perturbation theory which holds only for small values of χ_r. This is sufficient for most applications, since χ_r is small in the interior of uniformly rotating stars, even if they rotate with break up velocity. In the case of differential rotation with high angular velocities in the interior of the stars, our estimates are no longer valid. But even in the case for which $\chi_r \approx 1$ throughout the star, they may give the order of magnitude of the circulation occurring.

1.2. μ-CIRCULATION

Let us assume a chemically homogeneous region in a star mainly consisting of hydrogen and let us put a blob of pure helium somewhere in that region outside the center. If at the beginning the blob is not in hydrostatic equilibrium it first will expand or contract until the pressure inside is the same as outside. For any quantity y we define by Dy the difference between the value of y in the blob and the value of y in the immediate surroundings. Hydrostatic adjustment then means that for the pressure P we have $DP=0$. Then in general the density will not yet be the same in the blob as outside and the blob will rise or sink until there is no buoyancy force, i.e. until $D\varrho=0$. For the sake of simplicity we assume a perfect gas inside and outside the blob. Then as a consequence of the different mean molecular weights μ we have $D\mu>0$ and therefore $DT>0$. This means that the matter in the blob will be hotter than the surroundings. Therefore there is no thermal equilibrium. The blob will now try to adjust its temperature to the neighbourhood. It therefore will cool off, its density will increase and the blob will sink (Figure 1), with a velocity which is controlled by the thermal adjustment time. It is given by (see A2):

$$v_\mu \approx \frac{\varphi H_P}{\delta\,(\nabla_{\mathrm{ad}} - \nabla)\,\tau^*_{\mathrm{KH}}} \frac{|D\mu|}{\mu}, \tag{6}$$

where $\varphi=(\partial \ln\varrho/\partial \ln\mu)_{P,T}$ follows from the equation of state. For the perfect gas $\varphi=1$. H_P is the pressure scale height while τ^*_{KH} is the thermal adjustment time scale (Kelvin-Helmholtz time scale) of the blob. It is given by (see for instance Kippenhahn, 1969)

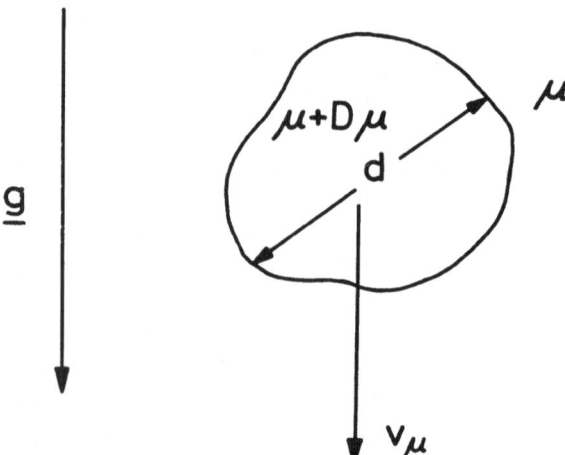

Fig. 1.   A blob of molecular weight $\mu + D\mu$ $(D\mu > 0)$ in surroundings with molecular weight μ in hydrostatic equilibrium undergoes a slow downward motion which is controlled by the thermal adjustment of the blob.

$$\tau_{KH}^* \approx \frac{3c_P \varkappa \varrho^2 \zeta d^2}{8acT^3} \tag{7}$$

where d is a measure for the size of the blob while ζ is a dimensionless factor which depends on the geometry. If V and O are volume and surface of the blob respectively, then one has $\zeta = V/Od$.

The time of thermal adjustment as it is defined in Equation (7) increases with the size of the blob. For blobs with a size comparable or even bigger than H_P, for $\varkappa, \varrho, T$ in Equation (7) mean values over the blob have to be taken. If the size becomes comparable with the radius of the star then τ_{KH}^* approaches the Kelvin-Helmholtz time scale of the star: $\tau_{KH}^* \to \tau_{KH}$.

1.3. MAGNETIC BUOYANCY

We now consider a magnetic blob, that is a subregion in a star which is filled with magnetic field but whose magnetic field lines are entirely in that blob; i.e. no field lines connect the inside with the outside. Finzi and Wolf (1968) used the word 'magnetic tangle' for such an element. Again if there is hydrostatic equilibrium between the magnetic tangle and the surroundings, we have $DP = 0$, $D\varrho = 0$. But in the blob the total pressure P consists of the gas pressure P_G and the magnetic pressure P_M. Therefore and if we again assume the perfect gas equation of state we find $DT < 0$. This time the blob is cooler than its surroundings and it will try to adjust itself thermally and will rise. The upward velocity is this time given by (see A3)

$$v_M \approx \frac{H_P}{\delta (\nabla_{ad} - \nabla) \tau_{KH}^*} \frac{P_M}{P_G}, \quad (P_M = -DP_G) \tag{8}$$

$c_P, \varkappa, a, c$ again have the usual meaning, while $\alpha = (\partial \ln \varrho / \partial \ln P)_T$.

1.4. NON-CONSERVATIVE ANGULAR VELOCITY DISTRIBUTION

There is another, more complicated, case in which hydrostatic equilibrium and thermal radiative equilibrium are not compatible. Let us assume that in a rotating star the angular distribution ω is not a function of the distance s from the axis of rotation only, but varies on coaxial cylinders $s = $ const. If z is the coordinate in the direction of the axis of rotation, we have in this case $\partial \omega / \partial z \neq 0$. Then the centrifugal acceleration

$$\mathbf{c} = \omega^2 \mathbf{s} \tag{9}$$

is a non-conservative vector field; i.e. $\nabla \times \mathbf{c} \neq 0$. The equation for hydrostatic equilibrium in the steady state is

$$\frac{1}{\varrho} \nabla P + \nabla \phi = \mathbf{c}, \tag{10}$$

R. KIPPENHAHN

where ϕ is the gravitational potential. If the temperature were constant on surfaces of equal pressure then the density would also be constant there and $\nabla P/\varrho$ would be a conservative vector field

$$\nabla \times \left(\frac{1}{\varrho}\nabla P\right) = 0\,; \tag{11}$$

then the left hand side of Equation (10) would be conservative, but the right hand side would not. Therefore hydrostatic equilibrium demands that the temperature is not constant on pressure surfaces. Therefore a special temperature distribution is necessary to maintain hydrostatic equilibrium. This temperature distribution is not that of thermal radiative equilibrium and consequently meridional motions ensue. This is a type of motion which comes from the Goldreich-Schubert-Fricke instability (Goldreich and Schubert, 1967; Fricke, 1968) of all angular velocity distributions with $\partial\omega/\partial z \neq 0$.

1.5. Limit towards small length scales

In the formulae (6) and (8) the velocity of the blob is higher for smaller blobs. This is due to the shorter time scale for thermal adjustment. For smaller scales these formulae are no longer valid since other effects become important, for instance friction. The smaller the blob, the more important the friction. A rough estimate for the lower limit of sizes of blobs can be obtained if one assumes that for the smallest sizes the Stokes velocity becomes comparable with the blob velocity. One then finds that for the order of magnitude of the smallest blobs (see A4)

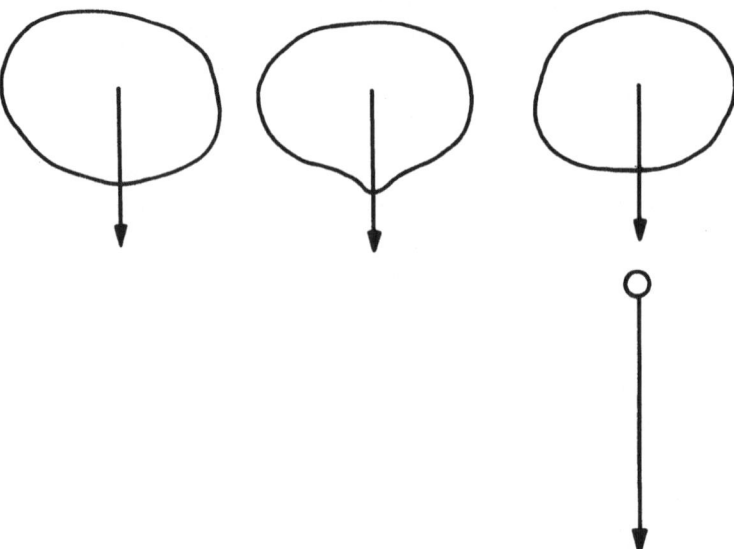

Fig. 2.   Since smaller blobs sink faster there is a tendency of fragmentation.

$$d \approx \sqrt{\frac{v_R c}{g_r}}, \tag{12}$$

where v_R is the local kinematic radiative viscosity and g_r the local gravitational acceleration. As a consequence, big blobs undergo fragmentation during their motions (see Figure 2).

Sometimes when viscosity is small the velocity for small blobs becomes of the order of the velocity of sound before friction effects set in. Then the assumption used in deriving the velocities is no longer valid.

For magnetic tangles d cannot be too small, in order to avoid a decay of the magnetic field before the element has moved over an appreciable distance.

In the case of non-conservative angular velocity distribution discussed in sub-Section 1.4 the motions which occur will also be small scale motions, since these modes can transport energy more effectively than large scale motions. Indeed in a normal star one should expect sizes of the order of 1 km. Some simple estimates of velocities and time scales of this type of motion have been given by Kippenhahn (1969).

2. Applications

2.1. MIXING IN CHEMICALLY HOMOGENEOUS REGIONS

Since on the main sequence the nuclear time scale is roughly 100 times longer than the Kelvin-Helmholtz time-scale with $\delta \approx 1$, $\nabla_{ad} - \nabla \approx 0.15$ one can estimate that τ_{EV} will be comparable with the time scale τ_{nucl} of nuclear evolution if $\bar{\chi}_r > 0.015$. Indeed for main sequence stars earlier than F5 the Eddington-Vogt time scale is shorter than the main sequence life-time. This has already been found by Sweet (1950). But there is observational evidence that even early type main sequence stars undergo unmixed evolution. For mixed evolution the stars would move to the upper left of the HR diagram while consuming their hydrogen but we are certain that they normally become red giants which is typical for unmixed evolution. Mestel (1953) found the way out of this difficulty. That there is no mixing of hydrogen and helium during the main sequence stage – not even for the fastest rotators – is due to the fact that with the same time-scale a gradient of molecular weight μ is being built up which prevents circulation, as we will later see. But the regions in these stars which remain chemically homogenous will indeed be mixed by the circulation. The flow pattern is rather complicated. Let us start say with rigid body rotation. For this type of motion the Eddington-Vogt circulation is a large scale motion rising at the poles and sinking at the equator. But since friction can be neglected each mass element conserves its angular momentum. Therefore after a while there will be a deviation from solid body rotation and ω will not be constant on cylinders $s = $ const. any more. As a consequence small scale motions will occur since the centrifugal acceleration is not conservative. Therefore a new, rather complicated ω-distribution will be established. The Eddington-Vogt circulation of this complicated circulation pattern will be even more complicated. One therefore should expect a kind of irregular small scale motion comparable with convection but smaller in scale and much slower. Probably the formula (1) will still give a good estimate for that type of small scale motion which sometimes is called *random circulation*. This circulation will mix angular momentum throughout the star. Although we do not know what the mean properties of that randomly fluctuating circulation field will be, we would expect that large differences in angular velocity would be smeared out although solid body rotation is not a solution.

Random circulation would also mix chemical elements as long as they do not effect the molecular weight considerably. This might be of some importance for the distribution of lithium through the envelopes of the stars. Recently Paczyński (1973) has suggested that the anomalous ratio of carbon to nitrogen in some early type stars can be explained by mixing. In his picture carbon is being transformed into nitrogen above the convective core in a region where helium is not enriched and therefore the molecular weight is practically constant. Random circulation could bring carbon from the outer layers into that region in which it is destroyed and consequently a large fraction of the carbon in the envelope can be transformed into nitrogen. Demarque and McClure (1973) have given arguments which favour this hypothesis.

Also in degenerate cores of evolved stars circulation might be important. Let us for instance take a degenerate carbon core in an evolved star before the onset of carbon burning. If this core were strictly isothermal, which means energy is neither created nor swallowed in the core, then we would have $L_r = 0$ for the core and v_{EV} would therefore vanish. But while the core mass is increasing due to helium burning at its surface, the core contracts and contraction energy is released, giving rise to a non-vanishing L_r and therefore a non-vanishing circulation. Neutrino-processes on the other hand take energy away and therefore energy has to be transported from one part of the core to the other, giving rise again to a non-vanishing L_r and therefore to circulation. It has been shown by Kippenhahn and Möllenhoff (1974) that the contraction energy released and the energy carried away by neutrinos in a rotating core cannot compensate each other. Therefore there will always be a conductive flux through the core, giving rise to a non-vanishing Eddington-Vogt circulation. By extrapolating the expressions for the circulation velocity derived for slow rotation to the case of $\chi_r \approx 1$ Kippenhahn and Möllenhoff (1974) have estimated that in rapidly spinning cores circulation can redistribute angular momentum within the time-scale of evolution. But this is only the case if the cores are spinning so rapidly that they are very flat and that χ_r is of the order 1. This might be of importance if one discusses the effect of rotation on the conditions for the onset of carbon burning (Sackmann and Weidemann, 1972).

Kippenhahn and Möllenhoff (1974) have also estimated the time-scale for mixing in rapidly rotating white dwarfs. Although white dwarfs can be roughly described as isothermal bodies due to their cooling, L_r does not vanish and if $\bar{\chi}_r = 1$ then the mixing time-scale is comparable with the cooling time. This might be of impor tance for the discussion of rapidly rotating white dwarfs with masses far above the

Chandrasekhar limit. The random circulation which then would occur would redistribute the angular velocity in a way similar to friction.

2.2. μ-BARRIERS

If, say, a helium core in a hydrogen rich envelope has been formed, then, in a transition region, the mean molecular weight μ increases inwards. If the Eddington-Vogt circulation tries to build up a circulation which, say, rises at the pole and sinks at the equator, then helium rich material is brought up along the polar axis while hydrogen rich material is brought inwards at the equator. This creates a situation where μ-currents try to reestablish spherical symmetry of the μ-distribution and act against the Eddington-Vogt circulation. For a rough estimate we describe the situation given in Figure 3 by a blob of the diameter $d \approx r$ where r is the core radius, we assume

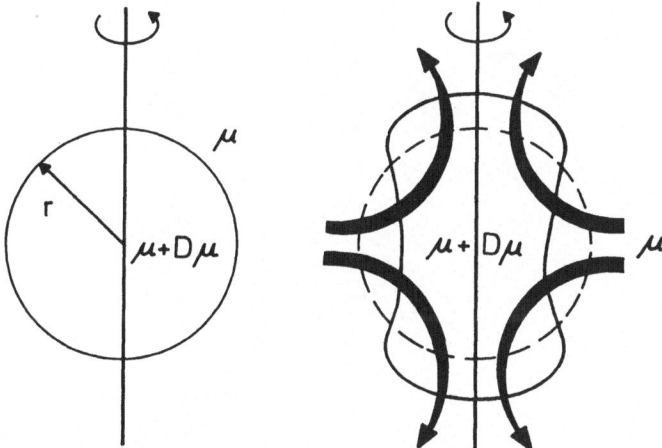

Fig. 3. In the case of a core of molecular weight $\mu + D\mu$ surrounded by an envelope of molecular weight μ meridional motions (black arrows) produce a non-spherical μ-distribution. Then μ-currents occur which try to reestablish spherical symmetry.

$\tau_{KH}^{*} \approx \tau_{KH}$ and we consider the matter in the equatorial plane as the 'surroundings' of the blob. In this very crude picture we can determine v_{μ} from Equation (6). No mixing will occur if $v_{EV} < v_{\mu}$. This gives (with $\varphi = 1$):

$$\chi_r < \left(\frac{2H_P}{r}\right) \frac{|D\mu|}{\mu}. \tag{13}$$

The factor $2H_P/r$ may be of the order 1, and since we have pure helium in the blob and a hydrogen rich mixture ($X_H = 0.7$, $X_{He} = 0.3$) outside, $D\mu/\mu$ is roughly 0.74. Thus one finds that there will be no mixing as long as $\chi_r < 0.74$. Meridional circulation will occur in the homogenous regions outside and inside the transition region. The region of variable molecular weight acts as a barrier, it prevents the Eddington-Vogt circulation from penetrating and mixing. For higher nuclear burning the differences in molecular weight which are being built up become smaller. For the transition from helium to a carbon-oxygen mixture one already has $D\mu/\mu \approx 0.27$. Therefore penetration of these μ barriers is more probable for higher nuclear burning. But even then rotation would have to be rather fast.

2.3. FORMATION OF μ-BARRIERS

In the foregoing we have discussed the problem of mixing through a μ-barrier which is already there – but in the star μ-barriers are being built up from homogenous regions by nuclear burning. There is no μ-barrier at the beginning. The question is whether mixing is strong enough to prevent the formation of a μ-barrier. This problem has first been investigated by Mestel (1957). At the beginning there is only Eddington-Vogt circulation. If a blob rises along the polar axis is still has the original chemical composition. After the time τ_{EV} it might sink towards the equatorial plane (Figure 4).

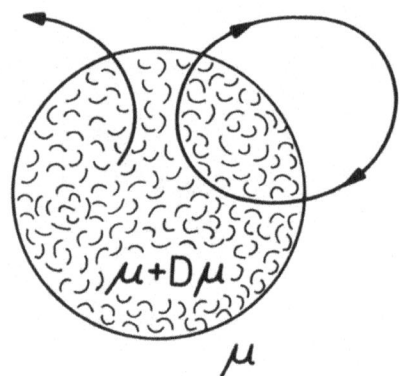

Fig. 4.   Meridional circulation in the central region of a star with convective core. Since the molecular weight in the core increases with time the molecular weight varies along the stream lines of the meridional velocity field. Again a non-spherical μ-distribution occurs which gives arise to μ-currents.

At the same time new material is coming up at the pole, enriched in ashes. If we define the rising material as the blob and the sinking material as its surroundings then we have ($\tau^*_{KH} \approx \tau_{KH}$)

$$v_\mu \approx \frac{\varphi H_P}{\delta\left(\nabla_{ad} - \nabla\right)\tau_{KH}} \frac{|D\mu|}{\mu}, \tag{14}$$

where again $D\mu$ is the difference in molecular weight between the blob and the surroundings. $D\mu$ has now to be distinguished from $\Delta\mu$ which is the difference of molecular weight before the onset of nuclear burning and after nuclear burning has been completed. One has

$$D\mu \approx \frac{\partial\mu}{\partial t}\tau_{EV}, \qquad \frac{1}{\mu}\frac{\partial\mu}{\partial t} = \frac{1}{\tau_{nucl}}\frac{\Delta\mu}{\mu} \tag{15}$$

where τ_{nucl} is the time scale of nuclear burning. Therefore $D\mu = \tau_{EV}\Delta\mu/\tau_{nucl}$. With this expression for $D\mu$ Equation (6) gives an estimate for the velocity with which the μ-currents try to reestablish spherical symmetry of the μ-distribution. No mixing will occur if $v_{EV} < v_\mu$. In order to estimate v_{EV} in the neighbourhood of the interface we find from Equation (1)

$$v_{EV} = \frac{V_{ad}}{\delta(V_{ad} - V)} \frac{L_r\chi_r}{M_r} \frac{3}{4\pi rG\varrho} = \frac{V_{ad}}{\delta(V_{ad} - V)} \frac{\chi_r r}{\xi\zeta\tau_{KH}^*}, \tag{16}$$

where we have introduced the dimensionless quantities ξ and ζ by

$$\xi = 4\pi r^3\varrho/3M_r, \quad \zeta = GM_r^2/rL_r\tau_{KH}^*.$$

For the characteristic time connected with these motions near the interface we define

$$\tau_{EV} = \frac{r}{v_{EV}} = \frac{\delta(V_{ad} - V)}{V_{ad}} \frac{\xi\zeta\tau_{KH}^*}{\chi_r}. \tag{17}$$

Here occurs a difficulty since $V_{ad} - V$ vanishes at the interface between a convective core and a radiative envelope and Equation (16) gives infinite velocities. Mestel (1953) has clarified the situation. In our problem it is sufficient to make clear that for $V_{ad} - V$ in Equation (17) a value has to be taken which is typical for the whole area through which matter coming out from the core is moving. From Equations (6), (16) and (17) we now obtain the condition that no mixing will occur if

$$\chi_r^2 < \frac{H_P}{r} \varphi\xi^2\zeta^2 \frac{\tau_{KH}^*}{\tau_{nucl}} \frac{\delta(V_{ad} - V)}{V_{ad}^2} \frac{\Delta\mu}{\mu}. \tag{18}$$

For any given stellar model the right hand side of Equation (18) can be determined and critical values of χ_r can be obtained. In order to give a rough idea we use some typical values and obtain from Equation (18) $\chi_r < 0.1(\Delta\mu/\mu)^{1/2}$ and conclude that between a helium core and a hydrogen-helium envelope $(\Delta\mu/\mu=0.74)$ no mixing will occur as long as $\chi_r < 0.09$ at the interface. For a carbon oxygen core in a helium envelope $(\Delta\mu/\mu=0.27)$ one obtains $\chi_r < 0.05$. As an example we can take an early type main sequence star in uniform rotation which is fully rotating – that is rotating at an angular velocity so high that centrifugal force and gravitational force cancel each other at the equator. The value of χ_r at the interface between convective core and envelope is then about 1/30 and no mixing will take place (Mestel, 1953). It should be kept in mind that in deriving and in using the condition (18) many dimensionless factors have been dropped in order to give some very simple estimates. For any given stellar model the criterion (18) can be derived in a more careful way following our procedure.

It seems surprising that main sequence stars are not too far from mixing and they might even mix if the angular velocity were to increase inwards sufficiently. On the other hand there is the well established fact that main sequence stars evolve to the right in the HR diagram which is typical for an unmixed evolution. Probably Equation (18) still exaggerates the danger of mixing. Indeed, in deriving this equation it was assumed that the Eddington-Vogt circulation is a large scale motion, where the characteristic lengths are given by the radius of the star. But as we have already seen Eddington-Vogt circulation by itself, starting from rigid body rotation, would very soon produce random circulation. Therefore, if in a main sequence star, helium enriched material would be brought out by circulation, say, at the polar region, then it would not move with the velocity v_{EV} but a random walk process would determine the diffusion of helium into the outer layers. Thus, the velocity by which helium is brought outwards is slower, and from this one might expect that even values of χ_r which are higher than the limit given by Equation (18) would still not cause mixing.

As we have seen in the case of the already fully developed μ-barrier for higher nuclear reactions the μ-effect becomes less and less important. For highly developed stars it therefore depends on the angular velocity of the cores whether mixing can take place or not. Unfortunately, there is practically no information whether the material in the very interior of a star conserves its angular momentum, whether the core which contracts during the evolution will rotate faster and faster or whether angular momentum is being exchanged between core and envelope. Since μ-barriers cannot be penetrated by circulation the only mechanism which brings angular momentum across the μ-barriers are magnetic fields with the appropriate topology. Fricke and Kippenhahn (1972) have used the angular velocities of pulsars in order to estimate that the cores of evolved stars at least rotate faster than the envelopes. But it is not clear whether the cores are really spinning rapidly.

2.4. MAGNETIC MIXING

If, as suggested in the last section, magnetic fields would transfer angular momentum from the cores to the envelopes, then rather strong magnetic fields are necessary, even if the initial poloidal field were small. Differential rotation between core and envelope would form a fairly strong toroidal field in the transition region and the question arises whether magnetic tangles are being formed, which would rise into the envelope. In the rotating convective cores of the stars one might expect dynamo effects similar to those discussed by Parker (1970), Krause and Rädler (1971), Köhler (1973) in connection with the solar cycle. One therefore can ask whether magnetic fields created in the very interior of the stars cannot mix the stars and maybe even bring nuclear material from the very interior to the surface. Gurm and Wentzel (1967) and Finzi and Wolf (1968) have discussed the effect of magnetic mixing. Gurm and Wentzel found that the velocity of the rising blobs should be sufficient to bring material from the core to the surface of a main sequence star in a time shorter than the time of nuclear evolution if the ratio of magnetic to gas pressure exceeds the value 10^{-4} (in order to get sufficient buoyancy) and if the diameter of the tangles is smaller than 10^{10} cm (in order to get a sufficiently short time of thermal adjustment). Finzi and Wolf even suggest that blue stragglers observed in the HR diagrams of globular clusters might be stars which undergo mixed evolution due to magnetic mixing.

The problem of magnetic fields in the very interior is also of importance for the problem of the solar neutrinos. Magnetic fields can reduce the observable solar neutrino flux considerably if they are strong enough. On the other hand strong magnetic fields in the very interior of the Sun would rise to the surface due to magnetic buoyancy. One has to keep in mind that the enrichment in helium in the central region of the Sun provides μ-currents which act against magnetic mixing. Equating the two expressions (6) for μ-currents and (8) for the velocity of magnetic blobs indicates that the magnetic fields would stay in the central region as long as

$$\varphi \frac{D\mu}{\mu} > \frac{P_M}{P_G}. \tag{19}$$

For μ and $\mu + D\mu$ we take the values for the envelope and the center of the present Sun and find $D\mu/\mu = 0.3$. With the central pressure of 2.2×10^{17} we find that the magnetic fields up to 1.3×10^9 G can be kept in the very interior of the Sun by the heavier molecular weight there.

If the magnetic field in the interior is stronger than that given in inequality (19), then tangles containing heavier material can move upwards bringing the heavier material into the outer regions. But this transport mechanism is not very effective. If at the beginning the magnetic buoyancy exceeds the extra weight of the heavier material, then while the tangle is rising it is expanding, the magnetic pressure is being reduced in such a way that P_M/P is getting smaller. After a finite path the tangle will come to an equilibrium position where the reduced magnetic buoyancy just balances the extra weight of the heavier material (see A5). In this position the tangle will not only be in hydrostatic but also in thermal equilibrium with its surroundings. Then with the decay time of the magnetic field the blob will sink again.

3. Nonspherical Instabilities in Circularly Symmetric Stars

Up to now we have always discussed cases where the star shows some deviation from spherical symmetry and where hydrostatic equilibrium and thermal radiative equilibrium cannot be fulfilled at the same time. Now we discuss spherically symmetric stars where indeed both equilibria can be fulfilled but where small deviations from symmetry will grow, giving rise to meridional motion which might mix the star. In accordance to the foregoing procedure we assume perturbations which are sufficiently slow so that hydrostatic equilibrium is fulfilled and the instability is of purely thermal origin. This type of instability is normally called *secular instability*. We do not mean that the fully developed motions which occurs from this type of instability must be slow and that hydrostatic equilibrium must always be fulfilled. But since we are only discussing the onset of this instability, we can work with hydrostatic equilibrium, which simplifies the problem extremely and which also gives us the possibility to use some of the formalisms given in the preceding chapters.

3.1. CARBON FINGERS

If neutrino losses are taken into account the helium flash of a 1.3 $M_\odot$ star occurs in a shell (Thomas, 1967). Therefore the carbon is being formed outside the center while the central region still contains the helium with the original low carbon content. After the helium flash helium burning continues to take place in a shell. Therefore a configuration occurs which is given in Figure 5. Consequently, we now discuss the case of a layer enriched in helium, which is above a helium layer as indicated in Figure 6. If one makes a small perturbation of the interfaces, then a drop of carbon which is hanging into the carbon region will, after reestablishing hydrostatic equilibrium, be hotter than the helium at the same level and we have the problem of μ-motion as discussed in sub-Section 1.2. Therefore the carbon enriched blob will sink even further and one has a situation similar to the salt finger experiment. The carbon will mix with the helium region below. From Thomas's models one obtains the following numerical values (in cgs units):

$$H_P = 6 \times 10^6, \qquad d = 24, \qquad \tau^*_{KH} = 1.5$$
$$\frac{1}{\mu}\frac{d\mu}{dt} \approx 10^{-5},$$

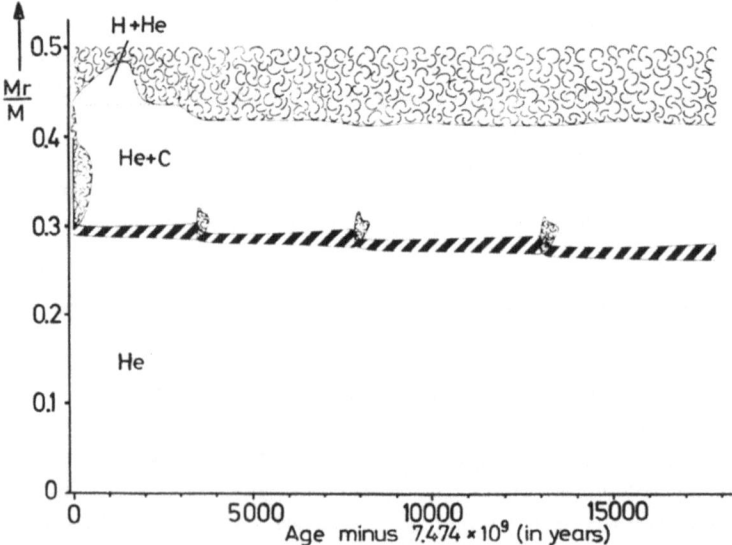

Fig. 5.   Helium burning in a shell after the helium flash of a star of 1.3 $M_\odot$ (Thomas, 1967). The cloudy regions indicate convection. The hatched region indicates shell burning. The regular variations of the shell including the small, short living convective zones just above the shell are due to thermal pulses. In the shell a layer is being built up where the molecular weight is higher than in the layers below.

where d again is determined by Equation (12). After the first second the velocity of the carbon enriched blobs will therefore be some 10 cm s^{-1}. Elements which start later will have higher velocity and will pass the earlier ones. This mixing problem has not yet been treated in detail. Thomas (1967) has assumed for his calculations complete mixing of the carbon enriched shell with the interior.

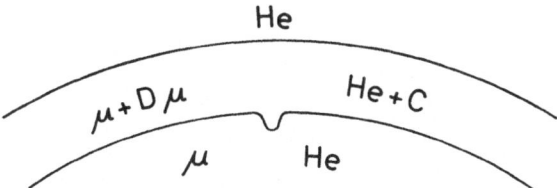

Fig. 6. The situation described in Figure 5 now in a simplified picture. The carbon enriched shell is secularly unstable and carbon fingers will move downwards.

3.2. CIRCULATION AND SHELL BURNING

In a non-rotating star pressure and density are constant on equipotentials, i.e. they are constant on spheres and from the equation of state, say, in the case of the perfect gas one can conclude that T/μ is also constant on spheres.

Let us now consider nuclear shell burning. Above the shell is the fuel, below are the ashes of higher molecular weight. The transition from fuel to ashes is continuous and therefore there is a transition region where molecular weight increases inwards. We now consider a slow perturbation of the type as it is indicated in Figure 7. We discuss the conditions on a sphere S which is right in the shell. Let A be an area where material is moving outwards through this sphere. There, material enriched in ashes, with a higher molecular weight, passes through S, while in the region B where matter is moving inwards, material enriched in fuel, with a lower molecular weight, penetrates the sphere. Now since T/μ is constant on S, the temperature in the area A is higher while in area B the temperature is lower. The effect of chemical composition diminishes the nuclear reactions in area A and increases the nuclear reaction rate in area B, but the variation of temperature does exactly the opposite. The variation of chemical composition tries to damp the circulation while the effect of temperature tries to enhance the motion. In order to find out whether or not shell burning is stable against this type of perturbation, one has to do a more detailed stability analysis. The first attempt in this direction has been done by Mestel (1957), later by Kippenhahn (1967), and then by Rosenbluth and Bahcall (1973), and by Richstone (1973). As one would already expect from our heuristic consideration, the higher the temperature dependence of nuclear reactions, the more likely the shell is to be unstable. This indeed comes from the detailed study of the stability problem, but one also has to take into account the effects of radiative transfer.

Some stability criteria have been given by Kippenhahn (1967). Biermann (1968) investigated the stability of shell burning in Red Giants, with the result that the shells seem to be stable. Recently Richstone (1973) has studied stellar models near and during the onset of helium flash, and he found that all models investigated which were spherically stable were also stable with respect to non-spherical perturbations. Some flash models were unstable but the e-folding time of the non-spherical instability was longer than the e-folding time of the spherical instability. It is therefore doubtful whether this type of circulation can have any influence on the evolution of the stars.

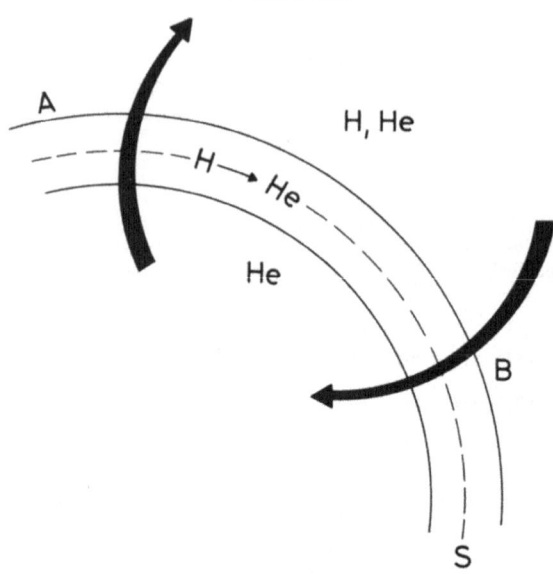

Fig. 7. In a spherically symmetric star with shell burning a non-spherical perturbation (black arrows) brings ashes into the shell in region *A* and at the same time *T* increases while in region *B* new fuel is brought into the shell while the temperature decreases.

3.3. PLUME MIXING

Recently Scalo and Ulrich (1973) have postulated a new mixing process which might

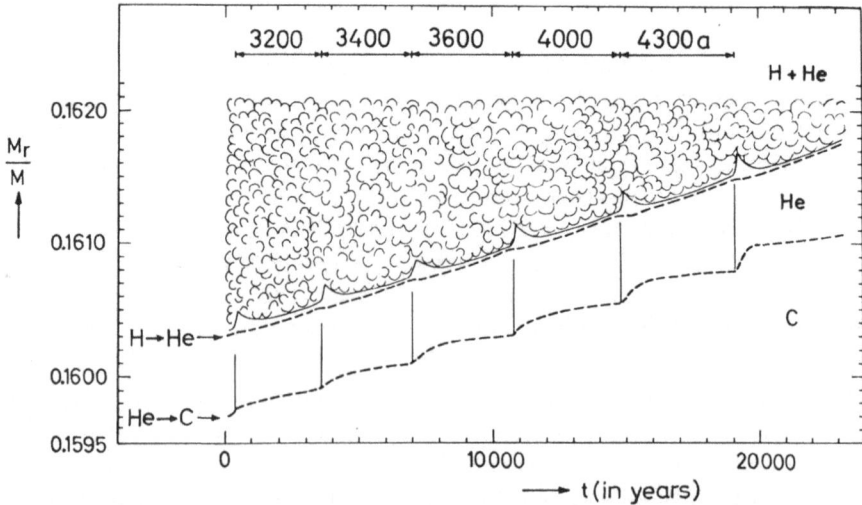

Fig. 8. Thermal pulses in a star of 5 $M_\odot$ (Weigert, 1966). The cloudy region gives the outer convective zone. The broken lines indicate the two shell sources eating outward. In intervals of about 4000 years thermal pulses occur forming short-living convective zones (vertical lines above the helium burning shell) which approach the hydrogen burning shell. The Scalo-Ulrich plume mixing is supposed to occur as soon as the top of such a convective zone penetrates into the hydrogen burning shell.

occur in an evolved star if carbon enriched material is brought up into a hydrogen rich outer region. This might be expected after many cycles of thermal pulses (Weigert, 1966) as indicated in Figure 8. During such a pulse a carbon enriched mixture is brought into the envelope by the convective zone which forms just above the helium burning shell during the pulse.

Due to the irregular convective pattern the material at the bottom of the envelope is heated in a non-spherically symmetric way, causing meridional circulation in the envelope. This motion would then bring hydrogen into the convective region where it would be mixed with the carbon enriched material (Figure 9). Since the CNO-cycle is controlled by the C^{12} content the energy production due to the CNO-cycle would be strongly enhanced. Under happy circumstances this extra energy could be used to enhance the circulation. Scalo and Ulrich give a non-linear model for this type of motion and discuss the consequences of this type of mixing on chemical abundances. The results look very promising.

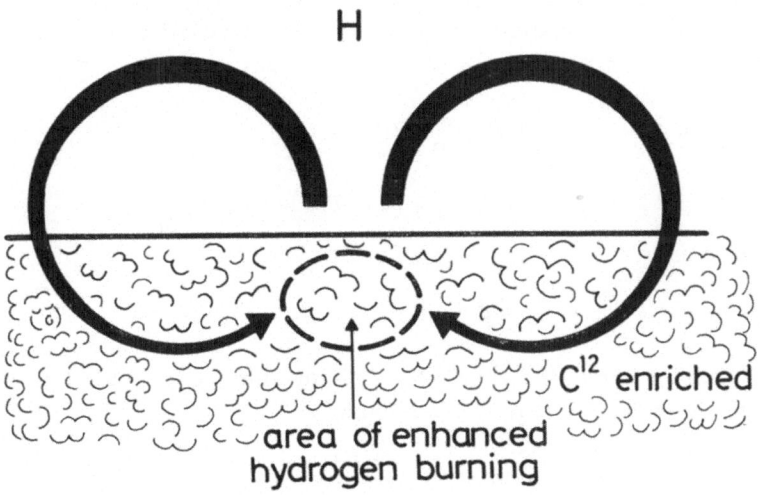

Fig. 9.   Scalo-Ulrich plume mixing. Enhanced CNO burning in the convective, C^{12} enriched zone drives circulation.

One would like to see this model of plume mixing worked out in detail. Maybe one should try to learn more of the driving mechanism by a stability analysis of a purely spherically symmetric model. In such a model non-spherical perturbations similar to those discussed in the preceding section should be applied. But the difference between this and the former case would be that mixing occurs at a certain depth. Such an approach could possibly help to find whether the excess energy released by the CNO-cycle would really be fed into the circulation, driving the motion rather than damping it, which in principle is also possible.

On the other hand model calculations beyond the phases covered by Weigert are necessary in order to find whether the convective zone can really penetrate into the hydrogen burning shell.

Acknowledgement

Part of this work was financed by the Deutsche Forschungsgemeinschaft.

Appendix

A1. ESTIMATE FOR v_{EV}

In a system of coordinates in which the matter of the star is at rest the first law of thermodynamics gives

$$\mathbf{V} \cdot \mathbf{F} = \varrho \varepsilon - c_P \varrho \, \frac{dT}{dt} + \delta \, \frac{dP}{dt}. \tag{A1.1}$$

Here, $\mathbf{F}$ is the radiative flux and ε the nuclear energy generation, while all the other symbols either have their usual meaning or have been explained in the foregoing text. If one now makes the transition to a system of coordinates in which the stellar matter moves with the velocity $\mathbf{v}$ by using the relations

$$\frac{d}{dt} = \frac{\partial}{\partial t} + \mathbf{v} \cdot \mathbf{V}, \quad \mathbf{V} T = \frac{T}{P} \mathbf{V} \mathbf{V} P$$

$$\nabla = d \ln T / d \ln P, \quad \nabla_{\text{ad}} = (d \ln T / d \ln P)_{\text{ad}} = \frac{P \delta}{c_P \varrho T} \tag{A1.2}$$

one obtains

$$\mathbf{V} \cdot \mathbf{F} = \delta \, \frac{\nabla_{\text{ad}} - \nabla}{\nabla_{\text{ad}}} \, \mathbf{v} \cdot \mathbf{V} P + \varepsilon \varrho. \tag{A1.3}$$

In a slowly rotating star $\mathbf{v} \cdot \mathbf{V} P \simeq v_r \, dP/dr \approx - v_r g \varrho$ and if we restrict ourselves to regions where there are no nuclear reactions we obtain

$$\mathbf{V} \cdot \mathbf{F} = - \delta \, \frac{\nabla_{\text{ad}} - \nabla}{\nabla_{\text{ad}}} \, g \varrho v_r. \tag{A1.4}$$

That in rotating stars there is normally no solution with $v_r = 0$ comes from the fact that $\mathbf{V} \cdot \mathbf{F}$ does not vanish.

As an estimate which is good for slowly rotating stars one can say (Baker and Kippenhahn, 1959, 144–145)

$$\mathbf{V} \cdot \mathbf{F} \approx \frac{3 L_r}{4 \pi r^3} \chi_r \tag{A1.5}$$

and from Equation (A1.4) one obtains

$$v_{EV} = |v_r| \approx \frac{\nabla_{\text{ad}}}{\delta \left(\nabla_{\text{ad}} - \nabla \right)} \frac{L_r \chi_r}{M_r} \frac{3}{4 \pi r G \varrho}. \tag{A1.6}$$

A2. μ-MOTION

If a blob of material in a star is carried with the velocity v either upwards ($v>0$) or downwards ($v<0$), its temperature is determined by the equation

$$\frac{\partial \theta}{\partial t} = (\nabla - \nabla_{ad}) \frac{v}{H_P} - \frac{1}{\tau_{KH}} \theta \tag{A2.1}$$

(Kippenhahn, 1969) where $\theta = DT/T$ and where the time τ_{KH}^* of thermal adjustment is given by Equation (7). If the velocity does not change very much during the time τ_{KH}^* then a quasi-steady state is achieved:

$$\theta = - (\nabla_{ad} - \nabla) \frac{v}{H_P} \tau_{KH}^*. \tag{A2.2}$$

For the case of different molecular weights inside and outside the blob, we obtain from $DP=0$, $D\varrho=0$

$$-\delta \frac{DT}{T} + \varphi \frac{D\mu}{\mu} = 0, \tag{A2.3}$$

where $\delta = -(\partial \ln \varrho / \partial \ln T)_{P,\mu}$, $\varphi = (\partial \ln \varrho / \partial \ln \mu)_{P,T}$ (δ, $\varphi = 1$ for an ideal gas). Consequently, we have

$$\theta = \frac{DT}{T} = \frac{\varphi}{\delta} \frac{D\mu}{\mu}. \tag{A2.4}$$

Combining Equation (A2.2) and (A2.4) yields

$$v_\mu = |v| = \frac{\varphi H_P}{\delta (\nabla_{ad} - \nabla) \tau_{KH}^*} \frac{|D\mu|}{\mu}. \tag{A2.5}$$

A3. MAGNETIC BUOYANCY

In order to derive the formula for magnetic buoyancy we again start with Equation (A2.2). We consider the size of the blob to be small compared to H_P. We may therefore assume the gas pressure P_{Gi} in the blob and P_{Ge} in the surroundings as being constant over the blob's diameter. P_M is a mean value of the magnetic pressure in the blob. We then define a dimensionless quantity ζ by

$$P_{Gi} = (1 - \zeta) P, \qquad P_M = \zeta P, \tag{A3.1}$$

where $P = P_{Gi} + P_M = P_{Ge}$ is the total pressure. We assume ζ to be small compared to unity.

$$DP_G = P_{Gi} - P_{Ge} = - P_M = - \zeta P \tag{A3.2}$$

and from the equation of state

$$\frac{D\varrho}{\varrho} = \alpha \frac{DP_G}{P_G} - \delta \frac{DT}{T} \tag{A3.3}$$

with $\alpha = (\partial \ln \varrho / \partial \ln P)_T$ ($\alpha = 1$ for an ideal gas). Since in hydrostatic equilibrium $D\varrho = 0$ we conclude from Equation (A3.2), (A3.4)

$$\theta = \frac{DT}{T} = -\frac{\alpha}{\delta} \frac{P}{P_G} \zeta = -\frac{\alpha}{\delta} \frac{P_M}{P_G} = \frac{\alpha}{\delta} \frac{DP_G}{P_G}. \tag{A3.4}$$

And if we combine Equation (A2.2) and (A3.4) we find

$$v_M = \frac{\alpha H_P}{\delta (\nabla_{ad} - \nabla) \tau_{KH}^*} \frac{P_M}{P_G}. \tag{A3.5}$$

A4. Size of the Fastest Elements

We consider a spherical blob of diameter d with $D\mu > 0$. If we assume pressure equilibrium ($DP = 0$) and thermal adjustment ($DT = 0$) then $D\mu/\mu = D\varrho/\varrho \neq 0$ and a buoyancy force

$$K = gVD\varrho = \frac{32\pi}{3} g\varrho d^3 \frac{D\mu}{\mu} \tag{A4.1}$$

acts on the blob where V is its volume. It would then move with the Stokes velocity v_{St} defined by

$$3\pi f \eta v_{St} d = K, \tag{A4.2}$$

where f is a dimensionless factor of order 1, while η is the viscosity. Thus, we obtain

$$v_{St} = \frac{32}{9f} \frac{gd^2}{v} \frac{D\mu}{\mu}, \tag{A4.3}$$

where v is the kinematic viscosity $v = \eta/\varrho$. This was *without thermal adjustment*. With thermal adjustment *without friction* it would move with

$$v_\mu = \frac{\varphi H_P}{\delta (\nabla_{ad} - \nabla) \tau_{KH}^*} \frac{|D\mu|}{\mu}. \tag{A4.4}$$

If we now express v in units of the kinematic *radiative* viscosity v_R

$$v_R = \frac{2}{15} \frac{aT^4}{c\varkappa\varrho^2} \tag{A4.5}$$

and if we put $v_{St} = v_\mu$, we obtain for the diameter of the blobs where friction becomes important

$$d \approx \left(\frac{45\nabla_{ad}\varphi f}{8\delta^2 (\nabla_{ad} - \nabla)} \frac{v}{v_R} \right)^{1/4} \sqrt{\frac{v_R c}{g_r}}. \tag{A4.6}$$

Ignoring the factor in front of the square root we obtain the estimate (12).

A5. A Magnetic Blob Lifting Heavier Material

We assume a rising magnetic blob with $D\mu > 0$. In this case

$$\frac{P_M}{P_{Gi}} > \frac{D\mu}{\mu}.$$ (A5.1)

As the element rises along a certain path P_M/P_{Gi} changes to $(P_M/P_{Gi}) + d(P_M/P_{Gi})$ while $D\mu/\mu$ remains constant, if the element is moving through a chemically homogeneous region. Thus, if $d(P_M/P_{Gi}) > 0$ then the effective buoyancy will become larger during the upward motion, while for $d(P_M/P_{Gi}) < 0$ the upward motion will slow down. The variation of P_M/P_{Gi} during the upward motion is given by

$$d\left(\frac{P_M}{P_{Gi}}\right) = \frac{dP_M}{P_{Gi}} - \frac{P_M}{P_{Gi}}\frac{dP_{Gi}}{P_{Gi}}.$$ (A5.2)

During the upward motion the density in the blob (and in the surroundings) changes according to

$$\frac{d\varrho}{\varrho} = \alpha\frac{dP_{Gi}}{P_{Gi}} - \delta\frac{dT_i}{T_i} = (\alpha - \nabla\delta)\frac{dP_{Gi}}{P_{Gi}},$$ (A5.3)

where $\nabla = d\ln T/d\ln P$ gives the ratio of the relative variations of temperature to pressure in the surroundings. While the density is changing during the motion, the element expands and the frozen in magnetic field changes. We define a dimensionless quantity ψ by

$$\frac{dP_M}{P_M} = \psi\frac{d\varrho}{\varrho}$$ (A5.4)

and we obtain

$$d\left(\frac{P_M}{P_{Gi}}\right) = [\psi(\alpha - \nabla\delta) - 1]\frac{dP_{Gi}}{P_{Gi}}\cdot\frac{P_M}{P_G}.$$ (A5.5)

During an upward motion $dP_{Gi} < 0$ and therefore

$$\psi(\alpha - \nabla\delta) \leqslant 1$$ (A5.6)

is the condition that the rising element will not slow down. For an ideal gas $\alpha = \delta = 1$, and we obtain from Equation (A5.6):

$$\psi \leqslant \frac{1}{1 - \nabla} \approx \tfrac{4}{3},$$ (A5.7)

where on the right we have taken $\nabla = 0.25$, a value typical for radiative regions in stars.

The simplest case of a magnetic tangle would be a cylinder of radius R with a magnetic field parallel to its axis. As it rises $\varrho \sim 1/R^2$, $B \sim 1/R^2$ and $P_M \sim 1/R^4$ and therefore $P_M \sim \varrho^2$ or $\psi = 2$ from which we can conclude that the upward motion will slow down. For a spherical tangle the magnetic pressure falls off with falling density even faster, ψ is therefore even larger than 2 and the rising tangle is decelerated even faster.

References

Aller, L. H.: 1961, *The Abundance of the Elements*, Interscience Publ. Inc., New York.
Baker, N. and Kippenhahn, R.: 1959, *Z. Astrophys.* **48**, 140.
Biermann, P.: 1968, Diplomarbeit.
Demarque, P. and McClure, R. D.: 1973, *Monthly Notices Roy. Astron. Soc.* **164**, 5.
Eddington, A. S.: 1925, *Observatory* **48**, 73.
Finzi, A. and Wolf, R. A.: 1968, *Astrophys. J.* **153**, 865.
Fricke, K. J.: 1968, *Z. Astrophys.* **68**, 317.
Fricke, K. J. and Kippenhahn, R.: 1972, *Ann. Rev. Astron. Astrophys.* **10**, 45.
Goldreich, P. and Schubert, G.: 1967, *Astrophys. J.* **150**, 571.
Gurm, H. S. and Wentzel, D. G.: 1967, *Astrophys. J.* **149**, 139.
Kippenhahn, R.: 1963, in L. Gratton (ed.), *Star Evolution*, Acad. Press. New York, p. 330.
Kippenhahn, R.: 1967, *Z. Astrophys.* **67**, 271.
Kippenhahn, R.: 1969, *Astron. Astrophys*, **2**, 309.
Kippenhahn, R. and Möllenhoff, C.: 1974, *Astrophys. Space Sci.*, in press.
Köhler, H.: 1973, *Astron. Astrophys.* **25**, 467.
Krause, F. and Rädler, K. H.: 1971, in *Ergebnisse der Plasmaphysik und der Gaselektronik*, Band II, R. Rompe and M. Steenbeck, Berlin, p. 3.
Mestel, L.: 1953, *Monthly Notices Roy. Astron. Soc.* **113**, 716.
Mestel, L.: 1957, *Astrophys. J.* **126**, 550.
Paczyński, B.: 1973, preprint.
Parker, E. N.: 1970, *Ann. Rev. Astron. Astrophys.* **8**, 1.
Richstone, D. O.: 1973, preprint.
Rosenbluth, M. N. and Bahcall, J. N.: 1973, *Astrophys. J.* **184**, 9.
Sackmann, I.-J. and Weidemann, V.: 1972, *Astrophys. J.* **178**, 427.
Scalo, J. M. and Ulrich, R. K.: 1973, *Astrophys. J.* **183**, 151.
Sweet, P. A., 1950, *Monthly Notices Roy. Astron. Soc.* **110**, 548.
Thomas, H.-C.: 1967, *Z. Astrophys.* **67**, 420.
Vogt, H.: 1925, *Astron. Nachr.* **223**, 229.
Weigert, A.: 1966, *Z. Astrophys.* **64**, 395.

ON THE MIXING OF THE MATTER IN SEMICONVECTIVE
REGIONS OF MASSIVE STARS

(Abstract)

A. V. TUTUKOV

The Astronomical Council of the Academy of Science, U.S.S.R.

There are two possibilities concerning the problem of convective stability of the region of varying molecular weight μ. Schwarzschild and Härm (1958) assumed that in semiconvective zones of massive stars the radiative temperature gradient is equal to the adiabatic one:

$$\nabla_r = \nabla_a.$$

Sakashita and Hayaski (1961) suggested the use of the Ledoux criterion for the convective stability:

$$\nabla_r < \nabla_a + \nabla_\mu, \tag{2}$$

where: $\nabla_\mu = (\beta/4 - 3\beta)\,(\mathrm{d}\ln\mu/\mathrm{d}\ln P)$.
Kato (1966) showed that the zone where the conditions

$$\nabla_a < \nabla_r < \nabla_a + \nabla_\mu \tag{3}$$

is satisfied is vibrationally unstable. This instability leads to the partial mixing of matter.

Dudorov and Tutukov (1972) investigated the mixing problem solving the linearized equation of motion of an element in a medium with varying molecular weight and with a radiative temperature gradient ∇_r satisfying conditions (3). The roots of the characteristic equation and rates of growth of perturbations of different size elements were found analytically for $\nabla_r - \nabla_a \ll \nabla_\mu$.

It appeared, that elements with the size $\sim 10^7$ cm have the largest increment of perturbation growth under conditions typical for semiconvective zones of massive stars. The amplitude of vibration is limited by the turbulent friction. The turbulence starts when the amplitude becomes $\sim 10^6$ cm or about 10^{-5} of the radial dimension of the semiconvective zone. The turbulence leads to a partial mixing of the matter. The necessary mixing velocity is secured during the hydrogen-burning stage if $\nabla_r - \nabla_a$ is about 10^{-5}.

References

Dudorov, A. and Tutukov, A.: 1972, *Nauch. Inform. Astron. Council Acad. Sci. USSR* **21**, 3.
Kato, S.: 1966, *Publ. Astron. Soc. Japan* **18**, 374.
Sakashita, S. and Hayashi, C.: 1961, *Prog. Theor. Phys.* **26**, 942.
Schwarzschild, M. and Härm, R.: 1958, *Astrophys. J.* **128**, 348.

DISCUSSION FOLLOWING PAPER BY KIPPENHAHN

Schwarzschild to Kippenhahn: May we ask Dr Kippenhahn for his personal estimate at this time as to whether 'plume' mixing will occur and be effective if in the evolution of a star an interface occurs between a carbon rich misture at the inside and a hydrogen rich mixture at the outside?

Kippenhahn: The position is not entirely clear. It will be necessary to take account of the mixing which occurs during the development of the instability.

What experience have you had of such interfaces?

Schwarzschild: In our calculations, mild hydrogen contact with carbon was reached in low mass stars.

Sugimoto: In the computation of the flickering helium shell burning by Prof. Schwarzschild and Härm, the effect of radiation pressure was not included. Since the entropy of radiation is rather large, the convection in the unstable helium shell might not reach the bottom of the hydrogen distribution. What do you think of this point?

Schwarzschild: Dr Sugimoto remembers correctly: our early model sequences for helium shell flash cycles were computed with neglect of radiation pressure. It is therefore not impossible that our results regarding the penetration of the convective tongue, caused by a helium shell flash, into the region containing hydrogen, may be affected negatively by the proper inclusion of radiation pressure.

Sugimoto: I think that there is certainly a need of computations which take into account the entropy of radiation.

Schwarzschild to Kippenhahn: Did your early calculations of $5 M_\odot$ evolution include radiation pressure?

Kippenhahn: Yes.

Tayler (ed.), Late Stages of Stellar Evolution, 42. *All Rights Reserved.*
Copyright © *1974 by the IAU.*

NON-EXPLOSIVE MASS LOSS

N. J. WOOLF

University of Minnesota, Minn., U.S.A.

Abstract. The various mechanisms whereby stars might lose matter are considered, together with the observational evidence of the mass loss rates associated with these mechanisms. The results are shown in a diagram giving the fractional mass that a star loses by various mechanisms as a function of initial mass.

1. Introduction

Since the last major conference to discuss mass-loss (Hack, 1969), there has been an abrupt change in our appreciation of the role that red giant stars play in stellar evolution. Mass ejection has become clearly visible through infrared observations of recently condensed circumstellar dust (Gillett *et al.*, 1968; Gehrz and Woolf, 1971), and through radio observation of molecules in these circumstellar clouds (Wilson and Barrett, 1972, Wilson *et al.*, 1972, 1973). And indeed it has become clear to the observers that a substantial fraction of the matter contained in stars is returned to the interstellar medium by this process.

In this connection it is interesting that the interstellar gas seems very depleted in elements that would form solid condensates above about 1000K, but not those that would condense at cooler temperatures. The Na/Ca problem of interstellar abundances that is a part of this has long been known. Unless most matter that returns to the interstellar medium can make solid condensates under circumstellar conditions, the interstellar gas abundances would not arise. Now moderately hot flows like those in planetary nebulae are known to be able to condense dust that did not arise in a previous red giant phase (Gillett *et al.*, 1973). But the situation may well be different for supernova ejecta. If so, these interstellar matter observations would be stating that the mass flow from supernova explosions is only a small part of the total return of mass to the interstellar medium. And it justifies us concentrating here on non-explosive mass loss.

This paper is divided into two main sections, each with sub-divisions. In the first we consider processes which might cause matter to be ejected, and decide which should be considered potentially important for stellar evolution. In the second section we consider the evolution of stars of various mass ranges, and attempt to infer when matter leaves them, and how much of the mass goes.

2. Mass Ejection Mechanisms

Some stars make a great display while they trickle off a little matter back to interstellar space. Others without much ado seem to return most of what originally formed them. Here we are attempting to find which phases have a mass ejection rate that significantly affects stellar evolution. It is also an attempt to find out which mechanisms operate the ejection process, so that their fuller implication for all areas of stellar evolution might be considered.

Tayler (ed.), Late Stages of Stellar Evolution, 43–53. All Rights Reserved.

Mass ejection significantly affects stellar evolution when the rate at which highest energy content fuel is depleted in the interior of a star is slower than the rate at which it is removed from the exterior. Thus if the exterior is hydrogen, ejection rates exceeding 1.4×10^{-19} g erg^{-1} radiated are significant. When the exterior is helium, ejection rates ten times larger are needed to be significant.

These rates should be considered in terms of the problem of a star reducing its mass enough to become a white dwarf, as appears to have happened to at least one star of the Pleiades (Luyten and Herbig, 1960). A 6 $M_\odot$ star crosses the Hertzsprung gap with perhaps 20% of its original mass in a burned core. To avoid acquiring more than 1.4 $M_\odot$ in its core, it will have to shed $\sim$4.6 $M_\odot$, while only burning $\sim$0.2 $M_\odot$. This must be a very efficient process.

2.1. WAYS FOR A STAR TO REDUCE

The possible means for a star to reduce its mass are:
   (i) Use of radiation pressure to drive off upper atmospheric layers,
   (ii) Feeding momentum into the outer layers through travelling waves,
   (iii) Feeding energy into outer layers via turbulent or convective motions,
   (iv) Release of energy that comes from the star being in a metastable state,
   (v) Rotational shedding of matter at the equator by the star reducing its moment of inertia,
   (vi) And giving the mass to a companion star.

Method (vi) is not helpful, since it only delays the day of reckoning. However it may be that binary stars can use their angular momentum to efficiently 'spill' mass while it is being transferred and this should be explored.

Method (v) is believed to operate for Be stars, and has been suggested (Limber, 1964) to operate for WR stars. A discussion by Auer and Woolf (1965) shows that rotational shedding will not allow a star to lose a lot of mass. Stellar model predictions would have a Be star losing perhaps 0.5% of its mass, while observations may suggest ten times this amount. Neither amount is very significant.

Explosive mass ejection, (iv), does occur, and it has also been suggested that the hydrogen ionization and/or molecule formation energy of a red giant might lead to fast or slow dispersal of the outer layers. We shall consider mainly the observational evidence for slow mass loss, and the alternative schemes for causing it. We shall also consider the observed mass loss in planetary nebulae, which may be caused this way, but not supernova or nova explosions.

Thus we are left to consider, (i) radiation pressure driven winds, which have been discussed for red giants by Weymann (1962), Hoyle and Wickramasinghe (1962), and Gehrz and Woolf (1971).

Then there are travelling waves (ii) which should only be important for pulsating stars, and can so be observationally distinguished.* And there is (iii) mass loss via macroscopic motions, a process best seen in the solar corona.

---

* It is possible that on a small scale, travelling waves may propagate 'microtrubulence', but we consider this under (iii).

2.2. EVIDENCE FOR MASS LOSS

The evidence that mass loss does occur can be briefly summarized.

(1) White dwarfs are found at relics of stars at least 2.5 $M_\odot$, and perhaps as much as 6 $M_\odot$

(2) Stellar spectra show that envelopes are receding from some stars.

(3) Stars are found with extended optical, infrared or radio emitting envelopes surrounding them.

(4) There are some weak dynamical and stellar model indications that some more evolved stars now have less mass than less evolved stars in the same clusters (Caputo *et al.*, 1973).

(5) The chemical composition of some stars seems to require that both mass loss and mixing have occurred. However this is always very difficult to distinguish from cases where mixing alone has happened to a highly evolved star.

Many of these phenomena are discussed in an article by Weymann (1963) and the book edited by Hack. Compositional changes are intriguing but tend to be inconclusive. Some rare low mass stars seem to develop almost pure helium envelopes (Hill, 1965). Nitrogen to carbon ratios seem highly variable, even in otherwise 'normal' blue stars (Walborn, 1971). And the red carbon stars seem to either be caused by a combination of normal mixing and mass loss (Thompson, 1972) or mixing via jets or plumes (Ulrich and Scalo, 1972) or both.

Some attempts have been made to total the fuel consumed by stars of various masses (see e.g. Hills and Dale, 1973). But for low mass stars, the fuel consumption is dominated by a short period of time spent as a high luminosity red giant, and this is hard to estimate observationally. One's best chance to pin down the lifetime of crucial phases are to use in addition to cluster membership, field c.p.m. pairs, total numbers of field stars of these types, and data on motions.

2.3. RADIATIVE MASS LOSS

If a star radiates a continuum, atoms in the outer atmosphere are in a reduced radiation environment because of the spectral lines absorbed by closer in atoms of the same kind, and even if the radiation field were present, the reduced excitation would leave only resonance lines potentially able to absorb momentum. If however the outer layers are disturbed, then the atoms may experience the full force exerted by the continuum on their resonance lines, in the Doppler line width from their disturbed motion. The momentum absorbed is at most $(L/c)\cdot(v/c)$, and it partially becomes momentum $v\,dm/dt$ for outflowing matter with a mass loss rate dm/dt. Thus even if there are N resonance lines, the maximum mass loss rate is given by:

$$\frac{dM}{dt}_{(line)} < \frac{NL}{c^2}. \tag{1}$$

If however a continuum can be used to absorb or scatter the radiation, this rate can be increased by a factor c/Nv, giving a rate

$$\frac{dM}{dt}_{(continuum)} < \frac{L}{cv},$$

(2)

where v is the outflow velocity.

But now the radiation field may be able to couple momentum across the envelope, and (2) is no longer a true limit. But in practice, considering plausible models one infers that (2) is unlikely to be in error by more that a factor ~ 10.

The line radiation process has been assumed to the responsible for mass ejection in early type supergiants (Morton, 1969). For such stars, even if $N=10$, one finds a rate of only 10^{-20} g erg^{-1} which does not affect stellar evolution.

For late type stars, envelopes have long been known (e.g. Deutsch, 1956), and for these stars solid particles that form in the upper atmosphere can create continuous absorption and/or scattering. The envelopes are seen to expand at typical velocities of 15 km s^{-1}, giving a possible mass loss rate from (2) of $\sim 2 \times 10^{-17}$ g erg^{-1}. If a star is in a suitable state to create enough solids for a long enough time, such a rate can dominate the future course of evolution.

2.4. TRAVELLING WAVES

Since Schwarzschild's (1938) study of δ Cephei, it has seemed that oscillating stars might feed an appreciable fraction of their energy of oscillation into travelling waves. This might then possibly lift the outer layers of the star and eject them. Such ejection could be one of the means of damping the oscillations. At present it seems best to explore this possibility observationally. We know that solid particles seem able to form remarkably easily, and therefore one of the best tracers of mass ejection is infrared emission.

In 1970 Gehrz and Strecker found that AC Her, an RV Tauri star had a huge excess emission, and so Gehrz and Woolf (1970) observed other pulsating stars to see if this was a common phenomenon. We found that the rare RV Tauri stars do indeed have enormous excess emission, and Gehrz in his thesis estimated mass ejection rates of 5×10^{-6} $M_\odot$ yr^{-1} to 6×10^{-5} $M_\odot$ yr^{-1}. But other pulsating stars showed little or no effect.

To be sure, Mira and SRa variables have envelopes around them, but their infrared radiation from their envelopes seemed to correlate with spectral type rather than with pulsation amplitude. And for this reason, and also the comparison of mass ejection rate to luminosity, we believe the radiation pressure effect dominates there. For these stars the typical ejection rate is about 4×10^{-18} g erg^{-1}, a factor smaller than given by (2).

Small infrared excesses were found for the longest period cepheids, SV Vul and RS Pup ($M_v \sim -6$), and the second of these has the outer parts of its shell visible (Havlen, 1971). From the infrared excess, the inferred current mass ejection rate is $\sim 10^{-6}$ $M_\odot$ yr^{-1}, however the nebula density distribution and density seem to refer to a 10 to 100 times larger mass ejection rate over the past 10^5 to 10^4 yr. Possibly this refers to a preceding red giant phase. The current rate of mass ejection would not be significant over the limited lifetime as a cepheid.

For very high mass stars, perhaps above 60 $M_\odot$, pulsational mass loss may also occur. Here there is a possible reason for the dense infrared envelopes around η Car and CIT + 10420, and perhaps the shell around P Cygni where the mass ejection rate seems to be $\sim 10^{-4} M_\odot$ yr^{-1}. Even for a 100 $M_\odot$ star, living a few times 10^6 yr, such rates are decisive. But the total lifetime of such phases needs investigation. Some of the most massive stars may spend their lives shrouded in dusty shells that they are ejecting (Hoyle *et al.*, 1973) and they may never be seen as the luminous stars that they really are.

For W Virginis stars there is good spectroscopic evidence for believing that matter is ejected, but the infrared observations are not yet good enough to give the rate. These stars are probably too rare to be important as major sites of mass ejection.

2.5. CONVECTIVE MASS EJECTION

This process is well known in the Sun, where convected energy heats up the corona to such a temperature that the particles have escape velocity from their thermal energy. It is possible that the ejection rates in lower main sequence stars might explain their lithium depletion (Weymann and Sears, 1964), but total mass depletion is insignificant.

One might attempt to theoretically estimate rates of mass ejection by this process, as has De Jager (1959) who decided it was not significant. Or we can use some kind of observations to imply a limit to it as did Weymann (1963). And we should expect that the more luminous the star, the greater the mass flow might be. Now return again to the long period cepheids. Stars of similar luminosity and temperature but non variable do not show excess infrared emission. Suppose that they all have coronae due to convection. Now if the coronal mass ejection is larger than the pulsational mass ejection, the pulsational mass ejection will be absorbed into the main mass flow, and there should be no infrared excess. Since we observe the infrared excess it seems likely that the convective mass loss is smaller than $10^{-6} M_\odot$ yr^{-1}, and as such it seems unlikely to be very important.

Again for the most luminous G and K supergiants there seems to be a mass ejection process that causes dust to form (Humphreys *et al.*, 1971). And since it does not obviously refer to a pulsation driven process, one wonders whether it might be this mechanism. However as one examines cool stars of higher and higher luminosity, one finds that there appears to be a continuous sequence linking all these stars ejecting mass. It is clearly seen at earliest types M6III, M5II, M1Iab, G8Ia and G0Ia$^+$. Thus it seems to all be part of a radiation pressure driven process.

One major problem for stellar wind mass loss is that one initially predicts that it would be most effective for red giant stars. But these stars condense dust that acts as a coolant for the outer layers, while simultaneously activating its own mass loss process.

N. J. WOOLF

3. Ejection for Stars of Different Mass Ranges

3.1. STARS BELOW 2 $M_\odot$

For these stars there is clear evidence that red giant mass loss is crucial. We shall be concerned with determining the mass loss rates and the duration of its operation. The stars at the tip of the red giant branches are radiating about 2×10^{37} erg s^{-1}. And using inequality (2), we find a maximum ejection rate could be 6×10^{-6} $M_\odot$ yr^{-1} for an ejection velocity of 15 km s^{-1}. We shall infer somewhat smaller rates from the observations.

Tsuji (1971) has attempted to use optical observations to derive mass ejection rates for two S type Mira stars, R Cyg and R And. He found the rate would be 3×10^{-8} $M_\odot$ yr^{-1} if all the matter were neutral. But this had to be increased by a large but very uncertain factor to allow for ionization in the envelope. His final estimate was a rate between 7×10^{-6} and 1.5×10^{-5} $M_\odot$ yr^{-1}.

R Cyg has been observed from 3–11 μ in the infrared by Gillett *et al.* (1971), and it has infrared excess emission which is entirely typical of other giant stars of similar temperature. An 11 μ peak emission and another one near 18 μ are attributed to emission by silicate particles formed in the envelope, and from this we can obtain a second estimate of mass ejection rates. We assume that all matter that might condense as silicates (1/250 by mass for a normal composition mixture) does in fact condense. The distance where the silicates condense is estimated from effects of optical depth on the silicate emission bands. The amount of matter is estimated from the silicate band strength, and the ejection velocities are taken from optical and radio data. By this means Gehrz and Woolf (1971) estimated ejection rates for typical disc and halo long period variables (Mira and SRa) to be about 1.4×10^{-6} $M_\odot$ yr^{-1}.

Alternative ways of calculating mass ejection rates come from the observation of OH and H$_2$O masers in the envelopes of some stars. Here there may be some dispute about the physical conditions that give rise to masers. None the less, the rates calculated this way also seem to be about 10^{-6} $M_\odot$ yr^{-1}. For all these stars the rates calculated depend upon the surface area of the star. Now a crude theoretical estimate for red giant stars would have that at a given surface temperature, the luminosity and hence surface area is proportional to the mass. Thus since the masses of these stars only vary by perhaps a factor 3, their sizes must be similar, and the use of a single rate to describe all the stars may be appropriate.

The only LPVs for which both lifetimes and mass ejection rates can be estimated are the 150–200 day Miras which are found in some globular clusters. In a study of such stars in the field, T. J. Pepin and I (unpublished) found that these stars also had excess infrared emission, which seemed to come from closer in denser shells. And when we had corrected for a heavy element deficiency of a factor 3, it seemed that the matter ejection rates were similar to those for the more usual longer period Miras and SRa stars. For the rich cluster 47 Tucanae there are three Mira stars, and I estimate that there is about 1 evolved star per 5×10^4 yr of a given phase. Thus these Mira stars may only last about 1.5×10^5 yr, ejecting only about 0.2 $M_\odot$ while they do so. However for such stars there is little mass expected to be available between an initial mass of perhaps 0.9 $M_\odot$, and a final white dwarf of perhaps 0.6 $M_\odot$.

Other estimates for typical Miras suggest that they last far longer. Thus at high galactic latitudes there are about 30 Mira stars for every dense planetary nebula. The expansion lifetime of such a planetary nebula is about 2×10^4 yr, giving the Mira stars about 6×10^5 yr each under the assumption that all the low mass stars go through both phases. The average Mira stars would be ejecting about 0.8 $M_\odot$ each. Regardless of the precise figures it is hard to avoid the conclusion that red giant mass-loss is more important than planetary nebula mass-loss.

Alternatively we can estimate the total mass ejection from stars that populate high latitudes. I have used infrared absolute magnitudes (which should be more constant than visual absolute magnitudes) to estimate the space density and scale height of the LPV's seen at high latitudes. I find a scale height of 600 pc, and a local space density of 8.0×10^{-8} stars pc^{-3}. Then the total matter returned to the galactic plane from these stars is 1.4×10^{-10} $M_\odot$ pc^{-2} yr^{-1}. Deutsch (1969) has estimated that stars of less than 2 $M_\odot$ eject 2×10^{-10} $M_\odot$ pc^{-2} yr^{-1}. The near coincidence of the two rates suggests little need to search for further mass ejection phases for these stars. To our estimate for matter from LPV's, we should perhaps add a further 20% to allow for planetary nebulae, and another 10% to allow for mass ejection while the star is a somewhat warmer red giant than the LPV's.

For Population II stars, because of the reduced heavy element abundance, radiation pressure driven mass loss will be less effective. Indeed it seems that when the heavy element abundance is reduced by more than a factor of about 10, the process should abruptly stop because of the star's gravitational field. This has the consequence that we may find an abrupt change in the evolution of globular cluster stars at about this chemical composition. There have been observational indications of globular cluster star mass loss for M3 (Woolf, 1964) but not M13 (King, 1962). And one wonders whether the above process might not cause the difference. In M3 the coolest star does show TiO bands. It is an SRd variable that may be a Population II extension of the LPV's. Gehrz in his thesis found that stars of this type such as SX Her did indeed seem to be ejecting matter.

Population II mass loss cutting off have important consequences for the evolution of our Galaxy. If mass loss did not occur, then many more stars should have become supernovae, and the heavy element enrichment of the Galaxy could have been speeded up (Gilman, private communication).

3.2. STARS OF 2 $M_\odot$ TO 6 $M_\odot$

If stars in this mass range do become white dwarfs, they must lose between 50% and 80% of their initial mass. Questions arise (a) can we show that mass loss has occurred on this vast scale and (b) can we find in which phases it occurred?

Let us start by adopting a common current assumption, that the nuclei of planetary nebulae are hot white dwarf stars with a thin fringe of non degenerate matter around them. Now planetary nebulae are known with A or B type companions. For NGC 3132

N. J. WOOLF

the companion appears to be an A3V star (Brown *et al.*, 1970), and the combined planetary nebula and carbon star UV Aur also has a B8.5 companion. These cases seem to be different from NGC 1514 which seems to have a horizontal branch star companion (Greenstein, 1973). Then these planetary nebulae imply that main sequence late B stars have been able to form white dwarfs, and the white dwarf suspected Pleiad is also consistent with this. Thus it seems that mass loss *has* occurred on this vast scale.

Now not only UV Aur, but other carbon stars too have known A or B type companions, so that some of these must also have been formed in this mass range. Indeed one carbon star appears to have a B5 companion. Also the distribution of long period S stars would be consistent with many of them too coming from these stars. In the CIT 2μ survey there are many 'infrared stars', very cool stars that are mainly M stars, but some are C stars. These have a flattened distribution. And in the original attempt to identify the source of radiation, a number of these were misidentified as nearby A and B stars. This suggests that some of the A and B stars may well be physical companions. Thus there seem to be red giants that might be sites of mass ejection. Unfortunately there seems no way of estimating whether there are enough red giants. A check of cool stars for membership of galactic clusters in the right mass range could give an answer.

The more massive stars of this group form short period cepheid variables, but these do not seem to be ejecting much matter. Equally, although most of these stars rotate rapidly, it seems unlikely that rotational mass ejection is important. We are left with the conclusion that mass ejection is important for these stars, but without further information we cannot partition the responsibility between planetary nebulae and red giants.

3.3. Stars of 6 $M_\odot$ to 20 $M_\odot$

A typical group of stars of this kind might be the Per I association. There are here 22 M supergiants of class Iab, but there are no known S, C or infrared stars that might belong to this association. (Though again we must remember W CMa and its B5 companion.) The estimated initial mass is about 15 $M_\odot$. And for such stars the lifetime as a red giant is 5 or possibly 15×10^5 yr (Barbaro *et al.*, 1971), depending on the importance of neutrino processes.

Now the mass ejection rate of the Per I stars has only been observed for a few stars in that association, but for the other stars, comparable objects have been observed elsewhere, and we can crudely estimate the total mass ejection. About 10% of the time as a red giant is spent with a rate $\sim 3 \times 10^{-5}$ $M_\odot$ yr^{-1}, about 30% is spent with a rate $\sim 5 \times 10^{-6}$ $M_\odot$ yr^{-1}, and the remaining 60% is with a rate $\sim 10^{-6}$ $M_\odot$ yr^{-1}. The total ejection is in the range 2.5 to 7.5 $M_\odot$, with a preference for the smaller figure.

Stars in this mass range will have a small rotational mass loss near the main sequence and possibly a small loss as a cepheid, but there are no other phases at which substantial mass loss is expected. Of course there could be phases with mass ejection rates $\sim 10^{-4} M_\odot \text{ yr}^{-1}$, which last too short a time to find even one star, but even then it seems likely that we will have to conclude that stars in this mass range do not manage to shed more than perhaps 50% of their initial mass before some catastrophic phenomenon removes them from the HR diagram. The coming into importance of neutrino energy loss process in this mass range could make a crucial difference: neutrino pressure cannot sustain mass loss at low densities, whereas radiation pressure can. The mass ejection from S Persei is about $4 \times 10^{-18} \text{ g erg}^{-1}$ radiated, but if the neutrino luminosity is several times the radiative luminosity, the mass loss rate may not be able to affect the evolution.

3.4. STARS ABOVE 20 $M_\odot$

As stars become more massive, and radiation pressure dominates their structure, it becomes possible to lose mass before the interior becomes hot and neutrino processes take over. In this mass range we find Of and WR stars, there are O–B supergiants that are probably using line radiation pressure to eject matter (Morton, 1969), at higher luminosities there are stars like P Cygni, there are G and K supergiants with dusty envelopes, and there are some class Ia M stars, all busy ejecting matter.

But in this rather wide range of masses there is no easy way to discriminate between different initial mass stars. Luminosity is nearly proportional to mass, and evolutionary time scales are almost independent of it. With observed mass loss rates it would be surprising if the most massive stars of this group do not eject most of their original mass before becoming a supernova. For the least massive stars of this group, the conclusions for the 6–20 $M_\odot$ stars probably apply. It does not seem possible to estimate the mass where a transition occurs.

4. Conclusions

The mass loss processes that seem to affect stellar evolution are, radiation pressure driven, pulsation driven and single events caused by metastable states. In practice these seem to come down to radiation pressure driven mass loss for red giants, pulsation driven mass loss for the most massive and luminous stars and formation of planetary nebulae or supernovae. Other processes may be important for rare stars.

The oldest lowest mass stars lose relatively little. and indeed for stars with less than 10% of the solar heavy element abundance, the red giant mass loss process fails. For stars with more nearly Population I composition, there is a mass loss of $\sim 0.2 M_\odot$ in the red giant phase at a rate $\sim 1.4 \times 10^{-6} M_\odot \text{ yr}^{-1}$.

Somewhat more massive stars up to 2 $M_\odot$ appear to have increasing red giant mass ejection, with the phase lasting longer as more matter needs to be ejected. For these stars the red giant phase ejects typically 5 × the matter lost as a planetary nebula. Up to 6 $M_\odot$, it appears that stars still succeed in becoming white dwarfs, but the appropriate red giant statistics do not yet exist to demonstrate whether in this mass range the red giants or planetary nebulae are the most important sites of mass loss.

For stars of 6–20 $M_\odot$, it appears that red giant loss, the dominant process of ejection fails to reduce the initial mass by more than ~50%. And the star probably becomes a supernova while still having a substantial mass outside the core. For somewhat higher masses the star seems to be losing matter at a high rate most of the time. For stars above some high mass it seems likely that most of the matter is returned to the interstellar medium before the core implodes.

In Figure 1 there has been an attempt to schematically indicate the mass ejection

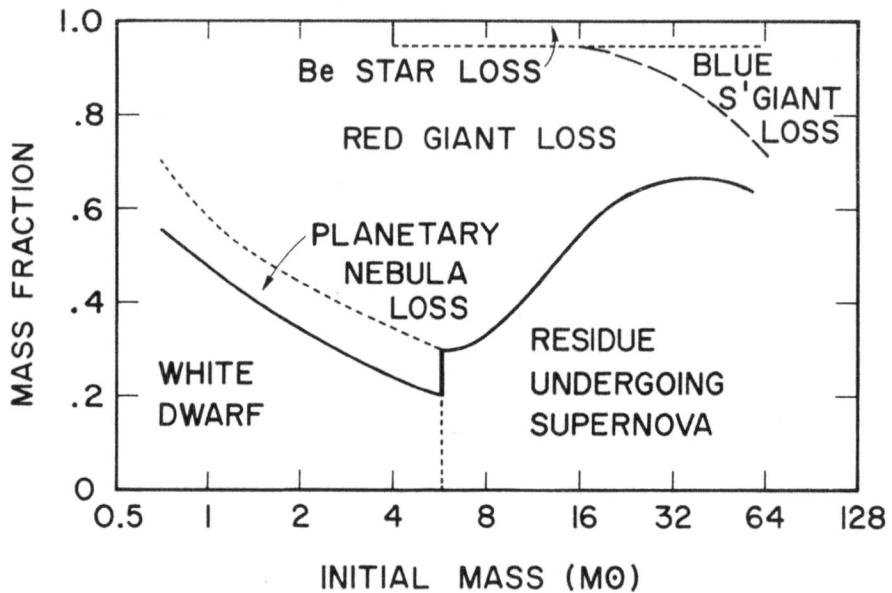

Fig. 1.   Mass ejection prior to a supernova (if that should happen) as a function of mass. Although the diagram is based on observations, the interpretation is both speculative and personal.

as a function of initial mass. The uncertainties about total mass loss increase above $6 \, M_{\odot}$, and the diagram for this portion indicates mainly that even in this mass range, ejection cannot continue to be ignored.

Acknowledgement

This research was supported by N.S.F. under grant GP32772.

References

Auer, L. H. and Woolf, N. J.: 1965, *Astrophys. J.* **142**, 182.
Barbaro, G., Chiosi, C., and Nobili, L.: in M. Hack (ed.), *Colloquium on Supergiant Stars*, Trieste Obs.
Brown, S., Higginbotham, N., and Lee, P.: 1970, *Publ. Astron. Soc. Pacific* **82**, 1372.
Caputo, F., Nata, A., and Castellani, V.: 1973, *Astrophys. Space Sci.* **22**, 213.
De Jager, C.: 1959, in *9th Colloque Int. d'Astrophysique*, Institut D'Astrophysique, Liège, p. 280.
Deutsch, A. J.: 1956, *Astrophys. J.* **123**, 210.
Deutsch, A. J.: 1969, *See* Hack 1969, p. 1.

Gehrz, R. D. and Woolf, N. J.: 1970, *Astrophys. J.* **161**, L213.
Gehrz, R. D. and Woolf, N. J.: 1971, *Astrophys. J.* **165**, 285.
Gillett, F. C., Low, F. J., and Stein, W. A.: 1968, *Astrophys. J.* **154**, 677.
Gillett, F. C., Merrill, K. M., and Stein, W. A.: 1971, *Astrophys. J.* **164**, 83.
Gillett, F. C., Forrest, W. J., and Merrill, K. M.: 1973, *Astrophys. J.* **183**, 87.
Greenstein, J. L.: 1973, *Astrophys. J.* **173**, 367.
Hack, M. (ed.): 1969, *Mass Loss From Stars*, D. Reidel Publ. Co., Dordrecht Holland.
Havlen, R. J.: 1971, *Astron. Astrophys.* **16**, 252.
Hill, P. W.: 1965, *Monthly Notices Roy. Astron. Soc.* **129**, 137.
Hills, R. J. and Dale, T. M.: 1973, *Astrophys. J.* **185**, 937.
Hoyle, F., Solomon, P. M., and Woolf, N. J.: 1973, *Astrophys. J. Letters* **185**, L89.
Hoyle, F. and Wickramasinghe, N. C.: 1962, *Monthly Notices Roy. Astron. Soc.* **124**, 417.
Humphreys, R. M., Strecker, D. W., and Ney, E. P.: 1971, *Astrophys. J.* **167**, L35.
King, I. R.: 1962, *Astrophys. J.* **136**, 784.
Limber, D. N.: 1964, *Astrophys. J.* **139**, 1251.
Luyten, W. and Herbig, G. H.: 1960, *Harvard Announcement Card*, 1474.
Morton, D. C.: 1969, in M. Hack (ed.), *Mass Loss From Stars*, D. Reidel Publ. Co., Dordrecht, p. 36.
Schwarzschild, M.: 1938, *Harvard Circ.* No. 431.
Thompson, R. I.: 1972, *Astrophys. J* **172**, 391.
Tsuji, T.: 1971, *Publ. Astron. Soc. Japan* **23**, 275.
Ulrich, R. K. and Scalo, J. M.: 1972, *Astrophys. J.* **176**, L37.
Walborn, N. R.: 1971, *Astrophys. J.* **165**, L67.
Weymann, R. J.: 1962, *Astrophys. J.* **136**, 476.
Weymann, R. J.: 1963, *Ann. Rev. Astron. Astrophys.* **1**, 97.
Weymann, R. J. and Sears, R. L.: 1965, *Astrophys. J.* **142**, 174.
Wilson, W. J. and Barrett, A. H.: 1972, *Astron. Astrophys.* **17**, 385.
Wilson, W. J., Schwartz, P. R., and Epstein, E. E.: 1973, *Astrophys. J.* **183**, 871.
Wilson, W. J., Schwartz, P. R., Neugebauer, G., Harvey, P. M., and Becklin, E. E.: 1972, *Astrophys. J.* **177**, 523.
Woolf, N. J.: 1964, *Astrophys. J.* **139**, 1081.

MASS EJECTION BY RADIATION PRESSURE

(Abstract)

ANNA ŻYTKOW

Institute of Astronomy, Polish Academy of Sciences, Warsaw, Poland

It was frequently suggested in recent years that the driving force for mass loss for the red supergiants may be the radiation pressure in continuum in the optically thick regions. There were many attempts to find selfconsistent models of time-independent mass outflow driven by this mechanism (Finzi and Wolf, 1971; Bisnovaty-Kogan and Nadyozhin, 1972; and many others); however, neither was satisfactory.

The aim of this paper is to indicate why radiation pressure in the high opacity regions should no longer be regarded as a promising mechanism in solving the problem of stationary mass ejection from the later type stars. The description of all the details may be found elsewhere (Żytkow, 1972, 1973).

In order to be able to treat the outflow as time-independent an element of matter undergoing acceleration should reach the velocity exceeding the escape velocity on a time-scale very short compared to the time required to eject the whole hydrogen rich envelope that can be expelled. In other words the reservoir of matter to be expelled should be much larger than the amount of matter actually undergoing acceleration. Let us mention that in the case of radiation pressure in continuum the basic assumption of time-independence was always made a priori with a merely superficial discussion of its validity. In view of our results it is this assumption which requires most careful consideration.

Within the framework of *stationary* outflow and spherical symmetry the models of outflowing envelopes were constructed by means of numerical integrations of equations of conservation of energy, momentum and mass together with radiative energy transport equation, equation of state including the effects of H and He ionization and H_2 dissociation and proper opacity values. The solutions have to be regular in the singular point where the transition from a subsonic to a supersonic velocity takes place. In the photosphere the commonly used photospheric thermal condition $L = 4\pi\sigma r_{phot}^2 T_{phot}^4$ (hereinafter all the symbols have their usual meaning) has to be satisfied. Thus, for a star mass M the models of outflowing envelopes are fully determined by two free parameters, say: the luminosity L of the star and the mass flux F. Obviously in this two parameter family of solutions only these leading to the expansion velocities that satisfy at infinity the condition $v_\infty^2 \geqslant 0$ (for large radii $v_\infty^2 = \text{const} = GM/r - v^2/2$) are relevant. Deep below the singular point in the (almost) static region of the models the properties of solutions should permit fitting of these models to the static interior (double-shell source carbon core in the case of low mass stars, helium core in the case of massive stars).

The models were constructed for $1\,M_\odot$ and $30\,M_\odot$ stars. The transition from

Tayler (ed.), Late Stages of Stellar Evolution, 54–55. All Rights Reserved.
Copyright © 1974 by the IAU.

subsonic to supersonic flow always takes place deep below the photosphere in the layers whose optical depth τ is large. So the upper limit to the rate at which momentum may be carried outwards to infinity is approximately $L\tau/c$, as may be derived from the equations of mass and momentum conservation. Thus the obtained rates of mass loss are rather large, ranging (approximately) from 10^{20} to 10^{25} g s^{-1}. For a 1 $M_\odot$ star the solutions satisfying the conditions stated above could be found for $L \gtrsim 1.625 \times 10^4 \, L_\odot$, for a 30 $M_\odot$ star for $L \gtrsim 1.15 \times 10^6 \, L_\odot$. The models occupy well defined areas in the yellow supergiant regions on the HR diagram; within the scope of adopted assumptions for both masses the complete range of relevant L and F values was investigated.

Some properties of the models seem to suggest that these models may provide a realistic description of the phenomenon of mass loss. This line of approach should, however, be no longer maintained. The calculations show that although one formally can construct the models of stationary outflowing envelopes these models cannot be physically meaningful. The main problem is that the reservoir of (almost static) matter to be expelled is always comparable or even less than the amount of matter actually undergoing acceleration. So in the case of radiation pressure in continuum in the optically thick region one cannot get selfconsistent results. We would like to emphasize that it is not the whole star that constitutes the reservoir of matter to be expelled – we have at our disposal only these layers of the star that are above the hydrogen burning shell source, as in a realistic model of a mass losing star on one hand the matter would have to flow downwards into the core through the shell (in view of its burning in the shell source) and on the other hand the matter would have to flow outwards.

Our results suggest that if one wants to describe the mass loss phenomenon in terms of stationary outflow, one should look for mechanisms in which matter is mainly accelerated above or close to the photosphere. In this respect the most promising seems to be the model of mass loss in which the radiation pressure on dust is taken into account as suggested by Weymann (1962), Wickramasinghe et al. (1966), and discussed by Gilman (1972) and others.

References

Bisnovaty-Kogan, G. S. and Nadyozhin, D. K.: 1972, *Astrophys. Space Sci.* **15**, 353.
Finzi, A. and Wolf, R. A.: 1971, *Astron. Astrophys.* **11**, 418.
Gilman, R. C.: 1972, *Astrophys. J.* **178**, 423.
Weymann, R.: 1962, *Astrophys. J.* **136**, 476.
Wickramasinghe, N. C., Donn, B. D., and Stecher, T. P.: 1966, *Astrophys. J.* **146**, 590.
Żytkow, A.: 1972, *Acta Astron.* **22**, 103.
Żytkow, A.: 1973, *Acta Astron.* **23**, 121.

STELLAR ENVELOPES AT SUPERCRITICAL LUMINOSITY

(Abstract)

G. S. BISNOVATYI-KOGAN

Institute of Applied Mathematics, USSR Academy of Sciences, Moscow, U.S.S.R.

Analytical solutions are obtained for static envelopes in radiative equilibrium for a dependence of opacity $\varkappa$ on temperature T given by

$$\varkappa = \varkappa_0, \quad T > T_1, \quad L/L_{c0} < 1$$
$$\varkappa = \varkappa_1 \, (T/T_1)^n, \quad T < T_1, \quad L/L_{c1} > 1 \,,$$

where L_{c0} and L_{c1} are critical luminosities corresponding to $\varkappa_0$ and $\varkappa_1$.

It is shown that there is an exact boundary between the regions of existence of static radiative envelopes and stationary outflowing envelopes, for which numerical solutions were obtained.

It is shown that convection makes the region of static envelopes broader and leads to the existence, for the same parameters, of both outflowing and static convective envelopes; i.e. to double-valued solutions. Such lack of uniqueness may be met during the computation of late stages of stellar evolution.

Tayler (ed.), Late Stages of Stellar Evolution, 56. All Rights Reserved.
Copyright © 1974 by the IAU.

DISCUSSION FOLLOWING PAPER BY WOOLF

Schramm to Woolf: How accurate is the value 6 $M_\odot$ which you presented for the division between stars eventually forming white dwarfs and stars becoming supernovae?

Woolf: There are several ways of estimating this number. One is the presence of a white dwarf in the Pleiades, where stars of 6–7 $M_\odot$ could just have evolved. The frequency of supernovae, assuming that they are stars more massive than those which become white dwarfs suggests 6 $M_\odot$. A similar value is obtained from the frequency of planetary nebulae, assuming that they do become white dwarfs. The uncertainty in the figure is however quite large; I estimate that it lies in the range 4 $M_\odot$ to 8 $M_\odot$.

Arnett to Woolf: Work by S. Falk and me suggests that to obtain light curves of type II supernovae we need an extended circumstellar shell of the sort mentioned by Prof. Woolf. It would be interesting to know whether these observed objects are in early or late evolutionary stages.

Mass loss seems to be a function of mass and chemical composition. Would Prof. Woolf like to summarize his ideas on this?

Woolf: There appear to be two groups of stars with regard to chemical composition. In those with heavy element abundances greater than 1/20 normal the formation of solid particles and the consequent radiation pressure on grains is very important. For lower heavy element abundance a different process is required. Similarly radiation pressure is less useful for the more massive stars but the precise dependence on mass is not clear.

Thomas: A group at Paris and Nice is studying the energetics of mass loss; in particular how is the energy stored in the region between the photosphere and the point where mass is lost. Eventually this study of atmospheric structure may provide good predictions of the rate of mass loss.

Woolf: The radiation pressure only becomes dominant in the envelope after the dust has formed and I agree that the energetics of the process are settled much lower in the atmosphere. However, I believe that the matter in the extended envelope is far too cool for it to escape, and this is why the formation of dust is crucial for mass loss.

However regardless of the physical process, the most important point for this conference is that the dust is a tracer for mass loss and allows us to estimate how much matter leaves a particular kind of star.

Thomas: I only wish to stress that mass loss will not be fully understood until the structure of the underlying star is also understood.

Paczyński: I would like to point out the existence of the so-called 'ring nebulae'

Tayler (ed.), Late Stages of Stellar Evolution, 57–58. All Rights Reserved.
Copyright © 1974 by the IAU.

which are known to exist around some single Population I Wolf-Rayet stars. NGC 6888 is the best studied case. The Population I W-R stars are believed to be 10 $M_\odot$ helium stars, which were originally the cores of 20–30 $M_\odot$ hydrogen stars. It looks like those massive stars lost all their hydrogen rich envelopes, presumably while being red supergiants. More details may be found in the proceedings of the IAU Symposium 'W-R and early type stars' (held in Buenos Aires, 1971) and in the proceedings of the conference held at Boulder in 1968 ('Wolf-Rayet Stars', K. Gebbie and R. N. Thomas, editors).

Faulkner: The comments so far have addressed themselves to mass loss after supergiant of late (second) giant stages. It is possible to distinguish mass loss between the first giant branch and horizontal branch (as currently required by theories), e.g. by coupling composition studies with mass loss studies?

Thomas: No, not from our point of view.

Woolf: I cannot distinguish.

REVIEW OF IAU SYMPOSIUM No. 59 ON
STELLAR INSTABILITY AND EVOLUTION,
CANBERRA, AUGUST 1973*

ICKO IBEN, Jr.

University of Illinois, Champaign-Urbana, Ill., U.S.A.

Abstract. Since the results of the Canberra Symposium are to be published, a complete review here would be redundant. I will therefore confine my written remarks to an enumeration of several topics which were discussed.

(1) Cepheid masses estimated in several different ways (using observational data in conjunction with theoretical evolutionary properties or in conjunction with theoretical pulsation properties) give discrepant results which may be interpreted to mean that stars lose considerable mass between the main sequence and Cepheid phases. One of the discrepancies disappears if the currently adopted distance to the Hyades is an underestimate by 0.2 mag. to 0.3 mag. A review by Van Altena at the IAU meeting in Sydney reveals that the convergent-point technique (which determines the current standard) gives a distance 0.2 mag. smaller than that given by all other techniques. It is therefore probable that Cepheids have essentially the same mass as their main sequence progenitors.

(2) The period-luminosity relationship for population I periodic variables now extends all the way to the main sequence, where a small pulsation amplitude has, until recently, prevented careful study. P-L relationships for Population I and Population II regular variables are now reasonably complete and the difference in slope and normalization between the two relationships is one of the most powerful discriminants between the two population types. Population I variables are represented by stars of many different ages and masses with a well defined mass-luminosity relationship ($L \propto M^4$) connecting them. Population II variables are all of low mass ($M \sim 0.5\ M_\odot$), with luminosity essentially uncoupled from mass.

(3) The evolutionary status of all Population II regular variables is understood in rough outline. For example: variables of intermediate period (~ 2 days) are evolving through the instability strip on a nuclear burning time scale, helium burning in one shell, hydrogen in another; long period variables are evolving through the strip on a Kelvin-Helmholtz time scale during excursions from the giant branch that are engendered by relaxation oscillations initiated in the double-shell source region.

(4) Understanding of detailed properties of Population II variables is still highly incomplete. Estimates of envelope helium abundance are a strong function of the mode of estimation. The quantitative influence of over-shoot and semi-convection at the edge of the convective core during core helium burning has not yet been satis-

---

* Supported in part by the U.S. National Science Foundation (GP-35863).

factorily demonstrated. A satisfactory understanding of the Osterhoof dichotomy is not yet at hand, although it is becoming increasingly probable that hysteresis in the domain of a transition region (between pulsation in the fundamental and in the first harmonic mode) may be responsible.

(5) Low-mass red giants that are in the double shell source stage lie on an extension of the normal giant branch that has a slope in the $\log T_e - M_{bol}$ plane much shallower than exhibited by the normal giant branch. This may be evidence for mass loss from very cool, luminous giants or may be an indication of a deficiency in stellar models.

(6) The formation of planetary nebulae may occur as a consequence of a final, large amplitude pulsation which terminates the Mira variable phase along the extention of the giant branch discussed under 5. Responsible for pulsation are properties of the star in the neighborhood of the hydrogen ionization zone, despite the fact that convection is important in this neighborhood. The final pulse may carry off all matter above the hydrogen-burning shell except for a very minute layer ($\sim 0.0003\ M_\odot$). The remnant is consistent with recent models of evolution for the central star of planetary nebulae. Of major interest is the fact that, in order to eject the nebula, it is not necessary to invoke either a large amplitude thermal pulse initiated in the helium-burning shell or radiation driven mass loss (the envelope-driven pulsation ejects matter before the critical luminosity for this process is reached).

(7) The observed properties of oscillations following the nova outburst are very strong evidence for non-radial pulsation. Observed periods are too long to be radial oscillations and an increase in period with decreasing radius (observed in one star) is a property unique to a non-radial g-mode.

(8) Statistical evidence makes it more and more likely that the progenitors of type I supernovae are binaries consisting initially of a star of intermediate mass ($3-8\ M_\odot$) orbiting a star of low mass ($<0.8\ M_\odot$) with initial periods in the range 1–6 yr. Following two phases of mass loss and mass transfer, the immediate supernova progenitor consists of a carbon-oxygen white dwarf orbiting about a red giant that is swelling beyond its Roche-lobe.

DISCUSSION FOLLOWING PAPER BY IBEN

Kippenhahn to Iben: Was there anything new about Breger's correlation between pulsational stability, rotation and metal content? Was there any explanation?

Iben: There is indeed a correlation between rotation rate and pulsation and between metallicity indicators and pulsation. For example, it seems that rotation prevents pulsation. I do not remember the sense of the correlation between metallicity and rotation. The correlations are strong since $\frac{2}{3}$ of the stars in the 'observational' instability strip are nonvariable. No detailed explanation was given.

Faulkner: On the topic of the distance to the Hyades, we have known since the mid-1960's that the best observed nearby dwarfs and sub-dwarfs, when corrected for line-blanketing effects, define a curve in the M_{Bol}, $\delta(U-B)$ diagram which passes $0.^m2$ above the Hyades position (i.e. the origin) in this diagram. One consequence of correcting the distance to the Hyades by $0.^m2$ is that the Hyades will now lie on the curve already defined by stars 'corrected to the Hyades'. There are those who have felt that this is not an unreasonable requirement.

Arnett to Iben: (1) Doug. Keeley, now at Santa Cruz, obtained results similar to those of P. R. Wood quoted by Iben (see Keeley's Ph.D. dissertation Cal. Inst. Tech. 1968).

(2) How does θ_d^2 variation in $^{12}C(\alpha, \gamma)^{16}O$ cause changes in the position of a star in the HR diagram.

Iben: The time which a star spends in this particular evolutionary phase is changed.

Tayler (ed.), Late Stages of Stellar Evolution, 61. *All Rights Reserved.*
Copyright © 1974 by the IAU.

EVOLUTION OF STARS WITH $M \leqslant 8 M_\odot$

B. PACZYNSKI

Institute of Astronomy, Polish Academy of Sciences, Warsaw, Poland

Abstract. The late stages of evolution of stars that develop degenerate carbon-oxygen cores are discussed. Model computations indicate that the initial masses of such stars, M, are below $M_1 = 8 M_\odot \pm$
$\pm 2 M_\odot$. The low mass stars ($M < M_0$) lose their envelopes and become white dwarfs. The intermediate stars ($M_0 < M < M_1$) ignite carbon in their highly degenerate cores of $1.4 M_\odot$. Present observational and/or theoretical estimates of M_0 are very uncertain. Problems associated with mass loss
and with carbon ignition and burning are discussed.

1. Introduction

Stars with the initial (i.e., main sequence) masses below $8 M_\odot$ develop degenerate
carbon-oxygen cores after hydrogen and helium exhaustion in their centers (Weigert,
1966; Arnett, 1969; Rose, 1969; Paczyński, 1970, 1971a; Uus, 1970, 1973) provided
there is neutrino emission due to the universal Fermi interactions (UFI). This introductory paper is concerned with the final stages of evolution of such stars. It will
be assumed that the UFI neutrinos are emitted. Most unfortunately, there are practically no model computations published so far which would follow the evolution
of any star with $M < 8 M_\odot$ all the way from the main sequence to the exhaustion of
nuclear fuel, and in which the UFI neutrinos would be neglected. It should be stressed,
therefore, that at the present time it is not possible to test the existence of the UFI
neutrinos on the basis of comparison between the observed stars and the published
models. In particular, nuclei of planetary nebulae cannot be used for such a test as
long as realistic models without the UFI neutrinos are not available.

Final evolution of a model with $M < 8 M_\odot$ and degenerate carbon-oxygen core is
governed by the core mass. The larger the core mass, M_c, the larger the luminosity, L,
produced by the hydrogen and helium shell sources. Model computations can be
fitted with the following analytic formula

$$L/L_\odot = 59250 (M_c/M_\odot - 0.522),$$ (1)

which is good for $0.6 < M_c/M_\odot < 1.4$ (Paczyński 1971b). To maintain such a high
luminosity, matter must flow from the hydrogen rich envelope through the shell
source region into the degenerate core. Every gram of matter releases about 6×10^{18} erg
in this process, and therefore the rate of growth for the core mass is given as

$$\frac{dM_c}{dt} (M_\odot \text{ yr}^{-1}) \approx 10^{-11} L/L_\odot.$$ (2)

The envelope mass does not affect the nuclear burning and the luminosity of these
models as long as it is larger than a certain limit, $M_{e,\text{min}}$. Model computations

Tayler (ed.), Late Stages of Stellar Evolution, 62–69. All Rights Reserved.
Copyright © 1974 by the IAU.

(Paczyński, 1971b) give $M_{e,\min} = 3 \times 10^{-4} M_\odot$ for $M_c = 0.6 M_\odot$, and $M_{e,\min} = 10^{-6} M_\odot$ for $M_c = 1.2 M_\odot$. If the envelope is considerably more massive than $M_{e,\min}$ the star is a red supergiant and its effective temperature is in the range of 2000–3000 K. Given the luminosity and effective temperature the stellar radius may be calculated at once. Models show that red supergiants with $L > 10^4 L_\odot$ or so have highly superadiabatic temperature gradients within the hydrogen and helium ionization zones that cover all the envelope, and the matter density is practically constant throughout a given envelope.

Degenerate carbon-oxygen cores have a structure similar to that of white dwarfs. As the core mass increases with time the core contracts slowly. The temperature at the centre is determined by the balance between the adiabatic heating and the cooling due to neutrino emission. Carbon is ignited when the central density rises up to about 2×10^9 g cm^{-3} if the recent strong screening corrections are used (DeWitt et al., 1973; Graboske et al., 1973; Graboske, 1973). At that time the core mass is 1.38 $M_\odot$. Arnett (1969) suggested that carbon ignition leads to a detonation and a total disruption of a star. This problem was recently analyzed by Bruenn (1972) who gives a lot of references. It is possible that the neutrino energy losses due to the URCA process driven by convective motion stabilizes carbon burning and prevents a total disruption of a star (Paczyński, 1972; Couch and Arnett, 1973; Ergma and Paczyński, 1974) but this is not certain (Bruenn, 1973; Paczyński, 1974).

Let us suppose that the core mass does not grow up to 1.38 $M_\odot$ either because the total mass of a star was too small from the beginning, or the envelope was lost in a course of evolution. In this case the core becomes a carbon-oxygen white dwarf with a little of unburnt hydrogen left at the surface. On its way from the red giant to the white dwarf state the star evolves rapidly through the high luminosity–high temperature region of the HR diagram (Paczyński, 1970, 1971; Rose and Smith, 1970; Schwarzschild and Härm, 1973) which is occupied by nuclei of planetary nebulae. If the former stellar envelope is still present as a circumstellar matter at the time when the core evolves to the white dwarf state then the ultraviolet radiation from the core may ionize the circumstellar matter and a nebula may be seen. I think there can be little doubt that nuclei of young planetary nebulae are in a double shell source phase, as this is the only phase of evolution of a medium mass star when the luminosity in excess of $10^4 L_\odot$ is achieved.

Schwarzschild and Härm (1965) discovered that helium shell burning around a degenerate core is thermally unstable and violent thermal pulses are produced in the stellar interior. Nobody was able so far to follow with the detailed model computations all the thermal pulses in a given stellar model, as too much computing time would be necessary. The impact of these instabilities on the stellar evolution is not really understood and it will not be discussed in this paper.

In the following discussion I shall concentrate on the two problems: mass loss from stars and carbon ignition and burning. I believe the other important aspect of stellar evolution, the large scale mixing which may bring to the stellar surface the products of nuclear burning, will be discussed by other speakers.

2. Mass Loss

The problem of mass loss from low or intermediate mass stars is closely related to the origin of planetary nebulae. Seventeen years ago Shklovsky (1956) convincingly suggested that planetary nebulae are formed from the envelopes of red supergiants. This idea is generally accepted now. Considering the high luminosity of nuclei of planetary nebulae it is clear that when a nebula is ejected the star must be in a phase of hydrogen and helium burning in two shells. There is no generally accepted mechanism for ejection of stellar envelopes, but a large number of different suggestions has been made. I shall discuss briefly those which seem to be the most common or plausible.

Helium shell flashes were frequently proposed to be the cause of envelope ejection. Perhaps the shell flashes do not drive but rather stimulate the mass loss (Smith and Rose, 1972). The envelopes of red supergiants may be unstable due to their own structure, as hydrogen and helium ionization zones are very thick. As a result an adiabatic exponent is smaller than $\frac{4}{3}$, and the total energy of an envelope is positive. It has been suggested (Lucy, 1967; Roxburgh, 1967; Paczyński and Ziółkowski, 1968) that envelopes are dynamically unstable. Nonadiabatic perturbations were recently studied by Smith and Rose (1972), Scott (1973), Sparks and Kutter (1972), and others. However, there is a very serious problem with all these suggestions and models. It is known that thermal and dynamical time scales for the envelopes of red supergiants are of the same order of magnitude, while the convective time scale is shorter than either thermal or dynamical time scale (Fawley, 1973). Therefore, we have to consider nonadiabatic motion of the envelope, and an interaction between the radial motion and convection must also be taken into account. There is no theory of such interaction. All investigators were forced to make (explicitly or implicitly) some ad hoc assumptions about this interaction and the results cannot be credible.

There is no compelling observational evidence that an ejection of an envelope is very rapid. Consider a young and dense planetary nebula. The electron density is hardly larger than 10^6, and an ionized mass may be something like $10^{-2} M_\odot$. This corresponds to a nebular radius of 2×10^{16} cm. If the original expansion velocity was 7 km s^{-1} then the time scale for an ejection would be 10^3 yr, and the implied mass loss rate only $10^{-5} M_\odot$ yr^{-1}. This is not much more than an average mass loss rate which is observed in a typical Mira variable $-2 \times 10^{-6} M_\odot$ yr^{-1} according to Gehrz and Wolf (1971). It is possible that the steady mass loss that is observed to be taking place in red supergiants is a process responsible for an ejection of a whole envelope and a formation of a planetary nebula (Paczyński, 1971c). Radiation pressure on dust grains formed in the atmosphere may be the driving force (Weymann, 1962; Wickramasinghe *et al.*, 1966; Krishna Swamy and Stecher, 1969; Balamore and Lucy, 1972; Gilman 1972). No satisfactory model has been published so far but this approach seems to be promising.

It is frequently suggested that the driving force for mass loss may be the radiation pressure in the high opacity regions where hydrogen and helium are partly ionized.

Models with a stationary mass loss were considered by Finzi and Wolf (1971), Bisnovaty-Kogan and Nadyozhin (1972) and many others. Unfortunately, Żytkow (1972, 1973) has shown that in such models the transition from a subsonic to a supersonic flow takes place below the photosphere, where the gas density is high. As a result the mass loss rates obtained from the models are so large that the process cannot be stationary. Therefore, we come back to the problem of interaction of radial motion with convection, and this approach seems to be hopeless at present.

Little is known observationally about the range of masses of planetary nebulae, but the range is likely to be very large. It should be stressed that there is no observational evidence that all nebulae have identical masses, say $0.2\,M_\odot$ or $0.6\,M_\odot$, though $0.2\,M_\odot$ may very well be a median value. A typical mass of the ancestor star is believed to be $1\,M_\odot$ or $1.5\,M_\odot$, but nothing is known about an upper mass limit. Frequently, when a circumstellar nebula is discovered to have a mass of say $10\,M_\odot$ it is immediately said not to be a planetary nebula as its mass is too large.

Nuclei of planetary nebulae evolve towards the white dwarf stage. One may try to estimate the range of stellar masses for which the mass loss is important by studying white dwarfs in open clusters. Jones (1970) estimated that stars with $M \leqslant 2\,M_\odot$ were producing white dwarfs. One may also assume that stars above a certain mass limit M_0 explode as supernovae, while the less massive stars lose enough matter to become white dwarfs. With the known (?) rate of supernova explosions and the known birth rate function the value of M_0 may be deduced. I believe that different authors gave estimates that ranged from $3\,M_\odot$ up to $10\,M_\odot$. This simply reflects the uncertainty of the statistical approach.

Considering the present status of the theory of mass loss and the present state of interpretation of the observations, it is fair to say that we do not know what is the value of M_0, the limit that separates the low mass stars which produce white dwarfs (and presumably planetary nebulae), and the medium mass stars which can ignite carbon in their $1.4\,M_\odot$ degenerate cores. If there was no mass loss such cores could be produced by stars in the mass range of $1.4\,M_\odot - M_1$, where M_1 is believed to be about $8\,M_\odot$. This upper mass limit is derived from the model computations only, and these computations are sensitive to the efficiency with which the convective envelope can penetrate the helium core. My impression is that we have $M_1 = (8 \pm 2)\,M_\odot$, and I would like to stress the likely uncertainty which places M_1 somewhere in the range of 6–10 $M_\odot$. It is not impossible that in fact we have $M_1 < M_0$, and that there are no stars igniting carbon in their degenerate cores (Arnett, private communication)!

3. Carbon Ignition and Burning

Let us consider now carbon ignition in degenerate cores, assuming that there are stars in which this process is taking place. According to Arnett (1968, 1969) carbon ignition leads to a detonation and a supernova explosion, and no stellar remnant is left. On the other hand it has been suggested by Gunn and Ostriker (1970), on the basis of a statistical analysis of the observational data, that the ancestors of pulsars should be stars in the main sequence mass range of 4–10 $M_\odot$. This suggestion created a demand for finding a physical process that could prevent a total disruption of a star undergoing carbon ignition within the degenerate core (Arnett, 1971; Barkat et al., 1971, 1972; Buchler et al., 1971; Bruenn, 1971, 1973a, b; Colgate, 1971; Paczyński, 1972, 1974; Sackmann and Weidemann, 1972; Iben, 1972; Couch and Arnett, 1973; Ergma and Paczyński, 1974). Energy losses due to convectively driven URCA processes are likely to prevent carbon detonation. Simplified models of Ergma and Paczyński (1974) indicate that it may be possible to exhaust carbon nonexplosively within the inner 0.5 $M_\odot$ of the core. Perhaps the inner core may then collapse and produce a neutron star of about 0.5 $M_\odot$, while the outer 0.9 $M_\odot$ of the core would explode due to carbon detonation.

It is unfortunate that the available model calculations are crude and the results are uncertain. At the same time it is no longer clear that pulsars have to be produced from the medium mass stars. Arnett and Schramm (1973) suggest that pulsars are produced by massive stars $(8 \leqslant M/M_\odot \leqslant 70)$. These stars have about the same death rate as the medium mass stars $(4 \leqslant M/M_\odot \leqslant 8)$ (cf. Salpeter, 1959; Schmidt, 1963). This new suggestion is reasonable as Arnett demonstrated that all massive stars are developing 1.4 $M_\odot$ central cores.

Sometimes it was argued that carbon detonation models would produce too much iron peak elements in the Galaxy. This is not obvious, as the explosions of massive stars may produce mostly elements from carbon to silicon (Arnett and Schramm, 1973); the mass ejected per explosion is larger than it is in a medium mass star, while the number of explosions may be comparable. If we notice that the cosmic abundance ratio $Fe/(C+O+N)$ is about $\frac{1}{4}$ (Allen, 1955) then it becomes clear that the present day theory of late stages of stellar evolution combined with statistical arguments can, at the best, narrow the range of masses for which carbon ignition is disruptive. There is no obvious reason to believe that carbon detonations are incompatible with the observations. Perhaps studies of individual pulsars and supernova remnants are more promising in determining the origin of pulsars than the statistical analysis is.

4. General Discussion

In this paper I intended to demonstrate that the two most fundamental processes for the late stages of evolution of intermediate mass stars $(M < 8\ M_\odot)$ are not theoretically understood, and that the interpretation of existing observations does not provide us with the answers we need. The two processes are the mass loss and carbon ignition. I believe that more effort should be made to interpret the observations and to verify observationally existing models of double shell source stars. One obvious theoretical prediction is the luminosity function for red supergiants that are in a double shell source phase. Combining the Equations (1) and (2) we obtain for the rate of change of a bolometric magnitude

$$\frac{dM_{bol}}{dt} = -6.4 \times 10^{-7}\ \text{yr}^{-1}, \tag{3}$$

for $-7.0 < M_{bol} < -4.4$, which corresponds to $1.4 > M_{core}/M_\odot > 0.6$. If all stars went through this range of the carbon-oxygen core masses the luminosity function would be constant. Loss of an envelope stops the core growth and may produce a deficiency of the most luminous supergiants.

Analysis of luminous infrared objects and objects suspected of being young planetary nebulae may provide us with an information about the time scale on which an envelope of a red supergiant is lost. A deep photograph of NGC 6543 (Millikan, 1972) shows a large faint nebulosity, 5' in diameter, and the inner bright nebulosity, only 0.5 in diameter. This may indicate that the rate of mass loss was variable, small in the past and large at the final phases.

There is a very spectacular variable star, FG Sge, which is a nucleus of a planetary nebula (Herbig and Boyarchuk, 1968; Langer et al., 1973). This star probably under-goes a thermal pulse in its helium shell source (Paczyński, 1971b). As the luminous nuclei of young planetary nebulae are almost certainly in a double shell burning phase of evolution many of them should exhibit light variations on the time scale of years or decades. Theoretical timescales can be calibrated by means of model compu-tations in terms of core and envelope masses. Observations could be used to check the models and to derive masses of the nuclei of planetary nebulae.

Type I supernovae are believed to explode in galaxies where there are no massive stars. There are suggestions (see e.g. Finzi and Wolf, 1967) that presupernovae of type I are white dwarfs with masses very close, but smaller than the effective Chandrasekhar mass limit. Supposedly such white dwarfs could be produced some 10^9 or 10^{10} yr ago, when there were massive stars available. On a long time scale either the white dwarf mass was increasing, or the effective Chandrasekhar mass limit was decreasing. Such an evolution could finally lead to an explosion. The best candidates for the ancestors of massive white dwarfs are the intermediate mass stars. In this case the initial conditions for a presupernova model may be obtained from the model calculations for the evolution of an ordinary star (Paczyński, 1971a). It is essential to have better knowledge about the pycnonuclear reaction rates and other physical properties of dense matter at low temperature. Evolutionary calculations for hypothetical type I presupernovae could link the theory of intermediate mass stars with the theory of supernovae.

Interaction between the convection and URCA processes is essential for carbon burning in degenerate matter and a consistent picture of this interaction is needed in order to be able to predict theoretically the final products of the evolution of intermediate mass stars.

Acknowledgements

I am very grateful to Dr W. David Arnett for many interesting and illuminating discussions on the late phases of stellar evolution, and to Prof. Abdus Salam, the director of the International Centre for Theoretical Physics in Trieste for his hospitality at the Summer Session during which this review was partly prepared.

References

Allen, C. W.: 1955, *Astrophysical Quantities*, University of London, The Athlone Press.
Arnett, W. D.: 1968, *Nature* **219**, 1344.
Arnett, W. D.: 1969, *Astrophys. Space Sci.* **5**, 180.
Arnett, W. D.: 1971, *Astrophys. J.* **169**, 113.
Arnett, W. D. and Schramm, D. N.: 1973, *Astrophys. J. Letters* **184**, L47.
Balamore, D. S. and Lucy, L. B.: 1972, *Bull. Am. Astron. Soc.* **4**, 234.
Barkat, Z., Buchler, J.-R., and Wheeler, J. C.: 1971, *Astrophys. Letters* **8**, 21.
Barkat, Z., Wheeler, J. C., and Buchler, J.-R.: 1972, *Astrophys. J.* **171**, 651.
Bisnovaty-Kogan, G. S. and Nadyozhin, D. K.: 1972, *Astrophys. Space Sci.* **15**, 353.
Bruenn, S. W.: 1971, *Astrophys. J.* **168**, 203.
Bruenn, S. W.: 1972, *Astrophys. J. Suppl.* **24**, 283.
Bruenn, S. W.: 1973a, *Astrophys. J. Letters* **183**, L125.
Bruenn, S. W.: 1973b, *Astrophys. J.* **186**, 1157.
Buchler, J.-R., Wheeler, J. C., and Barkat, Z.: 1971, *Astrophys. J.* **167**, 465.
Colgate, S. A.: 1971, *Astrophys. J.* **163**, 221.
Couch, R. G. and Arnett, W. D.: 1973, *Astrophys. J.* **180**, L101.
De Witt, H. E., Graboske, H. C., and Cooper, M. S.: 1973, *Astrophys. J.* **181**, 439.
Ergma, E. and Paczyński, B.: 1974, *Acta Astron.* **24**, 1.
Fawley, W. M.: 1973, *Astrophys. J.* **180**, 483.
Finzi, A. and Wolf, R. A.: 1967, *Astrophys. J.* **150**, 115.
Finzi, A. and Wolf, R. A.: 1971, *Astron. Astrophys.* **11**, 418.
Gehrz, R. D. and Woolf, N. J.: 1971, *Astrophys. J.* **165**, 285.
Gilman, R. C.: 1972, *Astrophys. J.* **178**, 423.
Graboske, H. C.: 1973, *Astrophys. J.* **183**, 177.
Graboske, H. C., DeWitt, H. E., Grossman, A. S., and Cooper, M. S.: 1973, *Astrophys. J.* **181**, 457.
Gunn, J. E. and Ostriker, J. P.: 1970, *Astrophys. J.* **160**, 979.
Herbig, G. H. and Boyarchuck, A. A.: 1968, *Astrophys. J.* **153**, 397.
Iben, I. Jr.: 1972, *Astrophys. J.* **178**, 433.
Jones, E. M.: 1970, *Astrophys. J.* **159**, 101.
Krishna-Swamy, K. S. and Stecher, T. D.: 1969, *Publ. Astron. Soc. Pacific* **81**, 873.
Langer, G. E., Kraft, R. P., and Anderson, K. S.: 1973, *Bull. Am. Astron. Soc.* **5**, 313.
Lucy, L. B.: 1967, *Astron. J.* **72**, 813.
Millikan, A. G.: 1972, *Mercury* **1**, No. 5, 13.
Paczyński, B.: 1970, *Acta Astron.* **20**, 47.
Paczyński, B.: 1971a, *Acta Astron.* **21**, 271.
Paczyński, B.: 1971b, *Acta Astron.* **21**, 417.
Paczyński, B.: 1971c, *Astrophys. Letters* **9**, 33.
Paczyński, B.: 1972, *Astrophys. Letters* **11**, 53.
Paczyński, B.: 1974, *Astrophys. Letters*, in press.
Paczyński, B. and Ziółkowski, J.: 1968, *Acta Astron.* **18**, 255.
Rose, W. K.: 1969, *Astrophys. J.* **155**, 491.
Rose, W. K. and Smith, R. L.: 1970, *Astrophys. J.* **159**, 903.
Roxburgh, I. W.: 1967, *Nature* **215**, 838.
Sackmann, I. J. and Weidemann, V.: 1972, *Astrophys. J.* **178**, 427.
Salpeter, E. E.: 1959, *Astrophys. J.* **129**, 608.
Schmidt, M.: 1963, *Astrophys. J.* **137**, 758.
Schwarzschild, M. and Härm, R.: 1965, *Astrophys. J.* **142**, 855.
Schwarzschild, M. and Härm, R.: 1973, *Bull. Am. Astron. Soc.* **5**, 315.
Scott, E. H.: 1973, *Astrophys. J.* **180**, 487.
Shklovsky, I.: 1956, *Astron. Zh.* **33**, 315.
Smith, R. L. and Rose, W. K.: 1972, *Astrophys. J.* **176**, 395.
Sparks, W. M. and Kutter, G. S.: 1972, *Astrophys. J.* **175**, 707.
Uus, U.: 1970, *Scientific Inf.* **17**, 3.
Uus, U.: 1973, *Astron. Zh.* **50**, 297.
Weigert, A.: 1966, *Z. Astrophys.* **64**, 395.

Weymann, R.: 1962, *Astrophys. J.* **136**, 476.
Wickramasinghe, N. C., Donn, B. D., and Stecher, T. P.: 1966, *Astrophys. J.* **146**, 590.
Żytkow, A.: 1972, *Acta Astron.* **22**, 103.
Żytkow, A.: 1973, *Acta Astron.* **23**, 121.

THE URCA PROCESS IN CONVECTIVE CORES*

(Abstract)

S. TSURUTA

*NASA, Goddard Space Flight Center, Greenbelt, Md., U.S.A.***

and

A. G. W. CAMERON

Harvard College Observatory, Cambridge, Mass., U.S.A.

The importance of the URCA neutrino process in convective cores in late stages of stellar evolution was already pointed out in this Symposium. In pre-supernova stages, Paczyński (1972) regarded convective motion as a stellar vibration. We explored the possibility for a more realistic approach. As a first step, we assumed convective motion as a fully developed turbulence, which is represented by super-position of eddies of different velocities and sizes obeying the Kolmogoroff spectrum. Using the Monte Carlo method, we followed the path of a lump of matter, with emission of neutrinos (and antineutrinos) whenever appropriate (though electron captures inside and through beta decays outside the URCA shell), until a steady state is reached. The calculations were carried out for a typical pre-supernova model (similar to that of Paczyński).

The results are: (a) when the stellar luminosity $L \simeq 100 \, L_\odot$, the URCA neutrino luminosity L_ν^u is $\sim 10^{38}$ erg s^{-1}, somewhat less than but comparable with Paczyński's result, while (b) a significant increase in L_ν^u is noted for $L \simeq 10^6 \, L_\odot$. In both (a) and (b), the convective velocity is sufficiently fast, so that the corresponding Reynolds number is far greater than the critical value. Therefore, the assumption of fully developed turbulence should be perfectly valid. It may be noted that the random walk approach we adopted for the direction of motion is not valid in the presence of discontinuity in composition. However, already at $L \simeq 100 \, L_\odot$, the timescale for beta processes ($\sim 10^5$ s) is longer than the convective timescale ($\sim 10^4$ s), and mixing of elements is nearly complete. In view of the absence of better convective theories in degenerate cores, our results may prove to be useful.

Reference

Paczyński, B.: 1972, *Astrophys. Letters* **11**, 53.

* This paper was presented by S. Tsuruta.
** Present address, Astronomy Centre, University of Sussex, Falmer, Brighton, U.K.

DISCUSSION FOLLOWING PAPER BY PACZYŃSKI

Iben to Paczyński: We may not be worried if stars in the range 2 $M_\odot$ to 8 $M_\odot$ do not make pulsars but, if these stars are detonated and leave no remnant do we not make far too much iron peak matter?

Ostriker to Paczyński: I have some comments on the insights offered by statistical considerations concerning the likely endpoint of stars in the 3–8 $M_\odot$ range.

(1) I doubt if enough stars die in the mass range $8 < M/M_\odot < \infty$ to account for the number of pulsars.

(2) If all stellar cores from stars $3 < M/M_\odot < 8$ were explosively turned into iron peak elements, I think that far more iron would have been produced in the history of the Galaxy than is currently observed.

(3) These statistical arguments have an uncertainty greater than ± 0.3 to 0.5 in the logarithm so that (1) and (2) do not constrain the theory overly.

Paczyński: I believe that it is better to look at the iron problem in a different way. One should compare the amount of iron ejected in carbon detonation supernovae (if carbon detonates) with the amount of C, N, O produced in massive supernovae discussed by Arnett. In this way we could put an upper limit to the mass range of stars which may be allowed to detonate carbon without violating the observed Fe/(C+N+O) ratio, but I do not believe that we could exclude on observational grounds the possibility of carbon detonation. I am afraid that at present this is purely a theoretical problem and the theoretical answer is uncertain as I tried to emphasise in my review. Indeed, when mass loss is taken into account, it is even possible that the lower mass limit for carbon detonation could be higher than the higher mass limit!

Sugimoto to Paczyński: I would like to give two comments concerning the cause of mass loss. (1) I computed the instability of the helium-burning shell for the stage of C–O core mass very close to the Chandrasekhar limit (1.39 $M_\odot$). The peak of the energy generation by the helium burning is only $L_{He} \simeq 10^7 L_\odot$, which is too small to produce any dynamical effect. The reason why the peak is so low is that radiation pressure is so much larger than gas pressure. If we then keep the pressure constant at the helium-burning shell, while the entropy increases, the temperature rises but the density decreases so much that the nuclear energy generation decreases.

(2) Paczyński and Ziolkowski (1968) estimate the upper limit of mass below which the binding energy of a red giant envelope becomes negative and found 3–4 $M_\odot$, for the case of mixing length equal to pressure scale height ($l = H$). Provided the mixing length satisfies $0.7 < l/H < 2.0$, the value of the upper limit of mass varies slightly. Another mechanism of mass loss will be necessary, if the maximum mass of stars becoming planetary nebulae is to be appreciably larger than 4 $M_\odot$.

Tayler (ed.), Late Stages of Stellar Evolution, 71–72. All Rights Reserved.
Copyright © 1974 by the IAU.

Arnett to Paczyński: Great care must be taken in estimating the number of high mass stars which may become supernovae.

(1) The 'Salpeter' mass function is sometimes taken as $\psi \sim m^{4/3}$ and sometimes as his tabulated version (which has $\psi \sim m^{5/3}$ or so for $m \geqslant 10\ m_\odot$). More realistic values similar to the latter are actually used by myself and Talbot in some of our statistical arguments.

(2) One should use direct counts of OB stars rather than the mass function for this question of pulsar formation since it avoids the question of whose luminosity function to mass function conversion you use.

Schwarzschild to Paczyński: I fully agree with Dr Paczyński that the mass ejection from low mass supergiants, particularly the ejection of a planetary nebula, is most likely caused by an envelope instability, not by the interior shell flash instability. Nevertheless, the structure of the envelope varies during an interior flash cycle and there appears to be a particular phase in the interior cycle at which the envelope instability is particularly strong. At this phase the envelope has a fairly unusual thermal structure. If the main part of the ejection of the planetary nebula occurs at this phase, it appears from recent Princeton computations that the remnant star is left in such a special state that it makes its transition from a red supergiant to a blue nucleus very fast (~ 3000 yr); that is fast enough to properly illuminate the ejected nebula before the latter hopelessly disappears.

EVOLUTION OF STARS WITH $M \geqslant 8\,M_\odot$*

A. G. MASSEVITCH and A. V. TUTUKOV

Astronomical Council of the Academy of Science of the USSR, Moscow, U.S.S.R.

Abstract. A review of research in the field of evolution of massive stars during the last three years is presented. The analysis of computed stellar models in the helium-burning stage provides evidence for the existence of neutrino emission predicted by the theory of universal Fermi-interaction. The criterion for convective stability in the zone of variable molecular weight still remains uncertain despite several possibilities for a unique choice of the stability criterion that were recently suggested.

Evolutionary computations of massive star models in the carbon, oxygen, neon and silicon-burning stages provide opportunities to obtain models of presupernovae of different types and to investigate the cause of instability responsible for the implosion as a function of the initial mass of the star.

If the neutrino emission is included in computations the core of a star with $M \gtrsim 90\,M_\odot$ loses stability due to the positron-electron pair formation in the oxygen-burning stage. A star with $16\,M_\odot \lesssim M \lesssim 90\,M_\odot$ collapses because of photodissociation of the iron nuclei but a core of a star with $9\,M_\odot \lesssim M \lesssim 16\,M_\odot$ collapses due to neutronization of silicon or iron nuclei.

Attention is paid to the possibility of circulation mixing of matter within the boundary of a carbon-oxygen core, due to the fast rotation of the core in advanced stages of evolution, if the angular momentum is conserved in the course of evolution. As the molecular weight barrier is low in late stages of chemical evolution, it can not prevent the penetration of circulation flows outside the convective core.

Difficulties of the present theory of convection in stars are stressed and the necessity of a two-dimensional approach to the solution of the stellar constitution equations is emphasized.

1. Introduction

Since the last meeting in Brighton a considerable increase of number of papers on evolution of massive stars may be noted. The main purposes for drawing attention to this problem are: (a) the still uncertain evolutionary status of blue and red supergiants; (b) a possibility to verify the theory of weak interactions; (c) problems connected with the synthesis of heavy nuclei and their penetration into the interstellar medium; (d) attempts to construct a realistic model of a presupernova; (e) the whole scope of new problems connected with the discovery of pulsars and neutron stars and the development of X-ray astronomy, particularly with the possibility of detecting collapsed bodies. Due to the development of computing techniques a large number of evolutionary sequences of stellar models has been constructed recently in many countries.

The lower mass range of stars we deal with in this review is determined in the title The upper limit is not so well defined. Ledoux (1941), Schwarzschild and Härm (1959) showed on the basis of the theory of linear pulsations that main sequence stars with $M \gtrsim 60\,M_\odot$ will be pulsationally unstable. This upper limit has been confirmed later by Stothers and Simon (1967). Recent computations of an evolutionary sequence for

---

* This paper was presented by A. G. Massevitch.

Tayler (ed.), Late Stages of Stellar Evolution, 73–92. All Rights Reserved.
Copyright © 1974 by the IAU.

stellar models with $M=130\,M_\odot$ carried out by Appenzeller (1970a, b) indicate however that pulsational mass loss might not be strong enough to affect considerably the evolution of a star with $M \lesssim 200\,M_\odot$. Still up to now detailed computations of advanced evolution of large mass stars have been carried out only for the mass range $M \lesssim 64\,M_\odot$.

The main feature of the evolution of stars with large masses is a rather 'calm' (noncataclysmic) consequent exhaustion of nuclear fuel in convective cores up to the iron group, followed by a collapse of the core caused by the photodisintegration of Fe-nuclei at very high temperatures.

The rather smooth change from one nuclear fuel to another (contrary to stars of low mass where it is accompanied by a flash) is caused by the relatively small degeneracy of stellar matter in the cores of massive stars.

A recent comparison of theoretical tracks with observations carried out by Stothers (1972a) led him to the conclusion that best agreement may be achieved for a chemical composition $X=0.70$, $Z=0.03$ in good agreement with earlier results (Ruben and Massevich, 1966). Similar values for X and Z have been derived recently by Barbaro and Chiosi (1972). Still the problem of an adequate comparison of theory and observations remains open. This is mainly due to the fact that such a comparison requires uncertain theoretical transformation of observational data – data that themselves are for most cases incomplete. This is particularly true for the effective temperature scale of red giant stars in very late stages of evolution which are the topic of the present Symposium. It should be noted that the present theory of stellar evolution has reached a state when it is possible in several cases to predict observable phenomena. So, for example, on the basis of evolutionary tracks for late stages of stellar evolution ratios of isotopes in stellar envelopes may be deduced.

The main difficulties that have to be dealt with when computing evolutionary models of massive stars are first the uncertainty of the criterion of convective instability in the intermediate layer with varying molecular weight and, second, the neutrino emission predicted by the theory of weak interactions.

A detailed review of papers on evolution of stars with large masses can be found in Stothers and Chin (1969, 1970), Ruben (1969), Dalloporta (1971), and Massevich and Schustov (1973).

2. The Stability Criterion in the Intermediate Layer with Varying Molecular Weight

Schwarzschild and Härm (1958) showed first that in models for stars of large masses a convective unstable layer (a 'semiconvective' region) develops at the boundary of the convective core. A 'semiconvective' region is one in which the stability criterion is initially violated, but which becomes almost stable if mixing of matter reduces the opacity, e.g. by mixing core helium with envelope hydrogen. A similar region develops in small mass stars on the stage of core-helium-burning. There have been some controversies as concerning the stability criterion in the semiconvective zone. Schwarzschild and Härm assumed:

$$\nabla_r = \nabla_a. \tag{1}$$

Sakashita and Hayashi (1959, 1961) introduced for the region of varying molecular weight the Ledoux criterion:

$$\nabla_r = \nabla_a + \frac{\beta}{4 - 3\beta} \frac{d \ln \mu}{d \ln P}, \tag{2}$$

where β – is the ratio of gas pressure P_g to the total pressure P, μ – molecular weight.

An important step was made by Kato (1966), who discovered that the zone where the conditions:

$$\nabla_a < \nabla_r < \nabla_a + \frac{\beta}{4 - 3\beta} \frac{d \ln \mu}{d \ln P} \tag{3}$$

is fulfilled is vibrationally unstable. The instability is caused by radiative heat exchange between convective elements and the surrounding medium. Vibrational instability leads to a mixing of matter until the Schwarzschild criterion of stability is fulfilled. The timescale of mixing is much shorter than the evolutionary timescale. Gabriel (1968) and Auré (1971) studied the vibrational stability of massive stars in a linear approximation. They showed that due to the stabilizing action of the radiative envelope no pulsations arise. This result is in favour of the Ledoux criterion of stability. It is however doubtful if the problem investigated by Gabriel and Auré is fully adequate for the problem of mixing in the layers of varying molecular weight of massive stars.

Dudorov and Tutukov (1972) investigated in detail the mixing problem using the linearized equation of motion of a convective element in a medium with varying molecular weight and with a radiative temperature gradient satisfying condition (3). They obtained analytically the roots of the characteristic equation for $\nabla_r - \nabla_a \ll$ $\ll d \ln \mu / d \ln P$ and velocities of growth of perturbations of different size elements. Elements with linear dimension 10^7 cm have the largest increment of perturbation growth under conditions existing in the hydrogen-helium layers of neutral stability of massive stars.

The growth of the amplitude of perturbations is limited by turbulent friction. The maximal amplitude is 10^{-5} of the linear dimensions of the semiconvective layer. The turbulent motion leads to a partial mixing of matter in the layer. The necessary mixing velocity is secured during the hydrogen-burning stage if $\nabla_r - \nabla_a$ is about 10^{-5}.

However in spite of important results achieved in the above mentioned investigations, the problem of stability of intermediate layers with varying molecular weight remains unsolved. The view has been expressed that numerous evolutionary computations might provide a rather simple solution to the question of what criterion of stability has to be chosen but these hopes were not realized.

The results of an experiment, studying heat transfer in a liquid with varying salt concentration seem to speak in favour of Schwarzschild criterion (Spiegel, 1969). But the identity of conditions in varying salinity liquids with conditions in stellar interiors has not been yet proved. The main difference between the experiment and reality is that heat conductivity coefficient in a liquid does not depend on salt concentration while in the stellar interiors it is determined by chemical composition (in massive stars).

Thus, up to now it seems necessary to perform two versions of evolutionary computations differing only in assumptions on the stability conditions. It can not be ruled out that both stability conditions might take place in reality. If a strong enough regular magnetic field which prevents turbulence of the main scale motions exists in the zone of varying molecular weight and if the amplitude of pulsations is limited by the stabilizing effect of the radiative envelope, then no mixing will occur and the Ledoux criterion of convective neutrality is realized. For stars without a magnetic field the Schwarzschild criterion remains valid.

The procedure of constructing evolutionary models for massive stars is described in details by Varshavsky and Tutukov (1972). Figure 1 shows how the structure of a

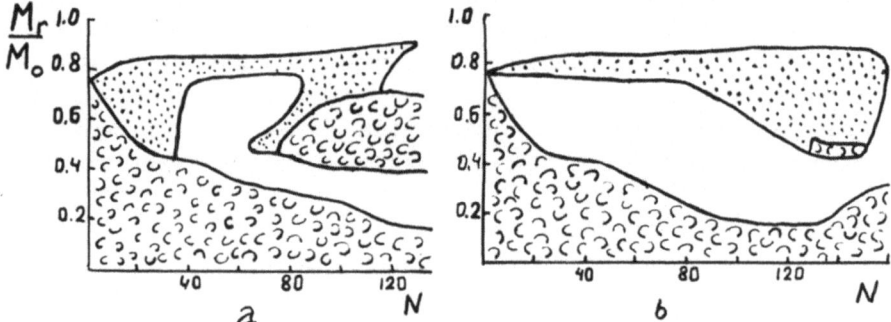

Fig. 1. Evolutionary changes in the structure of the 64 $M_\odot$ star (Varshavsky and Tutukov, 1972), N-number of the evolutionary model. On the left: $\nabla_r = \nabla_a$ in the semiconvective zone, on the right:

$$\nabla_r = \nabla_a + \frac{\beta}{4 - 3\beta} \frac{d \ln \mu}{d \ln p}$$

– convective zones;

– semiconvective zones.

star with 64 $M_\odot$ changes in the course of evolution if either the criterion given by Schwarzschild (Figure 1a) or by Ledoux (Figure 1b) is used. The intermediate convective layer develops in both cases as a result of high concentration of the energy sources in the hydrogen burning shell. The time of evolution in the hydrogen burning stage is for case 1b 10% shorter as compared with 1a (Tutukov, 1972). The most important difference between both sets of models shows in the resulting profiles of hydrogen distribution and in a consequent difference in hydrogen contents in the shell source. The distribution of hydrogen is for stars with $M \lesssim 50 \, M_\odot$ (when the Ledoux criterion is used) smooth and similar to that for a star with $M \lesssim 10 \, M_\odot$. The shell source develops in a region with low hydrogen content: $X \approx 10^{-2}$. If the Schwarzschild criterion is used the hydrogen distribution becomes discontinuous.

As a result of this the hydrogen content in the shell will be as large as 0.4 for a 16 $M_\odot$ star and about 0.2 for a 64 $M_\odot$ star (Varshavsky and Tutukov, 1972). The initial difference in the hydrogen profiles results in two different evolutionary tracks in the stage of helium burning.

Besides it should be noted that the hydrogen shell source situated at the discontinuity of the hydrogen profile will become unstable at early phases of helium burning for $M = 15 M_\odot$ (Stothers and Chin, 1972). The changes in luminosity caused by thermal flashes in the shell are too small to be used as a criterion for choice between the two stability conditions in the semiconvective region. The same instability was found also by Varshavsky (1972a) for 15 $M_\odot$ star and by Noels and Gabriel (1973) for 8 $M_\odot$.

3. The Stability Criterion and Evolution of Stars in the Helium Burning Stage

Starting with 1970, evolution of stars with large mass has been studied intensively in several countries (Chiosi and Summa, 1970; Paczyński, 1970; Simpson, 1971; Massevich et al., 1971; Robertson, 1971). A number of computations have been performed and new results obtained mainly dealing with the helium burning stage.

First, the lower limit of masses for Population I stars has been estimated for which discrepancies connected with different approaches to the stability criterion are developing in the helium burning stage. According to Barbaro et al. (1972) the lower limit is 13 $M_\odot$. Stars with $M \lesssim 13 M_\odot$ evolve, notwithstanding which stability criterion is used, like stars of medium masses, e.g. helium burning starts in the region of red supergiants and continues in the blue supergiants region. The upper limit for the phenomena considered is about 40 $M_\odot$. As Barbaro et al. (1971b) and Varshavsky and Tutukov (1973a) have shown, stars with masses larger than 40 $M_\odot$ have no blue supergiant phase with a lifetime comparable with the time of helium burning in the core: helium is almost completely exhausted in the phase of a red supergiant.

For a mass range $13 M_\odot \lesssim M \lesssim 64 M_\odot$ the following cases should be distinguished.

Case A. If the Ledoux criterion is used, helium burning start in the phase of a red supergiant. The following evolution is completely defined by the depth of penetration of the outer convective envelope into the intermediate layer of varying chemical composition in the stage preceding helium burning (Lauterborn, et al., 1971; Ziołkowski, 1972). Mixing at the inner boundary of the convective envelope causes a discontinuity in the profile of hydrogen distribution. As soon as the shell source developing outwards reaches the region of this discontinuity the star moves from the red supergiants region in the HR diagram to the region occupied by blue supergiants. This displacement occurs in a Kelvin timescale. Similar results have been obtained by Chiosi and Summa (1970) and Robertson (1972).

Case A1. If the convective envelope is not deep enough, the shell source will never reach the discontinuity in the hydrogen distribution during the time of helium burning. That means that the model of the star remains during its helium burning stage in the red supergiants region in the HR diagram (Paczyński, 1970; Varshavsky, 1972a, b; Varshavsky and Tutukov 1973a).

Case A1 can not be considered as the only possible way of evolution for stars of large masses because it can not explain the observed number of blue supergiants. It should be noted that in stars with $M \gtrsim 60\ M_\odot$ the occurrence of a discontinuity of hydrogen distribution caused by the development of a thin intermediate convective layer is possible even if the Ledoux criterion is used (Varshavsky and Tutukov, 1972). But as the shell source never approaches this discontinuity, the models remain in the red supergiant region.

Case B. When the stability criterion given by Schwarzschild is used, helium may be completely exhausted in the blue supergiants region (Case B1) or the star exhausts the main part of its helium fuel as a blue supergiant and only the remainder as a red supergiant (Case B2). Very massive stars, $M \gtrsim 40\ M_\odot$, notwithstanding the existence of an intermediate convective layer and a discontinuity in hydrogen distribution, will exhaust their core helium only in the red supergiants region (Case B3).

For Case A the ratio of lifetimes as blue and red supergiants τ_b/τ_r is highly dependent on uncertainties concerning efficiency of convection and the depth of its penetration. For Case B no such direct dependence has been noted. However we may note that the lifetimes in this case are functions of the profile of hydrogen distribution in the region where the shell source is located, of the absorption coefficient, the theory of convection, the initial chemical composition, possible mass loss and the initial mass of the model. A possibility of the existence of multiple solutions should not be excluded. For a model with $M = 32\ M_\odot$, Varshavsky and Tutukov (1973a) found that the transition from a blue to a red supergiant occurs in the Kelvin timescale (Case B has been investigated). The dependence of the ratio τ_b/τ_r on chemical composition has been studied by Robertson (1972) who found that a change of Z from 0.02 to 0.04 may change the ratio τ_b/τ_r by about three times. That means that even for a fixed stability criterion the lifetimes ratio cannot be determined uniquely. Stothers and Chin (1973a) showed that τ_b/τ_r is increasing with increasing Z, Y, $\varepsilon_{C+\alpha}$ and decreasing $\varepsilon_{3\alpha}$.

A thorough study of the behaviour of thermally stable models with varying chemical composition and different profiles of hydrogen distribution should shed more light on the influence of physical parameters (X, energy sources, convection, mass loss, rotation etc.) on models obtained at late stages of stellar evolution. Kozlowski (1971) computed models for $10\ M_\odot$ in the stage of helium burning with hydrogen profiles corresponding to Case A. He showed that for a certain rather small range of helium core masses two thermally stable solutions exist one for a blue and the other for a red supergiant. Frantsman (1973) and Frantsman et al. (1973) studied the influence of mass loss on the location of $15\ M_\odot$ and $30\ M_\odot$ models in various stages of helium burning. The hydrogen distribution profile was taken according to Case A. They showed, that for a He-content in the core about unity a continuous sequence of models with decreasing mass of the H-envelope can be obtained in the blue supergiants region. They did not succeed in obtaining a similar sequence for a much smaller He-content in the core (e.g. for a more evolved star).

The authors conclude that the necessary condition for obtaining a model of a blue supergiant is the activity of the hydrogen shell source.

The influence of the hydrogen distribution profile on the evolution of massive stars in the helium burning stage has been studied by Barbaro $et\ al.$ (1971b). They introduced two characteristic times determining the prehelium burning evolution.

(1) $\tau_{sh} = X_{sh} \cdot E/\varepsilon$, where X_{sh} is the H-content in the H-shell source, E – energy output by burning 1 g of hydrogen and ε – rate of energy generation in the shell.

(2) τ_d – time of diffusion of the radiation from the shell to the surface, which is about the Kelvin time scale for the envelope. If $\tau_{sh} < \tau_d$, as it usually is the case for small X_{sh}, the change of thermal conditions in the shell occurs at such a rate that the envelope starts to expand due to the increasing energy flux. As soon as the decreasing surface temperature reaches several thousands of degrees, a convective envelope develops and decreases τ_d to a value comparable with τ_{sh}. It should be noted that earlier computations of late stages of evolution, not taking into account convection in the outer layers lead usually to extremely large values of radii.

If the shell source developing outwards approaches the discontinuity in the hydrogen profile, e.g. caused by the outer convective envelope, τ_{sh} increases sharply and becomes larger than τ_d. As a consequence the model is transfered to the region of blue supergiants at a rate comparable to the Kelvin timescale for the envelope.

If in the model considered when leaving the main sequence $\tau_{sh} > \tau_d$ (because of a high H-content in the shell source), thermal stability of helium-burning model will be reached in the blue supergiants region. It should be noted also that the development of an intermediate convective layer and a layer of neutral stability will decrease the value of τ_d in this case.

All mentioned above is true only if the times τ_{sh} and τ_d are smaller than the characteristic time of complete He-exhaustion in the core. The activity of the H-shell source is another necessary condition as has been already mentioned. Briefly summarizing, for thermally stable stellar models with Population I composition the model will be a red supergiant if the H-content in the H-shell source is small and a blue supergiant if it is relatively large. The possibility that two stable solutions (for red and blue stars simultaneously) may exist for a range of X_{sh} values was proved by Kozlowski and Paczyński (1973).

It is interesting to note that Trimble $et\ al.$ (1973) found that stars with $M = 12\ M_\odot$ and 30 $M_\odot$ (for an abundance of heavy elements $Z = 0.001$) reach thermal stability in the helium burning stage in the blue supergiants region even if the Ledoux criterion of stability is applied. Similar results have been obtained by Varshavsky (1973) for models with 16 $M_\odot$ and 32 $M_\odot$ (for $Z = 0$). The decrease of opacity for small Z values is probably sufficient to decrease τ_d even if the H-content in the shell is low. Fricke and Strittmatter (1972) and Höppner and Weigert (1973) point out the importance of the gravitational field of the star core in defining of the position of a core-helium burning model in the HR diagram.

4. Blue and Red Supergiants and the Stability Criteria and the Hypothesis of Electron-Neutrino Interaction

The main uncertainties in the theoretically derived value of the ratio n_b/n_r from evolutionary computations of massive stars arise from the uncertainties concerning the stability criterion, efficiency of convection and the reality of weak interactions. From the observational point of view the uncertainties in deriving this ratio are caused by the difficulty to distinguish blue supergiants from main sequence stars in the stage of hydrogen exhaustion in the core and by uncertainties in the scale of effective temperatures and bolometric corrections for very hot stars. Theoretical values of n_b/n_r are discussed by Stothers and Chin (1969), Chiosi and Summa (1970), Simpson (1971), Robertson (1972), Barbaro *et al.* (1972b), Tutukov and Varshavsky (1973), and Robertson (1973).

As both the theoretical and the observational values appear to be rather uncertain, no final conclusion can be made at present concerning the stability criteria or the existence of weak interactions. This is one of the characteristic features of the development of the theory of stellar evolution. The values of such (important for stellar evolution) parameters as the opacity coefficient or the nuclear energy sources have been obtained either by numerical computations using known physical processes or by extrapolation of experimental results and cannot in general be verified in Earth conditions. The role of the number of various physical phenomena on stellar evolution such as magnetic fields, rotation, convection and semiconvective diffusion is at present not yet fully known. Observations provide an integral effect of all these possible factors and there are almost no possibilities at present to distinguish between them. There are only very few well established observational data that could be used as criteria for various theories of evolution. And regretfully almost for every positive argument there exists a negative one too.

The estimate for n_b/n_r obtained from observational results by Schild (1970) for the solar surroundings is about 1.6. Humphreys (1970) gives $n_b/n_r \approx 2.3$ for a mass range $10\,M_\odot \lesssim M \lesssim 20\,M_\odot$ and $n_b/n_r \approx 10$ for masses $M \gtrsim 20\,M_\odot$. According to Hartwick (1970) n_b/n_r decreases with the distance from the galactic centre from 10 for distances $R < 10$ kpc to about 3 at a distance of 12–13 kpc. These are the main results that can be derived from observations.

Evolutionary tracks for stars in the mass range $11\,M_\odot \lesssim M \lesssim 64\,M_\odot$ (Case B) are plotted on Figure 2 together with the Humphreys data.

Recent investigations have shown that it might be possible to discuss the problems concerning stability criteria and weak interactions separately. Let us summarize the results.

The Role of Neutrino Emission

(1) According to Stothers (1972), if neutrino emission is not taken into account, the ratio n_b/n_r never exceeds 3 due to the long time of depletion of carbon, oxygen, neon and silicon. Chiosi and Summa (1970) took into consideration the binary nature of

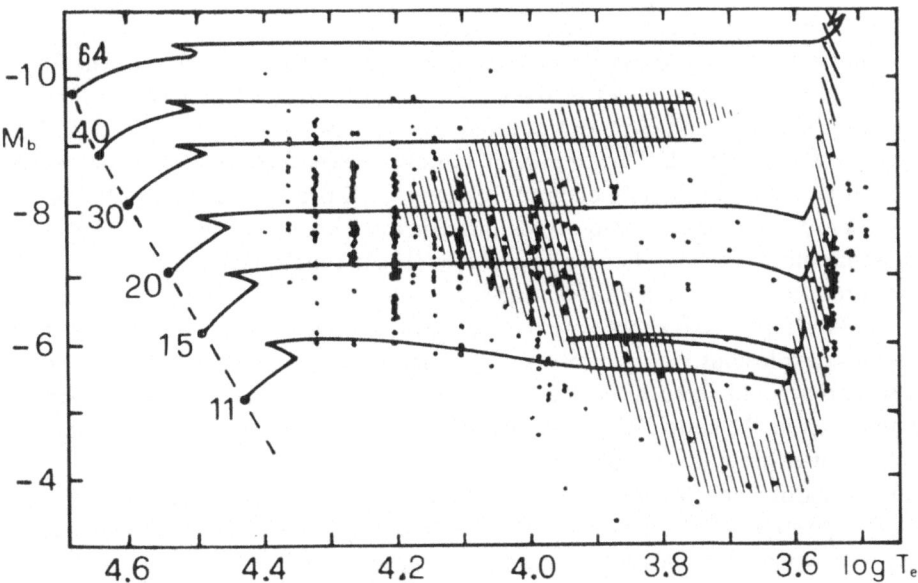

Fig. 2. Evolutionary tracks of 11, 15, 20, 30, 40 $M_\odot$ (Barbaro *et al.*, 1971a) and 64 $M_\odot$ stars (Varshavsky and Tutukov, 1972, 1973a) on the HR diagram. Version B. The position of supergiants after Humphreys (1970) are marked by dots. The hatched areas mark the position of core helium burning models.

a fraction of massive stars, which prevents them from evolving into red supergiants. If neutrino emission is absent then duplicity of massive stars allows us to increase n_b/n_r about one and a half times, but the ratio still remains less than 2 (Barbaro *et al.*, 1971). In the case neutrino emission is present $n_b/n_r \gtrsim 7$. This is considered as an argument in the favour of the existence of neutrino emission.

(2) As core carbon- and oxygen-burning inevitably occurs in the red supergiants region (Stothers and Chin, 1969; Varshavsky and Tutukov, 1973b), the absence of very high luminosity red supergiants provides evidence for acceleration of advanced evolution due to neutrino emission. However core helium burning in stellar models with $M \gtrsim 40 \, M_\odot$ (Case B) usually occurs in the red supergiants region, where the influence of neutrino emission is negligible. It is quite possible that these stars never reach the blue supergiants region (Barbaro *et al.*, 1971; Varshavsky and Tutukov, 1973b). Moreover Bisnovaty-Kogan and Nadyozhin (1972) have shown that all stars with $M \gtrsim 20 \, M_\odot$ lose rapidly the main part of their envelopes, while evolving in the red supergiants region. As a result of this mass loss the models are shifted into the Wolf-Rayet stars region of the HR diagram.

(3) Stothers and Chin (1969), Sugimoto (1970a), and Varshavsky and Tutukov (1973b) showed, that in absence of neutrino emission the outer convective zone in stars with $M \gtrsim 20 \, M_\odot$ penetrates subsequently into the hydrogen and helium shell sources during the core carbon burning stage and for less massive stars during the

core oxygen burning stage. Convective mixing leads to an enrichment of the stellar atmosphere by carbon, oxygen and other heavy species. The duration of the above mentioned evolutionary stages is about 5–20% of the core helium burning lifetime. But Stothers and Chin (1969) did not find any S or N stars in a sample of 50 blue supergiants investigated. This argument however needs further verification, as the position of S-stars in the HR diagram is not yet well defined (Motteran, 1971).

(4) In order to explain the observed abundance of iron, nickel, titanium, chromium and some other elements Fowler and Hoyle (1964) concluded, that these elements were produced at $T \approx 3.8 \times 10^9$ K and $\varrho \approx 3 \times 10^6$ g cm^{-3} during a time period: $3_{10}3 \lesssim \tau(\text{s}) \lesssim 8_{10}4$. Such short time intervals can be secured within the frame of the supernova explosions theory of elements formation only if pair-neutrino emission is taken into account (Ikeuchi et al., 1971, 1972a). However, the changes of chemical composition in the course of supernova explosion and the chemical composition of ejected matter are not very well known.

(5) Demarque and Mengel (1972) find that the agreement between stellar models and the position of the observed horizontal branch in globular clusters may be improved, if neutrino emission is taken into consideration.

Thus all the above mentioned considerations provide evidence in favour of the existence of neutrino emission, but a number of uncertainties in estimation of both the theoretical and the observed ratio n_b/n_r prevents some authors (Barbaro et al., 1972) from definite conclusions concerning the validity of universal weak interaction theory.

The problem of choice between two stability criteria for semi-convection remains very complicated even if neutrino emission is granted. This problem was discussed in detail by Ziołkowski (1972), Varshavsky and Tutukov (1973), and Robertson (1973). Let us describe the main points of this discussion.

(1) Blue supergiants occupy the effective temperature region $3.9 \lesssim \log T_e \lesssim 4.35$ in the HR diagram (see Figure 2). The effective temperatures of core helium burning models are $3.8 \lesssim \log T_e \lesssim 4.2$ for Case B and $4.15 \lesssim \log T_e \lesssim 4.25$ for Case A. (Ziołkowski, 1972). It was shown (Frantsman, 1973; Frantsman et al., 1973) that mass loss (up to 30% of the initial stellar mass) in the red supergiants region is able to extend the interval of effective temperatures of models sufficiently to reach an agreement with observations (for Case A). However no appropriate mechanism of this mass loss is proposed. Besides, there are some difficulties in obtaining models with $\log T_e \approx 4.3$ and $M_b \approx -9$ for Case B. At the same time, for Case A it is possible to compute models with still higher effective temperatures, if the penetration of the convective envelope is accurately enough taken into account. Thus, effective temperatures do not provide a possibility to make a definite choice between the two stability criteria, although Case A seems to agree better with observations if mass loss is taken into account.

(2) If the red supergiant evolutionary stage is preceding the blue supergiant stage, the relative abundance of the CNO-group elements should be changed: the ratio N/C increases during evolution. According to Ziołkowski (1972) N/C is about

5 for α Cyg, but other blue supergiants show normal N/C ratios. Besides several O-stars also have higher ratios N/C than normal stars. It may be possible, that the N/C ratio is influenced not only by convection but also by a number of other processes: e.g. circulation and diffusion.

(3) The variation of periods of long period cepheids (Ziołkowski, 1972). There are 3 long period cepheids in LMC and 3 in SMC. If stellar evolution corresponds to Case A, a part of the cepheids should show the increase of periods and the other part the decrease of periods. In Case B all cepheids should have increasing periods. But in spite of the fast rate of evolution of massive stars in the cepheid strip (it is crossed in some hundreds of years only), it is impossible to detect changes of periods as the periods are themselves very long. Nevertheless, this method appears a very promising one and it might allow in future to come to a definite conclusion concerning the direction of stellar evolution in the Hertzsprung gap.

(4) Stothers and Evans (1970) suggested to employ the statistics of binaries with one blue supergiant component. It is necessary to use short period binaries in order to be sure that the blue supergiant has not evolved into a red one. But the number of such binaries is too small to allow any definite conclusion (Ziołkowski 1972). Moreover even in Case A the primary component of a binary appears after mass loss as a blue supergiant in the core carbon and oxygen burning stages (Tutukov and Yungelson, 1973, see Figure 3). If neutrino emission is absent, lifetime in these evolutionary stages is long enough for the star to be observable.

(5) The evolution of remnants of primary components of massive binaries suffering mass exchange depends on the stability criterion used. As Barbaro et al. (1969) have shown for a 30 $M_\odot$ star evolving according to Case B, if mass exchange starts before the helium burning stage in the blue supergiant region, the effective temperature of the remnant does not exceed 4×10^4 K in core-helium burning stage. Tutukov and Yungelson (1973) studied the evolution of close binaries with mass exchange for primary stars with $10 M_\odot \leqslant M \leqslant 64 M_\odot$, using the Ledoux criterion for convective stability. Their evolutionary tracks are plotted on Figure 3. The effective temperatures of the remnants, which are usually regarded as models of Wolf-Rayet stars are of the order of 8×10^4 K. So it seems that there is a considerable difference between the two cases of evolution. The temperatures of Wolf-Rayet stars are not well known, but it is most agreed that they are very high: Cherepashchuk (1972) obtained $\sim 80\,000$ K for V444 Cyg. This result is an evidence in favour of Case A. But it is necessary to mention, that mass loss which is of great importance for Wolf-Rayet stars may increase their effective temperatures also for Case B.

(6) Different authors obtain different theoretical estimates for the ratio n_b/n_r for different stability conditions, e.g. Chiosi and Summa (1970) came to practically equal values of n_b/n_r for both stability conditions for a 20 $M_\odot$ star. To clarify this problem it is necessary to study the properties of models in thermal equilibrium and their dependence on the H-profile, chemical composition and input physics.

(7) Barbaro and Chiosi (1972) estimated the dependence of chemical composition on the galactocentric distance R of stars using a set of evolutionary tracks com-

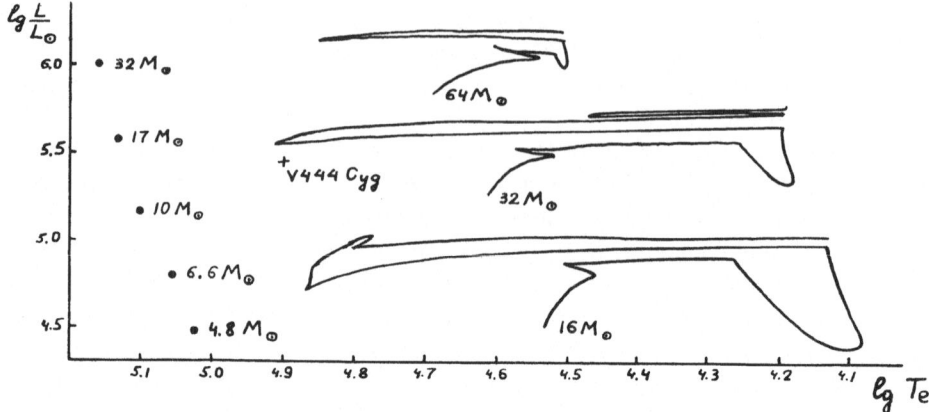

Fig. 3. Evolutionary tracks of primary components of massive binary stars on the HR-diagram.
● – models of homogeneous helium stars ($Y = 0.956$, $Z = 0.044$); + – position of WR-component
of V444 Cygni (Cherepashchuk, 1972).

puted with the Schwarzschild stability criterion. The obtained result is somewhat
striking at first sight. The abundance of heavy elements grows with increasing R:
$Z = 0.02$ for $R \approx 7.5$ kpc and $Z = 0.04$ for $R \approx 10.5$ kpc. This result may be a consequence of uncertainties in the value of n_b/n_r obtained from evolutionary computations. But it appears that the abundance of heavy species must not necessarily
monotonically increase with increasing age. Ikeuchi et al. (1973b) found that if
mass loss by stars with small masses is taken into account the heavy element abundances in the interstellar matter should have grown during the early history of the
Galaxy, as long as the amount of matter lost by supernovae exceeds the flow of
mass lost by stars of small masses. The maximal value of Z is attained after about
10^9 yr. After that the value of Z is decreasing, as the flow of matter from supernovae
becomes smaller than the flow from stars with small masses. This allows to explain
the chemical composition of stars with high metal contents studied by Van den Bergh
and Sackman (1964), Spinrad and Taylor (1969), and Taylor (1970). The results of
Barbaro and Chiosi can be also explained if we assume that the rate of chemical
evolution of the periphery of the Galaxy is slower than that in the central parts
owing to a lower star density. The absence of active mixing leads then to big values of
Z for those remote regions.

5. Chemical Composition of Stellar Matter After the Depletion of Helium

Conditions defining the chemical composition of stellar matter after core helium
depletion were examined by Uus (1970), Arnett (1972), Ferrari et al. (1972), and
Varshavsky and Tutukov (1973a). An analysis of the chemical composition variation
equations showed that the final composition is defined by temperature and density of
stellar matter in the course of helium burning. On Figure 4 are plotted constant carbon
abundance lines for the final configuration on the $\log T - \log \varrho$ diagram according

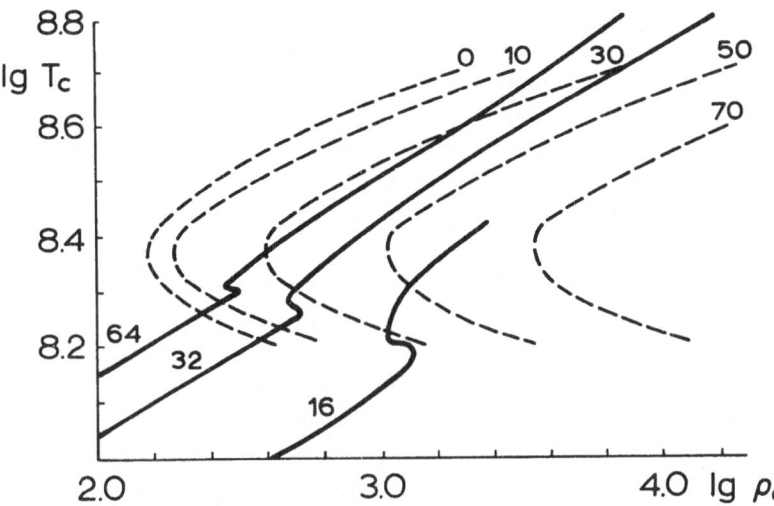

Fig. 4. Evolutionary tracks of stellar centres in the $\lg T - \lg \varrho$ plane (heavy lines). $M = 16, 32,$ 64 $M_{\odot}$. Constant final carbon abundance curves are marked (dashed lines) after Varshavsky and Tutukov (1973a). Numbers near the dashed lines show the percentage of carbon.

to Varshavsky and Tutukov (1973). The values of the concentrations can be obtained by the numerical solution of a very simple equation. Helium burning occurs mainly near the centre of the star, but the position of this centre on the $\log T - \log \varrho$ diagram varies. It follows from evolutionary computations, that the main part of helium is depleted in the upper part of the 'S-like' bend of the path in Figure 4. This allows us to explain the variation in composition of the final helium burning products with stellar mass. This method can be also used to estimate the variations of nuclear reaction rates. E.g. Austin *et al.* (1971) varied the rate of the 3α-reaction, and Weisser *et al.* (1972) – the rate of the $C + \alpha$ reaction. The variations of the reaction rates shift the set of constant carbon abundance lines in the $\log T - \log \varrho$ diagram and lead to slight deformations of their upper parts. Tutukov and Varshavsky (1973a) showed that due to the above mentioned variations the final carbon content changes from 0.45 to 0.2 for a 16 $M_{\odot}$ star and from 0.3 to 0.10 for a 64 $M_{\odot}$ star.

It is possible to estimate in the same way the neon formation efficiency. After helium depletion the abundance of core neon is 0.003 for a 16 $M_{\odot}$ star and 0.03 for a 64 $M_{\odot}$ star (Varshavsky and Tutukov, 1973a). Similar values were obtained also by Arnett (1972b). It is worthwhile to mention that for new neon-formation reactions rates given by Toevs *et al.* (1972) the estimated final neon abundance does not change significantly.

6. Core Carbon, Oxygen, Neon, Silicon and Nickel-Burning Stages of Evolution of Massive Stars

Direct computations of evolution of massive stars from the main sequence up to the iron core formation stage are heavily embarrassed by the necessity to deal with very complex models consisting of many layers with a number of nuclear burning shells. The large number of meshpoints necessary for precise computations also does not facilitate the task. One possible way to overcome some of those difficulties is to compute a number of models in thermal equilibrium for regions in the HR diagram representing late stages of evolution. In such a way Stothers and Chin (1969) computed core carbon, oxygen and neon-burning models for red supergiants region of the HR diagram.

The evolution of helium models, representing cores of normal stars up to the oxygen depletion stage was studied by Sugimoto (1970a, b) for 3 $M_\odot$ and 10 $M_\odot$ stars. Arnett (1972a, b, c) also studied carbon and oxygen burning starting from pure helium models. Core oxygen burning was studied also by Woosley *et al.* (1972), Rakavy and Shaviv (1967), Barkat *et al.* (1967), and Fraley (1968).

Varshavsky and Tutukov (1972, 1973a, b) computed evolutionary tracks for 32 $M_\odot$ and 64 $M_\odot$ stars starting from hydrogen burning on the main sequence up to oxygen depletion in the core. The computations were performed for both the Ledoux and Schwarzschild criteria of convective stability in the zone of variable molecular weight. Their results confirmed the conclusion of Stothers and Chin (1969) that the core carbon and oxygen-burning models appear as red supergiants in the HR diagram. Tutukov and Varshavsky also found that evolution after the start of carbon-burning does not noticeably depend on assumptions concerning stability conditions in the layer of variable molecular weight.

Evolution of carbon-oxygen cores with 1.5 $M_\odot$, 2.6 $M_\odot$, 5 $M_\odot$, 10 $M_\odot$, 30 $M_\odot$ up to formation of an iron core was studied by Ikeuchi *et al.* (1971, 1972a). Evolutionary tracks in the $\log T_c - \log \varrho_c$ plane for 5 $M_\odot$, 10 $M_\odot$ and 30 $M_\odot$ cores with neutrino emission taken into account are plotted on Figure 5. Regions of dynamical instability

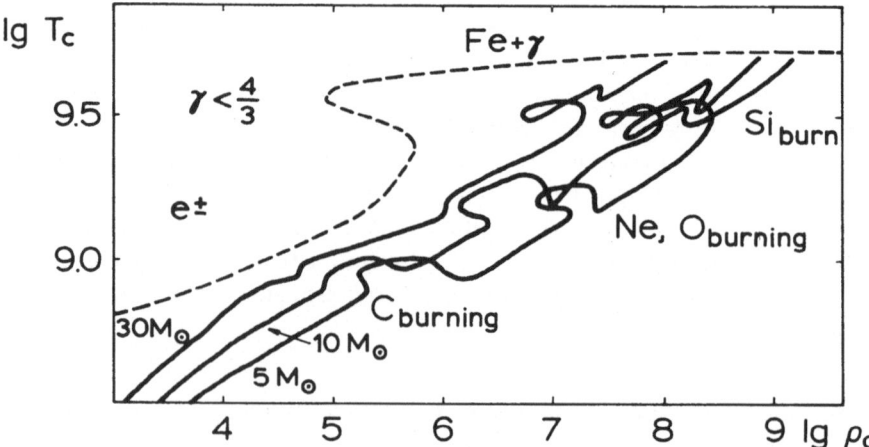

Fig. 5.   Evolution of central values of temperature and density of carbon-oxygen stars (Ikeuchi *et al.*, 1971, 1972a).

due to pair formation and iron photodissociation are also marked on the same figure. A comparison of results obtained by Varshavsky and Tutukov (1972, 1973a, b) and by Ikeuchi *et al.* (1971, 1972a) shows that carbon-oxygen cores with 2.5 $M_\odot$, 10 $M_\odot$ and 30 $M_\odot$ correspond to cores of main sequence stars with 16 $M_\odot$, 32 $M_\odot$ and 64 $M_\odot$. Thus, the two papers are supplementing each other in some respect. The main shortcoming of the carbon-oxygen cores evolution computations is the impossibility of taking into account the decrease of mass of the core caused by penetration of the outer convective zone.

Evolutionary changes of the structure of a 10 $M_\odot$ star are shown on Figure 6 (Ikeuchi *et al.*, 1971). The upper drawing corresponds to the case, when no neutrino emission is taken into account, the lower corresponds to evolution with pair- and

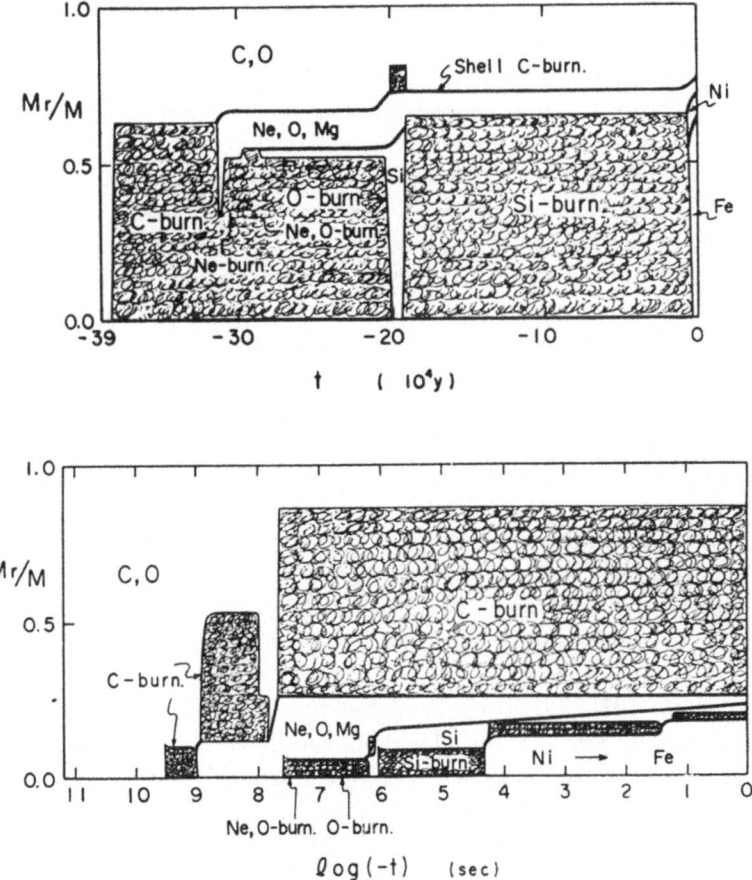

Fig. 6. Evolutionary changes of the structure of carbon-oxygen 10 $M_\odot$ star (Ikeuchi *et al.*, 1971). Upper drawing corresponds to evolution without neutrino emission, lower one – to evolution with neutrino emission.

photoneutrino emission. The main structural differences are the following: models with neutrino emission have small convective cores and thick convective nuclear burning shells, whereas models without neutrino emission have large convective cores and nearly always radiative shell sources. Neutrino emission accelerates the evolution: the carbon depletion time shortens about 10^2 times, the oxygen and neon depletion time $\sim 10^4$ times, the silicon depletion time $\sim 10^7$ times.

As Varshavsky and Tutukov (1973a) have found, only one or two nuclear burning shells nearest to the core are active, all outer shell sources are dying out in succession and getting absorbed by the convective envelope if neutrino emission is absent. For models with neutrino emission all shell sources are active during the computed evolutionary stages. Evolution without neutrino occurs nearly homologously, the point representing the centre of the star moves in the $\log T - \log \varrho$ plane along an almost straight line, for stars with a core $M_{CO} \geqslant 5 \, M_\odot$. Neutrino emission leads to higher density of the core and to a higher degree of degeneracy. The temperature needed for starting of burning of each successive nuclear fuel is larger by about 25% if neutrino emission is taken into account.

If pair- and photo-neutrino processes are taken into account, the computations become more complicated, as the energy flow in the interiors of the model changes by several orders on the way from one shell source to another. This implies the necessity to introduce a large number of new meshpoints in the nuclear burning layers. At the inner boundaries of the carbon, oxygen and silicon shell sources the energy flow is usually directed inwards, so that it compensates the neutrino energy losses from the interior parts of the star.

If neutrino emission is absent, the burning of nuclear fuel starts each time in the centre of the star. In models with neutrino losses taken into account, as have shown Ikeuchi *et al.* (1972) for $M_{CO} = 1.5 \, M_\odot$ and $2.6 \, M_\odot$ stars, nuclear burning starts in the layer surrounding the degenerate core. This creates conditions favourable for appearance of the Rayleigh-Taylor instability, but the mixing does not in a considerable way influence the evolution. The lowest mass of the chemically homogeneous core, necessary to start the following stage of nuclear burning is also increased if neutrino energy losses are present (Ikeuchi *et al.*, 1972). The lowest critical mass for carbon burning grows from $0.75 \, M_\odot$ to $1.06 \, M_\odot$, for neon burning – from $0.9 \, M_\odot$ to $1.1 \, M_\odot$, for oxygen burning – from $0.93 \, M_\odot$ to $1.15 \, M_\odot$, for silicon burning – from $1.4 \, M_\odot$ to $1.55 \, M_\odot$.

Ikeuchi *et al.* (1971, 1972a) distinguished two kinds of thermal flashes in shell sources occurring in all cases with and without neutrino emission: flashes in degenerate and nondegenerate layers. Particularly unstable are shell sources in carbon-oxygen stars with $M_{CO} = 1.5 \, M_\odot$ and $2.6 \, M_\odot$ if neutrino energy losses are taken into consideration. If the flashes do not result in mixing of layers with different chemical composition they do not considerably influence the evolution. The only effect is an increase of the burning time of the given nuclear fuel. Stothers and Chin (1973) found that the helium burning shell in a $15 \, M_\odot$ star is thermally unstable with a characteristic time of the instability growth about some tens of years.

Let us discuss the reasons for collapse of cores of different mass stars. If neutrino emission is absent, carbon-oxygen cores with mass $M_{CO} \gtrsim 30\ M_\odot$ (this corresponds to main sequence stars with mass $M \gtrsim 64\ M_\odot$) are collapsing owing to formation of electron-positron pairs (Rakavy and Shaviv, 1967; Barkat et al., 1967; Fraley, 1968). The explosion of the star in this case was studied by Fraley (1968). The evolution of carbon-oxygen cores with $1.4\ M_\odot \lesssim M_{CO} \lesssim 30\ M_\odot$ (corresponding to main sequence stars with masses $9\ M_\odot \lesssim M \lesssim 64\ M_\odot$) can be followed without neutrino emission still to the stage of formation of an iron core with $M_{Fe} = 0.5\text{–}0.7\ M_{CO}$. The iron core collapses owing to photodissociation of the iron nuclei.

If pair- and photoneutrino cooling is taken into consideration, all aforementioned estimates of limiting masses will be increased. Collapse caused by pair formation in carbon-oxygen cores will occur for $M_{CO} \gtrsim 40\ M_\odot$ (main sequence stars with masses $M \gtrsim 90\ M_\odot$). For collapse caused by photodissociation of iron nuclei the mass of the carbon-oxygen cores will be $2.6\ M_\odot \lesssim M_{CO} \lesssim 40\ M_\odot$ ($16\ M_\odot \lesssim M \lesssim 90\ M_\odot$ on the main sequence). The lowest mass of a collapsing iron core will be $\sim 3\ M_\odot$ for a carbon-oxygen star with $M_{CO} = 30\ M_\odot$ and $M_{Fe} \approx 1.3\ M_\odot$ for $2.6\ M_\odot \lesssim M_{CO} \lesssim 10\ M_\odot$. And finally carbon-oxygen cores $1.4\ M_\odot \lesssim M_{CO} \lesssim 2.6\ M_\odot$ ($9\ M_\odot \lesssim M \lesssim 16\ M_\odot$ on the main sequence) collapse owing to neutronization of silicon or iron (Ikeuchi et al., 1972). The upper limit $\sim 2.6\ M_\odot$ is not well defined as, because of the complexity of interaction of different energy sources, the evolutionary path of the stellar centre in the $\log T - \log \varrho$ plane is very sensitive to the input physics. In the course of computation of the evolution of carbon-oxygen stars with $M_{CO} = 1.5\ M_\odot$ and $2.6\ M_\odot$ Ikeuchi et al. (1972) have increased the neutrino energy loss rate 5 to 20 times as compared to values given by Beaudet et al. (1967). This might have led to an overestimate of the upper limiting mass: $M_{CO} \approx 2.6\ M_\odot$.

The influence of rotation, magnetic fields and mass loss on advanced evolution of massive stars has not yet been studied in details. It is evident however that even the loss of almost the whole hydrogen envelope in the red supergiant stage cannot change noticeably the evolution of the core. The role of rotation in advanced evolution has up to now attracted little attention. If we assume, that the local angular momentum is conserved in the course of evolution, the angular velocity of the core would increase considerably and the ratio of centrifugal force to gravitation α approaches unity. Under such conditions the velocity of circulation is strongly enhanced and the molecular weight barrier for late stages reactions becomes low (Varshavsky and Tutukov 1973b). As Varshavsky and Tutukov showed, if $\alpha_0 \gtrsim 0.06$ on the surface of the main sequence star, total depletion of carbon within the carbon-oxygen core becomes possible. If there exists a mechanism supporting the carbon-oxygen core on the verge of rotational stability, it is possible for all successive reactions to occur also within the original core borders.

The distribution of chemical elements of the presupernova will in this case be considerably changed. The evolutionary lifetimes will be also increased several times.

Conclusion

Summarizing we may state with some satisfaction that at present the evolutionary path of massive stars, starting from the zero age main sequence up to the occurrence of dynamical instability in the core of a star in a very late stage of evolution actually preceding a presupernova can be followed not only in a general outline but also in some details. The main results in this field mainly for advanced stages have been achieved in the last 3–4 years and the amount of work (particularly computations) implied is amazing. It has also to be stated that at present we are much better aware of the difficulties and 'weak points' of the present theory then perhaps 3–4 years ago and that in some aspects we even do not see a way out how to overcome or avoid these difficulties. This should by no means lead us to a pessimistic view for the future. It is not the first time that the theory of stellar evolution faces awkward problems for the solution of which new physical theories or quite new observational data have been needed. There were even worse situations in the past.

What really is very badly needed to improve our present views on advanced stages of evolution of stars of large masses is a better theory of convective envelopes (one possible solution may be a theory of nonlocal convection and time dependent convection) a relevant mechanism of mass loss, improved bolometric corrections and spectral or effective temperature of blue and red supergiants, improved nuclear reaction rates for heavy nuclei etc. A very important up to date problem is the study of the influence of rotation and magnetic fields on stellar evolution. This problem is of interest for the theory of evolution of stars of all masses. A two-dimensional approach to the solution of the stellar constitution equations seems to be the most promising way in this respect.

There is another problem that deserves thorough consideration. Recently a large amount of work has been carried out on studying supernovae explosions. This is the topic of one of the following reviews at this Symposium. Now the problem consists in that there remains a gap between the most advanced stage of evolution obtained by means of the present theory of stellar evolution with nuclear energy sources (a 'last' stellar model preceding a presupernova, as stated above) and the 'presupernova' model that is the starting point for the theory of supernovae explosions. The discrepancies between these two models are very large, particularly because the theory of supernova explosions starts usually with a very simplified model. We have tried to show above how very important for advanced stages are details concerning the structure of models at early stages of evolution: initial chemical composition, hydrogen profiles, intermediate convective layers, layers of varying chemical composition etc. The same is doubtless true for the initial presupernova model. There remains now to cover this gap and to achieve a fit between both these theories. This is by no means an easy task but an absolutely necessary step towards understanding of the latest stages of evolution of a star of large mass.

References

Appenzeller, I.: 1970a, *Astron. Astrophys.* **5**, 355.
Appenzeller, I.: 1970b, *Astron. Astrophys.* **9**, 216.
Arnett, W. D.: 1972a, *Astrophys. J.* **173**, 393.
Arnett, W. D.: 1972b, *Astrophys. J.* **176**, 681.
Arnett, W. D.: 1972c, *Astrophys. J.* **176**, 699.
Arnett, W. D.: 1973, *Astrophys. J.* **179**, 249.
Auré, J.-L.: 1971, *Astron. Astrophys.* **11**, 345.
Austin, S. M., Trentelman, G. F., and Kashy, E.: 1971, *Astrophys. J.* **163**, 79.
Barbaro, G., Giannone, P., Giannuzzi, M. A., and Summa, C.: 1969, in *Proc. of the 2nd Trieste Colloq. on Astrophys.* p. 217.
Barbaro, G., Chiosi, C., and Nobili, L.: 1971a, in *Proc. of the 3rd Colloq. on Astrophys.*, p. 313.
Barbaro, G., Chiosi, C., and Nobili, L.: 1971b, in *Proc. of the 3rd Colloq. on Astrophys.* p. 334.
Barbaro, G., Chiosi, C., and Nobili, L.: 1972, *Astron. Astrophys.* **18**, 186.
Barbaro, G. and Chiosi, C.: 1972, in *Proc. of the IAU Colloq.*, No. 17, Meudon, Section XV.
Barkat, Z., Rakavy, G., and Sack, N.: 1967, *Phys. Rev. Letters* **18**, 379.
Beaudet, G., Petrosian, V., and Salpeter, E. E.: 1967, *Astrophys. J.* **150**, 979.
Bisnovatyi-Kogan, G. S. and Nadyozhin, D. K.: 1972, *Astrophys. Space Sci.* **15**, 353.
Cherepashchuk, A. M.: 1972, *Astron. Tsirk.* **739**, 1.
Chiosi, C. and Summa, C.: 1970, *Astrophys. Space Sci.* **8**, 478.
Dallaporta, N.: 1971, in *Proc. of the 3rd Colloq. on Astrophys.*, p. 250.
Demarque, P. and Mengel, J. G.: 1972, *Nature Phys. Sci.* **239**, 55.
Dudurov, A. and Tutukov, A.: 1972, *Nauch. Inform. Moscow* **21**, 3.
Ferrari, A., Gillino, R. and Masani, A.: 1972, *Mem. Soc. Astron. Ital.* **43**, 731.
Fowler, W. A. and Hoyle, F.: 1964, *Astrophys. J. Suppl.* **91**, 201.
Fraley, G. S.: 1968, *Astrophys. Space Sci.* **2**, 96.
Frantsman, Ju. L.: 1973, *Nauch. Inform. Moscow* **26**, 62.
Frantsman, Ju. L., Popova, E. I., and Ziolkowski, J.: 1973, *Nauch. Inform. Moscow* **27**, 54.
Fricke, K. J. and Strittmatter, P. A.: 1972, *Monthly Notices Roy. Astron. Soc.* **156**, 129.
Gabriel, M.: 1968, *Astron. Astrophys.* **1**, 321.
Hartwick, F. D. A.: 1970, *Astrophys. Letters* **7**, 151.
Höppner, W. and Weigert, A.: 1973, *Astron. Astrophys.* **25**, 99.
Humphreys, R. M.: 1970, *Astrophys. Letters* **6**, 1.
Ikeuchi, S., Nakazawa, K., Murai, T., Höshi, R., and Hayashi, C.: 1971, *Prog. Theor. Phys. Kyoto* **46**, 1713.
Ikeuchi, S., Nakazawa, K., Murai, T., Höshi, R., and Hayashi, S.: 1972a, *Prog. Theor. Phys. Kyoto* **48**, 1870.
Ikeuchi, S., Sato, H., Sato, T., and Takeda, H.: 1972b, *Prog. Theor. Phys. Kyoto* **48**, 1885.
Kato, S.: 1966, *Publ. Astron. Soc. Japan* **18**, 374.
Kozlowski, M.: 1971, *Astrophys. Letters* **9**, 65.
Kozlowski, M. and Paczyński, B.: 1973, *Acta Astron.* **23**, 65.
Lauterborn, D., Refsdal, S., and Roth, M. L.: 1971, *Astron. Astrophys.* **13**, 119.
Ledoux, P.: 1941, *Astrophys. J.* **94**, 537.
Ledoux, P.: 1947, *Astrophys. J.* **105**, 305.
Massevich, A. G.: 1970, *Stellar Constitution*, Presidents Report, 1970, Draft Reports IAU, Brighton.
Massevich, A. G., Tutukov, A. V., Dluzhnevskaya, O. B., Varshavsky, V. J., Uus, U., Ergma, Eh. V., Popova, E. I., and Rodionova, G. G.: 1971, *Nauch. Inform. Moscow* **19**, 45.
Massevich, A. G. and Schustov, B. M.: 1972, *Physics and Evolution of Stars*, Astronomy Vol. 8, VINITI Reviews on Science and Technics, Moscow.
Motteran, M.: 1971, in *Proc. of the 3rd Colloq. on Astrophys.*, p. 292.
Noels, A. and Gabriel, M.: 1973, *Astron. Astrophys.* **24**, 201.
Paczyński, B.: 1970, *Acta Astron.* **20**, 195.
Pontekorvo, B. M.: 1959, *Zh. Exp. Theor. Phys.* **36**, 1915.
Rakavy, G. and Shaviv, G.: 1967, *Astrophys. J.* **148**, 803.
Robertson, J. W.: 1971, *Proc. Astron. Soc. Australia* **2**, 23.
Robertson, J. W.: 1972, *Astrophys. J.* **177**, 473.

Robertson, J. W.: 1973, *Astrophys. J.* **180**, 425.
Ruben, G.: 1969, *Nauch. Inform. Moscow* **14**, 3.
Ruben, G. and Massevich, A. G.: 1966, *Nauch. Inform. Moscow* **3**, 36.
Sakashita, S. and Hayashi, C.: 1959, *Prog. Theor. Phys. Kyoto* **22**, 830.
Sakashita, S. and Hayashi, C.: 1961, *Prog. Theor. Phys. Kyoto* **26**, 942.
Schild, R. F.: 1970, *Astrophys. J.* **161**, 855.
Schwarzschild, M. and Härm, R.: 1958, *Astrophys. J.* **128**, 348.
Schwarzschild, M. and Härm, R.: 1959, *Astrophys. J.* **129**, 637.
Simpson, E. E.: 1971, *Astrophys. J.* **165**, 295.
Spiegel, E. A.: 1969, *Comments Astrophys. Space Phys.* **1**, 57.
Spinrad, H. and Taylor, B. J.: 1969, *Astrophys. J.* **157**, 1279.
Stothers, R. and Simon, N.: 1968, *Astrophys. J.* **152**, 233.
Stothers, R. and Chin, C.-W.: 1969, *Astrophys. J.* **158**, 1039.
Stothers, R. and Evans, T. L.: 1970, *Observatory* **90**, 186.
Stothers, R.: 1972a, *Astrophys. J.* **175**, 431.
Stothers, R.: 1972b, *Astrophys. J.* **175**, 717.
Stothers, R. and Chin, C.-W.: 1972, *Astrophys. J.* **177**, 155.
Stothers, R. and Chin, C.-W.: 1973a, *Astrophys. J.* **179**, 555.
Stothers, R. and Chin, C.-W.: 1973b, *Astrophys. J.* **182**, 209.
Sugimoto, D.: 1970a, *Prog. Theor. Phys. Kyoto* **44**, 375.
Sugimoto, D.: 1970b, *Prog. Theor. Phys. Kyoto* **44**, 599.
Taylor, B. J.: 1970, *J. Suppl., Ser. Kyoto* **22**, 177.
Toevs, J. W., Fowler, W. A., Barnes, C. A., and Lyons, P. B.: 1971, *Astrophys. J.* **169**, 421.
Trimble, V., Paczyński, B., and Zimmerman, B. A.: 1973, *Astron. Astrophys.* **25**, 35.
Tutukov, A. V.: 1972, *Proc. of the IAU Colloq.*, No. 17, Meudon, Section XIV.
Tutukov, A. V. and Yungleson, L. R.: 1973, *Nauch. Inform. Moscow* **27**, 1.
Uus, U.: 1970, *Nauch. Inform. Moscow* **17**, 35.
Van den Bergh, S. and Sackman, I. J.: 1965, *Astron. J.* **70**, 133.
Varshavsky, V. I.: 1972a, *Nauch. Inform. Moscow* **21**, 25.
Varshavsky, V. I.: 1972b, *Astron. Zh. Akad. Nauk SSSR* **49**, 1055.
Varshavsky, V. I. and Tutukov, A. V.: 1972, *Nauch. Inform. Moscow* **23**, 47.
Varshavsky, V. I. and Tutukov, A. V.: 1973a, *Nauch. Inform. Moscow* **26**, 35.
Varshavsky, V. I. and Tutukov, A. V.: 1973b, *Nauch. Inform. Moscow* **27**, 73.
Varshavsky, V. I.: 1973c, *Nauch. Inform. Moscow* **27**, 96.
Weisser, D. C., Morgan, J. F., and Thompson, D. R.: 1972, in press.
Woosley, S. E., Arnett, W. D., and Clayton, D. D.: 1972, *Astrophys. J.* **175**, 731.
Ziolkowski, J.: 1972, *Acta Astron.* **22**, 327.

STABILITY PROPERTIES OF STARS
WITH CENTRAL He BURNING*

(Abstract)

M. GABRIEL, S. REFSDAL, and H. RITTER

Hamburger Sternwarte, University of Hamburg, F.R.G.

Stars with central He-burning have recently been studied in great detail by several groups. One of the most interesting results has been that several static $(\varepsilon_g = 0)$ solutions with the same chemical profile sometimes exist for such stars when $M \gtrsim 9\ M_\odot$ (Lauterborn *et al.*, 1971; Kozlowski, 1971). Which of these solutions the corresponding evolutionary models will choose depends on the evolutionary history of the star. Stability properties of such stars are therefore very important, since instabilities (if they exist) may trigger a transition from one branch of solutions to an other branch.

We have therefore carried out a complete secular stability analysis (including a search for complex eigenvalues) for the sequence of models discussed in Lauterborn *et al.* (1971, § 5). These models have $M = 9\ M_\odot$, homogeneous He-cores of mass M_c and the envelopes have step-profiles with $X = 0.302$ between $M_r = M_c$ and $M_r = M_1$, and $X = 0.602$ for $M_r > M_1$. M_1 was chosen constant $(M_1 = 0.203\ M)$ and M_c-values between $0.1890\ M$ and $0.1910\ M$ were considered. In agreement with earlier results we found 3 different static solutions for M_c-values between $0.1897\ M$ and $0.1908\ M$. The solution with the lowest value of T ($\log T \approx 3.63$), and the one with the highest value $(3.82 > \log T > 3.70)$ were always found to be secularly stable (all real eigenvalues are negative, and all complex eigenvalues have negative real parts). The models with an intermediate value of T_{eff} were, however, always found to be unstable (one real eigenvalue is positive). For M_c-values smaller than the triple solution region we found only solutions with low effective temperatures ($\log T \lesssim 3.62$) and for larger M_c-values only high effective temperatures ($\log T > 3.82$). These solutions are always found to be stable. The results mentioned above are consistent with results based on general arguments concerning changes of eigenvalues in turning points of a linear series (Gabriel and Ledoux, 1967; Paczyński, 1972). The present investigation gives, however, more information than one can obtain from such general arguments. The information on the complex eigenvalues can for instance be obtained from a complete secular stability analysis only, as we have done here.

The results obtained above indicate that the branch of intermediate solutions is never reached in evolutionary sequences, and that a transition from the branch with low T_{eff} to the branch with high T_{eff} (or vice versa) can only be initiated from a turning point $(\partial M_c / \partial T_{\text{eff}} = 0)$ of the linear series.

* This paper was presented by S. Refsdal.

References

Gabriel, M. and Ledoux, P.: 1967, *Ann. Astrophys.* **30**, 975.
Kozlowski, M.: 1971, *Astrophys. Letters* **9**, 65.
Lauterborn, D., Refsdal, S., and Roth, M. L.: 1971, *Astron. Astrophys.* **10**, 97.
Paczyński, B.: 1972, *Acta Astron.* **22**, 163.

THE EFFECT OF CHANGING THE INITIAL CHEMICAL
COMPOSITION ON THE EVOLUTION OF A 20 $M_\odot$ STAR*

(Abstract)

G. BARBARO, G. BERTELLI, C. CHIOSI, and E. NASI

Istituto di Astronomia, Universita di Padova, Italia

The evolution from the Main Sequence up to central He depletion of a 20 $M_\odot$ star with six different sets of initial chemical composition parameters has been followed. Semiconvective regions have been analysed according to the Schwarzschild-Härm criterion.

The comparison of the different evolutionary tracks seems to indicate that during central He-burning two main phases can be displayed: a first one, at the beginning of the burning ($Y_c \gtrsim 0.500$), in which the evolutionary effects are essentially due to the advancement of the H-burning shell, and a late one ($Y_c \lesssim 0.200$), whose behaviour can be accounted for only by the core chemical evolution, with the exception of a phase of secular instability.

This conclusion is supported also by computation of fictitious evolutionary sequences of models in thermal equilibrium both with constant central He content and with constant core mass, which respectively mimic the first and the late phase.

Theoretical distributions of supergiant stars in the different regions of the HR diagram are derived in the cases of both the Schwarzschild-Härm and the Ledoux criterion, and compared with observed distributions of galactic supergiant stars. The comparison seems to indicate that a better agreement is obtained when the neutrality condition adopted for semiconvective region is the Schwarzschild-Härm condition.

A more complete account of this work is to be published in *Astronomy and Astrophysics*.

* This paper was presented by C. Chiosi.

EVOLUTION OF MASSIVE CLOSE BINARIES*

(Abstract)

L. YUNGELSON and A. V. TUTUKOV

The Astronomical Council of the Academy of Science, U.S.S.R.

(1) The great interest to the evolution of massive binary stars ($M_1 \gtrsim 10\ M_\odot$) is to a considerable degree stimulated by a close connection between their advanced evolution and the origin of a part of W-R stars and of X-ray sources.

(2) In the present communication the results of computations of evolution of primary components of close binary systems with masses 16 $M_\odot$, 32 $M_\odot$, 64 $M_\odot$ are presented. The mass exchange in systems under consideration starts after hydrogen exhaustion in the primary core. The Ledoux stability criterion is used for handling of semiconvection.

(3) in all those systems the mass exchange lasts some thousand years only. The mass of remnants of primaries M_f after mass exchange is related to their initial mass M_i by an approximate relation $M_f/M_\odot = 10^{-0.96}(M_i/M_\odot)^{1.4}$. Mass exchange stops when the surface is reached by layers with a hydrogen content $X \approx 0.2$. After detaching from the Roche lobe the primary remnant moves rapidly to the region of the HR-diagram occupied by Wolf-Rayet stars. The range of masses of the remnants 5–35 $M_\odot$, their radii 1–7 $R_\odot$ and effective temperatures 50000–80000 K fit well the range of the same parameters for hot compact cores of WR stars. In order to fit the theoretical periods of binaries with WR components to observations it is necessary to assume a considerable mass and angular momentum loss from the system. Mass loss always occurs in such a way that first a WR star of the nitrogen sequence is formed. But the position of layers enriched by carbon allows the star to evolve into a carbon WR star by means of mass loss with a rate 10^{-6}–$10^{-5}\ M_\odot$ yr^{-1} during the time of He exhaustion in the core.

(4) In the course of further evolution in core carbon and oxygen burning stages it is possible for the star to fill the Roche lobe several times but the subsequent mass loss is small – only some tenth of a solar mass. It is possible to observe the star in those evolutionary stages only if the neutrino emission is absent.

(5) After the primary explodes as a supernova, the systems remain bound as always the less massive component explodes. If the explosion is spherically symmetric this system with a component – a neutron star or a 'black hole' receives velocity up to 100 km s^{-1}, high enough to bring the system to the distance up to 100 pc from the galactic plane before the second supernova explosion in the system. The accretion on the relativistic component may lead to X-ray emission. After the second component explodes as a supernova, the system, as a rule, disrupts, giving birth to two single neutron stars or 'black holes', which may appear as sources of radio of X-ray

---

* This paper was presented by L. Yungelson.

emission. The spatial distribution of pulsars along the Z-coordinate agrees well with the observed (Gunn and Ostriker, 1970).

Reference

Gunn, J. E. and Ostriker, J. P.: 1970, *Astrophys. J.* **160**, 979.

ON THE EVOLUTIONARY STAGE OF V448 CYG

(Abstract)

M. KUMSIASHVILY

Astronomical Council of the Academy of Science, U.S.S.R.

The light-curve of the massive binary star V448 Cyg has been investigated. The mass of the components and rate of mass exchange is estimated; the evolutionary stage of the star is determined. It appears that in V448 Cyg the mass exchange possibly occurs for the second time.

THE INVESTIGATION OF OPEN CLUSTERS BY MEANS
OF THEORETICAL HR-DIAGRAMS*

(Abstract)

O. B. DLUZHNEVSKAYA and A. E. PISKUNOFF

Astronomical Council of the Academy of Science, U.S.S.R.

An investigation of the initial mass function for 19 open clusters of different ages was carried out. The membership of stars in clusters was determined by means of UBV-photometry data, proper motions or radial velocities. The masses of cluster members were estimated using Paczyński's (1970) evolutionary sequences isochrones (Dluzhnevskaya *et al.*, 1971) and B.C. and T_e scales given by Johnson.

Assuming that the mass function may be expressed as $f(\mathfrak{M}) = a\mathfrak{M}^{-\alpha}$ the values of 'a' and 'α' were derived for each cluster. It is shown that 'α' is an increasing function of the cluster age, while 'a' correlates with the cluster mass.

This may be interpreted as an evidence of certain evolutionary changes in the initial mass functions for stars with masses larger than 8 $\mathfrak{M}_{\odot}$ in the solar vicinity.

References

Dluzhnevskaya, O. B., Musylev, V. V., and Rodionova, G. G.: 1971, *Nauch. Inform. Moscow* **21**.
Johnson, H. L.: 1966, *Ann. Rev. Astron. Astrophys.* **4**, 93.
Paczyński, B.: 1970, *Acta Astron.* **20**.

* This paper was presented by O. B. Dluzhnevskaya.

Tayler (ed.), Late Stages of Stellar Evolution, 99. All Rights Reserved.

INTEGRAL PARAMETERS AND THE TOTAL MASS
OF OPEN CLUSTERS

(Abstract)

A. E. PISKUNOFF

Astronomical Council of the Academy of Science, U.S.S.R.

Masses and ages of very distant open clusters for which the colour-magnitude diagram cannot be constructed may be estimated from their integral magnitudes and colours. The network of evolutionary tracks for Population I stars allows a study of evolutionary changes of integral properties of model clusters. Evolutionary tracks for stars with masses $0.8 \, \mathfrak{M}_\odot \leqslant \mathfrak{M} \leqslant 15 \, \mathfrak{M}_\odot$ ($X=0.7$, $Z=0.03$) by Paczyński (1970) have been used and isochrones have been derived by an interpolation method. The initial mass function assumed for cluster models is the Salpeter function for 58, 285 and 580 stars. The integral colours and magnitudes were calculated for cluster ages $t=0$ and $6 \leqslant \log t \leqslant 9$ and the ratio $\log \mathfrak{M}$/integral magnitude has been derived.

Parameters obtained for cluster models with different ages and integral masses have been compared with observational data for clusters with similar characteristics. Correlations $\log \mathfrak{M}$/integral magnitude for these clusters have been derived on the basis of observations. The agreement of observational data with theoretical results is rather good.

Reference

Paczyński, B.: 1970, *Acta Astron.* **20**, 47.

MASS LOSS FROM STARS AND THE LOCATION OF BLUE SUPERGIANTS IN THE HR DIAGRAM*

(Abstract)

Yu. FRANTSMAN and E. POPOVA

Radio Astrophysical Observatory, Latvian Academy of Sciences, Riga, U.S.S.R.

Continuing the discussion about the evolution of massive stars, about cases A and B of the computations of evolution, I want to say some words about the comparison of the observations and theoretical computations.

As it is known, on the HR diagram blue supergiants occupy the effective temperature region between $3.9 < \log T_e < 4.35$.

In case B, the Schwarzschild stability criterion, the star exhausts the main part of its He-fuel as a blue supergiant. The core He-burning models are placed in $\log T_e$ region between 3.8 and 4.2. This is in good agreement with observations. The computations for the case A, Ledoux criterion, give the effective temperature only in the narrow region $(4.15 < \log T_e < 4.25)$. This does not agree with the observational data. But these calculations of stellar evolution were performed assuming that mass loss does not occur.

The question is: if the mass loss from the star in the previous evolutionary stage, red supergiants, may give greater dispersion of the effective temperature for blue supergiants. Sequences of inhomogeneous equilibrium models of stars with a constant mass of the He-core and decreasing mass of the H-rich envelope are calculated for stars with central helium burning in the blue supergiant region. The initial masses of the sequences are 15 $\mathfrak{M}_\odot$ and 30 $\mathfrak{M}_\odot$. Such models may represent the stars, which lost some amount of mass in the red supergiant region.

The chemical composition distribution was taken according to case A, $X = 0.70$ in the envelope, $X = 0$ in the core. It was shown that continuous sequences of models with decreasing mass of the hydrogen-rich envelope can be obtained in the blue supergiant region.

The effective temperature of these models depends on the amount of mass lost in the red supergiant stage, or equivalently the mass of the hydrogen envelope.

Conclusion

If a massive star loses up 20–30% of its mass in the red supergiant stage it returns to the blue supergiant region in HR diagram where the blue supergiants really are observed.

Then, in this point of view, there is no discrepancy between observational data and computations in case A.

* This paper was presented by Yu. Frantsman.

Tayler (ed.), Late Stages of Stellar Evolution, 101. *All Rights Reserved.*
Copyright © 1974 by the IAU

SUPERGIANT STARS AS CHEMICAL
COMPOSITION INDICATORS*

(Abstract)

C. CHIOSI and E. NASI

Istituto di Astronomia, Universita di Padova, Italia

By investigating massive stars' theoretical evolution, it has been found that the fractions of time spent in the different effective temperature intervals in respect to the total core He-burning lifetime, which give the probabilities of finding the star in these temperature ranges, are very sensibly dependent on the initial chemical composition. The evolutionary tracks are computed under the assumption of the Schwarzschild-Härm criterion for semiconvective instability and neglecting the effect of mass loss from single stars, because no completely satisfactory theory of dealing with it, has so far been presented although it might be a very important effect.

The above lifetime fractions can be converted into percentages in terms of predicted number of stars expected in the different effective temperature intervals. The comparison of the theoretical percentages with the observed ones suggests the possibility of investigating whether differences in initial chemical composition exist among Population I stars.

Such an inhomogeneity in the chemical composition for young stars was already suggested by several authors, and moreover some features of the composite experimental HR diagrams of supergiant stars in our own Galaxy, Large and Small Magellanic Clouds (Humphreys, 1970; Brunet and Prevot, 1971; Osmer, 1973, respectively) seem to require it. From the above experimental sources we have derived the percentage distribution in the different temperature intervals by considering only those supergiants more luminous than -7^m in M_b and excluding also the Ib luminosity class, up to spectral type B2, stars for which it has been suggested, on the basis of their position in the HR diagram compared with evolutionary tracks, that they are in core H-burning phase.

An attempt is made to assign the average metal and hydrogen contents for supergiants stars in the three galaxies, and the following sequence is derived: $X_G < X_{LMC} < X_{SMC}$ and $Z_G > Z_{LMC} > Z_{SMC}$. This result is in quite satisfactory agreement with the indications derived by the period-colour relationships of Cepheids in the three galaxies.

A more complete account of this work will be submitted for publication to *Astronomy and Astrophysics*.

References

Brunet, J. P. and Prevot, L.: 1971, in M. Hack (ed.), *Colloquium on Supergiant Stars*, Trieste, p. 119.
Humphreys, R. M.: 1970, *Astron. J.* **74**, 602.
Osmer, P. S.: 1973, *Astrophys. J.* **181**, 327.

* This paper was presented by C. Chiosi.

DISCUSSION AFTER PAPER BY MASSEVITCH

Massevitch (in answer to comment by Chiosi): There are some difficulties in explaining the observed effective temperature range for blue supergiants in the case of both stability criteria. For the Ledoux criterion, an assumption of mass loss in the red supergiant region allows us to extend the interval of T_e sufficiently to reach an agreement with observations. On the other side, there are some difficulties in obtaining models with $\log T_e = 4.3$ and $M_{Bol} = -9$ if the Schwarzschild criterion is used. Thus effective temperatures do not provide a possibility to make a definite choice between the two criteria.

Schwarzschild: Mr Chairman, since the afternoon is quite advanced would it be proper to turn to some quite speculative points? I would very much like to ask Dr Paczyński whether he thinks that in the envelopes of massive supergiants instabilities exist which lead to a non-explosive major mass ejection, as seems to be the case for the less massive supergiants. Similarly, I would like to ask Dr Arnett whether he is ready to fill in for us the little gap Dr Massevitch referred to between the last systematically computed evolutionary model for a massive star and its final death.

Paczyński: I can only repeat that there are single massive Population I Wolf-Rayet stars surrounded by massive ring nebulae. This suggests that H-rich envelopes have been removed in a gentle fashion, not unlike the formation of a planetary nebula. Mass must have been lost in the red giant phase and the deficiency of red supergiants may be due to massive mass loss.

Schwarzschild: How does this happen?

Paczyński: Mass loss due to radiation pressure on dust grains can run at $10^{-5} M_\odot \, \text{yr}^{-1}$ at least. Alternatively ionization processes in a distended atmosphere may be involved.

Schwarzschild: You might need both mechanisms.

Paczyński: I believe that there is considerable misunderstanding associated with the term 'planetary nebula'. Originally, in order to get a distance scale, Shklovsky adopted an *average* mass for a nebula as a fraction of solar mass. Later Harman and Seaton and O'Dell used a similar method to find distances and again they used $0.2 \, M_\odot$ or $0.6 \, M_\odot$ as an *average* mass for a nebula. But it must be emphasized that while the concept of 'average mass' may be used to derive a distance it cannot be used as an indication that all planetary nebulae have indeed identical masses. There are no observational data that I know that would indicate that all nebulae are equally massive. In fact circumstellar ionized nebulae are known to have masses from $10^{-3} M_\odot$ up to $10 \, M_\odot$ or more (ring nebulae around single W-R stars). I think it is possible that radiation pressure on dust in the atmospheres of red supergiants is responsible for most of the mass loss, but perhaps the final mass loss is due to some

Tayler (ed.), Late Stages of Stellar Evolution, 103–104. *All Rights Reserved.*
Copyright © *1974 by the IAU.*

large scale instability of a whole envelope. Perhaps observations of young high density planetary nebulae and luminosous infra-red objects will help us to specify the real mechanism of mass loss.

Arnett: The gap mentioned by Dr Massevitch may not really be that large. My models start from helium burning and go all the way well into hydrodyamic core collapse. The cores are past the white dwarf maximum on their way to neutron stars. Work by me; Ivanova, Imshennik and Nadyozhin; and J. Wilson gives a fairly realistic picture of the collpase (if taken together). Falk and I have calculated supernova light curves from hydrodynamic models as have Ostriker and collaborators.

Bisnovatyi-Kogan: In the work by D. K. Nadyozhin and me (*Astrophys. Space Sci.* **15**, 353, 1972) the evolution of a 30 $M_\odot$ star was considered, taking into account the mass loss due to radiation pressure. Self consistent models with static core and outflowing envelope (3% of the total mass) were constructed with the boundary conditions far from the star being treated approximately. It was found that very intensive mass loss must occur leading to the loss of all the hydrogen envelope, with the remaining helium core ~ 10 $M_\odot$ becoming a W-R star. It was indicated that, if this picture is right, a single W-R star must have a massive and extended transparent envelope ~ 10–20 $M_\odot$. The observations of such envelopes about single W-R stars about which Paczyński has spoken seem to support our mechanism of W-R star formation and support the hypothesis of drastic mass loss. It seems at present that we have overestimated the mechanism of mass loss due to radiation pressure in the optically thick layer and that radiation pressure on dust grains must be an additional support to the mass loss. So, the mass loss from red supergiants must occur due to both these mechanisms acting together.

Woolf: The largest mass-loss rates observed for red supergiants are about 10^{-3} $M_\odot$ yr^{-1}. This occurs in very rare luminous stars of type type M5Ia$^+$. Such stars are both very massive and very rare. There do not seem to be enough objects, which would have to persist for about 10^5 yr, for this to be a major site of mass ejection for the most massive stars.

Appenzeller: Spectroscopic observations of some bright blue (and probably very massive P Cyg stars indicate mass loss rates in the order of 10^{-4} $M_\odot$ yr^{-1}. Thus, massive stars may also lose a large amount of mass during a P Cyg stage and I wonder if the bright single Wolf-Rayet stars mentioned above could simply be 'evolved' P Cyg stars.

Massevitch to Arnett: I should be very happy if there was not gap between the last directly evolved star and the supernova but there still seems to be one.

Frantsman to Chiosi: The problem of differences of chemical composition of SMC, LMC and Galaxy may be solved with better results if both the differences of parameters of supergiants and cepheids are taken into account. The evolution of supergiants depends on too many free parameters and from the observational data for supergiants only it is very difficult to draw any conclusions concerning the chemical composition.

MIXING BETWEEN THE CORE AND THE ENVELOPE
IN STARS WITH DEEP CONVECTION ZONES*

DAIICHIRO SUGIMOTO

Institute of Earth Science and Astronomy, College of General Education,
University of Tokyo, Komaba, Meguro, Tokyo, Japan

and

KEN-ICHI NOMOTO

Dept. of Astronomy, University of Tokyo, Tokyo, Japan

Abstract. The extensive surface convection zone of a red giant star becomes deeper and deeper, as the star evolves. In some cases it reaches the bottom of the hydrogen-rich envelope and even penetrates into the stellar core. From the standpoint of entropy distribution in the star, the structure of the convective envelope and the mechanism of its penetration into the core are summarized. Possibilities of the penetration are discussed for stars of masses in the range of 1–60 $M_\odot$ over their entire lifetimes of evolution. Effects of uncertainty in the mixing length theory of convection, neutrino loss and thermal instability of helium-burning shell are studied also. When the convection penetrates into the core, material of the core is brought up to the stellar surface and the star will be a peculiar star. It is concluded that there are possibilities to interpret the origin of peculiar stars with luminosities fainter than $7 \times 10^4 L_\odot$, but that the more luminous ones do not originate from the penetration of surface convection.

1. Introduction

Development of an extensive surface convection zone was found by Hoyle and Schwarzschild (1955) in their study of evolution of less massive stars through a giant branch of a globular cluster. They discussed that an envelope in radiative equilibrium cannot be consistent with a photospheric boundary condition in the case of red giant stars, and that it requires the development of the extensive surface convection zone.

This situation is understood more intuitively, though more roughly, as follows. We approximate the photospheric boundary condition by $\kappa_{ph}\varrho_{ph}R \simeq \frac{2}{3}$, where κ, ϱ and R denote opacity, density and the stellar radius, respectively, and the subscript 'ph' refers to the photospheric value. A condition that the mean density of the star is higher than the photospheric density is expressed as

$$M > (4/3)\,\pi R^3\varrho_{ph} \simeq (8/9)\,\pi R^2\kappa_{ph}^{-1},\tag{1}$$

where M denotes the mass of the star. This condition implies that there should be an upper limit to the stellar radius. It corresponds to the Hayashi limit in the HR diagram (Hayashi and Hōshi, 1961), when computed quantitatively.

Since the polytropic index in a convective region is always smaller than that in a radiative region, an envelope with a surface convection zone has a relatively small

---

* This paper was presented by Daiichiro Sugimoto.

Tayler (ed.), Late Stages of Stellar Evolution, 105–121. All Rights Reserved.
Copyright © 1974 by the IAU.

radius (Chandrasekhar, 1939; Hayashi *et al.*, 1962). Thus, the emergence of the convection zone is consistent with the limitation to the stellar radius.

In the present paper, we shall use the notions of the core and the envelope, dividing a star at the hydrogen-burning shell. As the star evolves, the central condensation of matter becomes stronger in the core, because the temperature and/or the electron degeneracy in the central region must increase. However, the hydrogen-burning shell hardly contracts, because its temperature must not be higher than the hydrogen-burning temperature $(2-10 \times 10^7 \, \text{K})$. Since the volume of the core does not change much, the density and thus the pressure near the hydrogen-burning shell become lower, while the central condensation proceeds in the core. In order to sustain the envelope against gravity, the bulk of the envelope should be brought to a region of weak gravity, i.e., of a large radial distance from the center.

Since the radius of the star is bounded as discussed above, the convection in the envelope becomes deeper and deeper in order to push the bulk of the envelope to the relatively outer portion of the envelope. It is our problem whether the convection in the envelope grows deep enough to penetrate into the core, and whether the convective mixing brings material of the core to the stellar surface.

In the next two sections, the mechanism of penetration of convection and nature of the convective envelopes will be summarized. In the later sections, existing stellar models will be discussed and compared one another. Some of new models will be incorporated also, which we computed recently and details of which are in preparation for publication. In the final section, concluding remarks will be given concerning the possible origin of peculiar stars.

2. Mechanism of the Penetration of Convection

It is convenient to discuss the convection zone in terms of entropy distribution through the star. In a chemically homogeneous region in radiative equilibrium, the entropy is increasing outward, while it stays constant in adiabatic convection and even decreases outward in superadiabatic convection. We shall approximate that the chemical composition and the energy flux L_r change discontinuously at the interface between the core and the envelope, where a hydrogen-burning shell lies (active or inactive). Quantities at the interface will be denoted by the subscript 1, adding e or i to indicate the external or internal side, if necessary. The distribution of entropy is illustrated in Figure 1. We shall use different definitions of entropy for the core and the envelope. Entropy defined for the core will be denoted by $s^{(\text{core})}$ and one defined for the envelope by s. Non-dimensional value of the entropies will be denoted by $\sigma \equiv (H/k) \, s$, where H and k denote the mass of hydrogen and the Boltzmann constant. Anyhow, we shall assume that there exists one-to-one correspondence between $s_{1i}^{(\text{core})}$ and s_{1e}, or $\sigma_{1i}^{(\text{core})}$ and σ_{1e} under continuity of pressure and temperature across the point 1. Practically, s is defined with the quantum statistical zero-point, so that it is continuous even in the ionization and dissociation zones. We shall use the definition of $s^{(\text{core})}$ as given in Sugimoto (1970a).

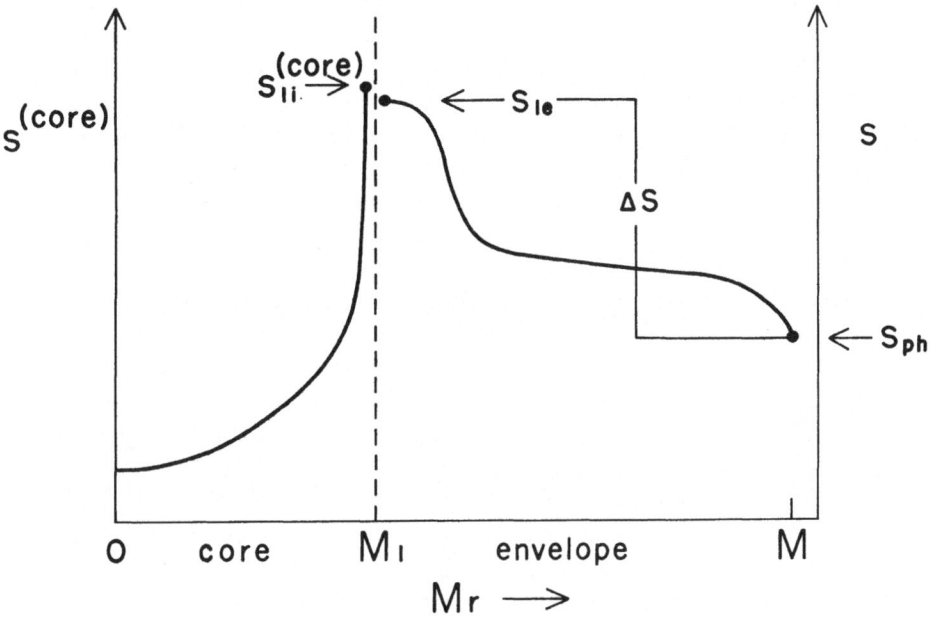

Fig. 1.   Illustration of entropy distribution in the star.

From the boundary conditions at the photosphere $\kappa_{\mathrm{ph}}\varrho_{\mathrm{ph}}R \simeq 2/3$ and $L = \pi acR^2 T_{\mathrm{eff}}^4$, where L and T_{eff} denote luminosity and effective temperature of the star, the entropy at the photosphere is computed. Near the photosphere of the red giant stars, κ_{ph} is so small that the radiation pressure is negligible in many cases. In such cases, the entropy at the photosphere is expressed as

$$s_{\mathrm{ph}} = (k/\mu H)\ln\kappa_{\mathrm{ph}}R^{1/4}L^{3/8} + \mathrm{const}, \tag{2}$$

where μ denotes the mean molecular weight. In the super-adiabatic envelope, entropy takes the maximum value $s_{\mathrm{ph}} + \Delta s$ at the bottom of the convective zone. The structure of the core including the stellar luminosity depends only slightly on the structure of the envelope (Paczyński, 1970a; Sugimoto, 1970a). The entropy s_{1e} determined from the $s_{1e}^{(\mathrm{core})}$ should satisfy the condition,

$$s_{1e} \leqslant s_{\mathrm{ph}} + \Delta s. \tag{3}$$

If the equality holds in the condition (3), the convection reaches just the bottom of the hydrogen-rich envelope. If the condition (3) is violated, the stellar structure is inconsistent. If we assume a smaller value of the core mass M_1, we can obtain a consistent structure. This implies that the convection should have penetrated into the core. Such a treatment has been made by Stothers and Chin (1969) and by Sugimoto (1970a, b, 1971). They fitted their core model to envelope at the hydrogen-burning shell.

In an actual star, the hydrogen content and L_r change continuously across the hydrogen-burning shell. The stellar structure near a thin nuclear burning shell is almost stationary, if the independent variables are taken to be M_r/M_1 and t, or r and t (Eggleton 1967), where M_r denotes mass contained in a sphere of radius r, and t denotes time of evolution. Such a nature was extensively used by Paczyński (1970b) and by Uus (1970a, 1971) to compute the model envelopes, into which the thin hydrogen-and helium-burning shells were incorporated. Though their envelope is described by ordinary differential equations, the so-called gravitational energy generation is included to a good approximation. They integrated such envelopes down to a point just interior to the helium-burning shell, where the core model is fitted to the envelope. Uus (1971, 1972c) discussed the penetration of convection into the hydrogen-burning shell in detail. We shall consider inward integration of the envelope. When we reach the region with appreciable hydrogen-burning, L_r is decreasing inward. Then, the radiative temperature gradient is diminished and the convection stops. Moreover, the gradient of the mean molecular weight in the hydrogen-burning shell makes this shell more convectively stable. This tends to inhibit the convection from penetrating into the core.

In some cases, however $s_{1i}^{(core)}$ and then s_{1e}, which are computed from the core model, become larger and larger, until the condition (3) is strongly violated (Sugimoto, 1970b). In principle, we can compute a structure for which pressure is continuous across the point 1 but the temperature jumps there. If $s_{1i}^{(core)}$ becomes so large that ϱ_{1i} is smaller than ϱ_{1e} in spite of the difference in the mean molecular weights, the Rayleigh-Taylor instability takes place. In view of the strong dependence of $\varrho_{1i}^{(core)}$ on the core mass (Sugimoto 1970b), the violation of the condition (3) may be considered to lead the mixing between the core and envelope, though detailed computation has never been made.

3. Structure of Envelope

In order to apply the condition (3), it is necessary to estimate the increase of entropy Δs in the envelope. At present there are no other means than the mixing length theory of convection (Böhm-Vitense, 1958), for which a parameter related with the efficiency of the convective energy transport must be assumed. Usually, it is described by $\alpha_p = l/H_p$ or $\alpha_\varrho = l/H_\varrho$, i.e., by the ratio of the mixing length l to the scale height of pressure H_p or of density H_ϱ.

When $\alpha_p = 1.0$, the superadiabaticity is so large that an inversion of density appears (Paczyński and Ziołkowski 1969, 1970a). The density inversion is not irrational by itself, because the hydrostatic equilibrium is established with this density inversion. The Rayleigh-Taylor instability takes place in this layer of density inversion, which corresponds just to the convective instability.

What happens, if we use α_ϱ with the density scale-height instead of the pressure scale-height? From the physical point of view, the mixing length must be smaller than the distance to the stellar surface and/or to the center of the star. Imposing this limitation, Uus (1972a) integrated some envelopes with $\alpha_\varrho = 0.5$ and 1.0, and obtained the density inversion. Moreover, the global nature such as the relations among Δs, luminosity and the core mass is essentially the same as those obtained for the pressure scale-height, if the value of H_ϱ is adjusted appropriately. Uus (1972b) has also made a non-local treatment of convection, but the global nature in the deep interior is hardly affected.

Thus, the most important uncertainty left to us is the choice of a value of α_p. In many computations of stellar evolution, α_p is assumed to lie in the range of 1–2, for which the computed giant branch can be fitted with observed HR diagram. However, such a comparison is made only for stars of relatively small luminosity as compared with the local critical luminosity,

$$L_{cr}(M_r) = \frac{4\pi Gc}{\varkappa} M_r, \tag{4}$$

at the bottom of the convective envelope (Nomoto and Sugimoto, 1972). In such stars Δs is relatively small and weakly dependent on α_p. Moreover, α_p may not be constant throughout the convection zone. Thus, we can not draw any definite conclusion concerning the value of α_p.

Nomoto and Sugimoto (1972) studied the dependence of Δs on the value of α_p as well as on the mass M, core mass M_1, and luminosity of the star. Their results are summarized as follows. The value of Δs is larger, as the ratios of the luminosity to the local critical luminosities, $L/L_{cr}(M)$ and $L/L_{cr}(M_1)$ are larger. Where Δs is large, its value depends strongly on the value of α_p. For $\alpha_p = 1.0$, the effect of super-adiabaticity is essential. For $\alpha_p = 1.5$, however, the envelopes are not much different from adiabatic one, unless $L/L_{cr}(M_1)$ is close to unity. Details may be found in graphs given by Nomoto and Sugimoto (1972).

4. Massive Stars

In this section we shall discuss the stars of mass greater than 8 $M_\odot$, for which electrons are non-degenerate in the carbon-oxygen core.

4.1. Phases before the exhaustion of carbon

Detailed Henyey-type computation including the hydrogen-rich envelope was made by Ziołkowski (1972) for the stars of masses 15 and 30 $M_\odot$ up to the carbon ignition, and for the star of mass 60 $M_\odot$ up to the exhaustion of helium. The convective envelope was treated in detail with $\alpha_p = 1.0$.

Distribution of hydrogen content in the star is illustrated in Figure 2. The intermediate zone, where the hydrogen content changes smoothly, has been produced in the core hydrogen-burning phase. Just after the helium ignition, the surface convection reaches the outer part of the intermediate zone. Then a part of helium in the intermediate zone is mixed into the convective envelope, as illustrated in Figure 2. The hydrogen concentration in the envelope, which was $X_e = 0.700$ initially, are decreased to 0.682, 0.630 and less than 0.604, respectively, for the stars of masses 15, 30 and 60 $M_\odot$.

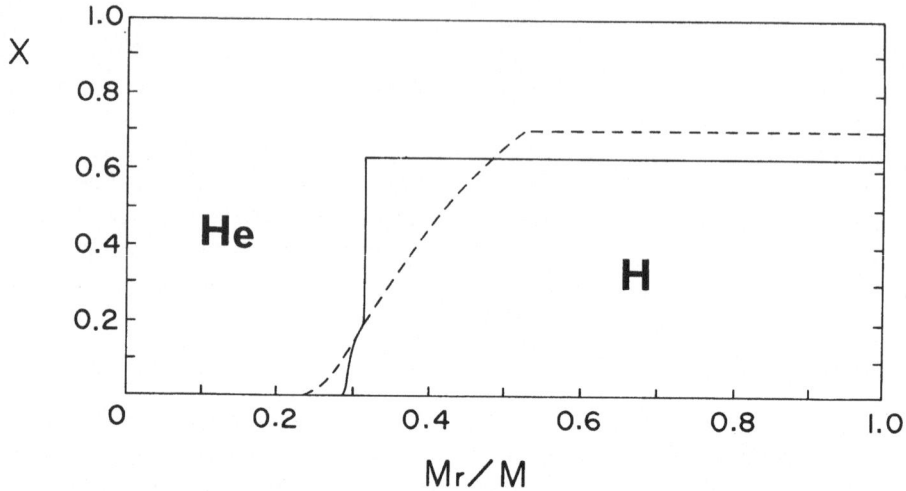

Fig. 2.   Distribution of hydrogen concentration X in the star changes from the dashed profile to the solid one. Computed by Ziolkowski (1972) for the star of 30 $M_\odot$.

In the main phase of the core helium burning, the surface convection retreats. From the helium exhaustion, the star becomes a red giant again and the surface convection becomes deeper. For the stars of 15 and 30 $M_\odot$, the convection does not reach the hydrogen-burning shell, before the ignition of carbon. For the star of 60 $M_\odot$, the bottom of convection may reach the active hydrogen-burning shell, though Ziołkowski did not compute such stages in detail.

As shown in Figure 2, the mass M_r at the inner edge of the intermediate zone increases as a result of the hydrogen-shell burning. Then, an almost discontinuous profile of the hydrogen distribution is realized. Stothers and Chin (1969) and Sugimoto (1970a, b) computed models, assuming such a discontinuous distribution of hydrogen. Both of them assumed adiabatic convection in the envelope. Stothers and Chin fitted such envelopes to homogeneous cores neglecting gravitational energy release. Sugimoto made Henyey-type computation of the core using the boundary conditions at the outer edge of the core,

$$\left(\frac{GM_r\varrho}{rP}\right)_{1e} = (n+1)_{1e} = \begin{cases} 4 & \text{(radiative)} \\ (n+1)_{ad} & \text{(convective)} \end{cases}, \tag{5}$$

where n and P denote the polytropic index and pressure, respectively, and the subscript 'ad' denotes the adiabatic value.

In the stars of 12 $M_\odot$ (Sugimoto, 1970b) and 15 $M_\odot$ (Stothers and Chin, 1969), convection does not penetrate into the core. In the star of 30 $M_\odot$ both results by Stothers and Chin, and by Sugimoto indicate that the convective envelope begins to penetrate into the core just before the ignition of carbon burning. As shown in Figure 3, helium of 1.2 $M_\odot$ is mixed from the outer part of the core into the convective envelope in early stages of the core carbon burning (Sugimoto, 1970b).

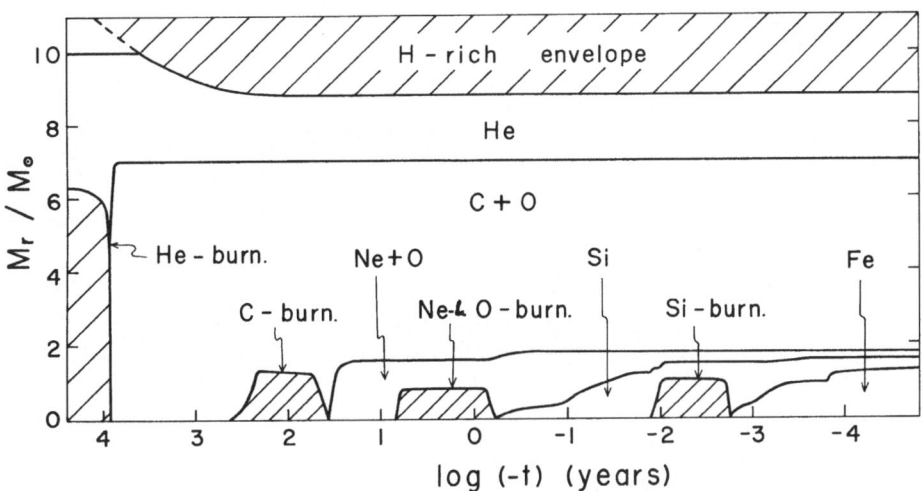

Fig. 3.   Chemical evolution of the star of 30 $M_\odot$ computed by Sugimoto 1970b) and by Sugimoto and Nomoto (1974). Neutrino loss is taken into account. Shaded regions are in convective equilibrium. A part of the hydrogen-rich envelope is omitted from the top of the figure.

Though the penetration of convection is somewhat overestimated in the treatment by Sugimoto (Nomoto and Sugimoto, 1972), it does not reach the carbon-oxygen zone by a large margin.

Nomoto (1974) computed the evolution of a 60 $M_\odot$ star with $\alpha_p = 1.0$, using boundary conditions which are essentially the same as Equation (5). They assumed a core mass of 30 $M_\odot$, which is somewhat larger than the core mass of 25.05 $M_\odot$ obtained by Ziołkowski (1972). A discontinuous change in the distribution of hydrogen was assumed. After helium has been almost exhausted in the central region, the hydrogen-shell burning remains inactive, entropy at the edge of the core becomes larger and larger, and then the convective envelope penetrates into the core. About 5 $M_\odot$ of helium and 2 $M_\odot$ of carbon and oxygen are mixed into the envelope.

Such a star may be a peculiar star. Lifetime of this star after the mixing of carbon and oxygen is as short as 3×10^3 yr, because extensive neutrino loss accelerates the evolution in later phases. The lifetime of the core helium burning is 2.5×10^5 yr, which the star spends as a red supergiant (Ziołkowski, 1972). Then, the ratio of the peculiar stars to the red supergiants will be 0.01.

4.2. LATER PHASES AND THE TIMESCALE OF MIXING

If the entropy distribution near the outer edge of the core is expressed as a function of M_r/M_1 and t, it remains almost stationary, even when M_1 is being reduced by mixing between the core and envelope. Then, the entropy of a mass element in the outer part of the core should have been greatly increased, before it is mixed into the envelope (Sugimoto, 1970a). This is accomplished by absorbing radiative heat flux from the interior, which takes a time of heat transfer $\tau_h (\Delta r = H_p)$ over unit scale-height of pressure. Thus, the rate of mixing is limited by

$$-\frac{d \ln M_1}{dt} < \frac{\Delta M/M_1}{\tau_h(\Delta r = H_p)},$$

(6)

where ΔM denotes mass contained in unit scale-height of pressure. Typical values are $\Delta M/M_1 \sim 10^{-3}$ and $\tau_h \sim 10$ yr. near the edge of the core, and the timescale of mixing $|dt/d \ln M_1|$ amounts to 10^4 yr.

After the carbon-burning phase, stellar evolution is greatly accelerated by copious emission of neutrinos. Typical timescale of evolution is 40 yr for the star of mass $30 \, M_\odot$, which is too short compared with the timescale of mixing. For the star of mass about $12 \, M_\odot$, the corresponding lifetime is 1500 yr. However, the convection does not reach the bottom of the envelope even in the phase of neon burning. Anyhow, the mixing is negligible after the carbon-burning phase of massive stars (Sugimoto, 1970b; Stothers and Chin, 1969).

4.3. EVOLUTION WITHOUT NEUTRINO LOSS

Hayashi *et al.* (1962) discussed an inconsistency between the observed number of red and yellow supergiants in star clusters and their lifetimes computed with neutrino loss. Now, this inconsistency seems to be resolved by the interpretation that most members of the red supergiants are in the phase of helium burning (Stothers and Chin, 1969; Stothers, 1972, and papers referred therein). Laboratory experiment of the direct interaction between electrons and neutrinos is under way by Gurr *et al.* (1972). Up to the present, they gave only an upper limit to the cross section, which is 1.9 times that of the universal Fermi interaction.

It is still meaningful to consider a limiting case, where the neutrino loss is completely neglected. Recently, Nomoto (1974) computed such a limiting case. Results for the star of $60 \, M_\odot$ with $\alpha_p = 1.0$ are shown in Figure 4. Since the timescale of evolution is long enough for mixing, the mass of the silicon core is reduced by mixing below the Chandrasekhar limit, as anticipated by Stothers and Chin (1969). For a smaller mass star, the invasion of the convective envelope is somewhat weaker, because the effect of superadiabaticity is large for the smaller mass star (Nomoto and Sugimoto, 1972). The core of $20 \, M_\odot$ star evolves into the iron core, before its mass is reduced below the Chandrasekhar limit. For the star of $12 \, M_\odot$, the core mass is not reduced below the Chandrasekhar limit, until the photo-disintegration of iron takes place and the core becomes unstable. The above results are based on the assumption of $\alpha_p = 1.0$. When $\alpha_p = 1.5$ is assumed, the iron core of $12 \, M_\odot$ star is reduced below the Chandrasekhar limit by mixing.

In the core of mass smaller than the Chandrasekhar limit, electrons become degenerate. Convection in the envelope extends down to the active hydrogen-burning shell, which surrounds the silicon or iron core. Hydrogen will be depleted uniformly in the envelope (Sugimoto, 1971). The lifetime of this phase of hydrogen depletion in envelope is estimated to be of the order of 10^7 yr, which is longer than

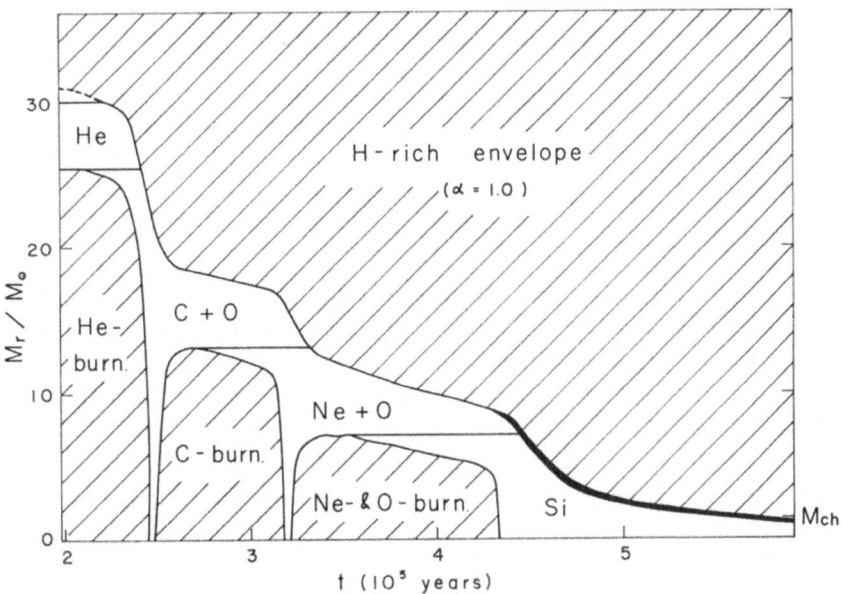

Fig. 4.   The same as Figure 3, but for the star of 60 $M_\odot$. Neutrino loss is neglected. Chandrasekhar's limiting mass is indicated by M_{Ch}.

the lifetime of helium burning by a factor more than 20. In this lifetime, the star has abundance peculiarity. Thus, if the neutrino loss were completely negligible, there would be as many peculiar stars as the massive main sequence stars (Stothers and Chin, 1969).

For the case of 12 $M_\odot$ star with $\alpha_p = 1.0$, the lifetime as a peculiar star is 5.7×10^4 yr, which extends from the penetration of convection into the carbon-oxygen zone up to the onset of dissociation of iron. The lifetime of this star from the contraction of carbon oxygen core up to the dissociation of iron is 7.2×10^5 yr. The lifetime as a red giant is somewhat longer than that, because the star spends a part of its helium-burning lifetime as a red giant. Then, the number of peculiar stars is somewhat less than 10% of normal red supergiants. In view of uncertainties involved, such a proportion of peculiar stars is not prohibitively high, because carbon stars and S-type stars consist 10^{-1}–10^{-2} of M-type giants (Iwanowska, 1966). We reach the following conclusion. The relative paucity of peculiar red supergiants is an evidence of the existence of the neutrino loss as discussed by Stothers and Chin (1969) and by Stothers (1972), unless the value of α_p is appreciably smaller than 1.0.

If we want to speak more strictly, the following comment should be amended. In all models computed above, the core mass is reduced below the Chandrasekhar limit after the oxygen-burning phase. Thus, the number of peculiar star is negligible, as long as the neutrino loss rate is larger than about 10^{-2} of the rate given by the universal Fermi interaction.

5. Stars of Intermediate Mass

In this section we shall discuss the stars of masses in the range of 3–8 $M_\odot$, which evolve into the carbon detonation supernovae (Arnett, 1969; Paczyński, 1970a; Bruenn, 1971). These stars have a carbon-oxygen core in which electrons are strongly degenerate due to the neutrino cooling.

5.1. Evolution toward the carbon detonation supernovae

When the central density becomes more than 10^6 g cm^{-3}, the convection becomes deeper and deeper. In the star of 5 $M_\odot$, a part of helium in the intermediate zone is mixed into the outer envelope (Weigert, 1966). For more massive stars the convective envelope penetrates into the helium zone. The core mass is reduced to about 1.0 $M_\odot$, if the mass of the helium core was larger than that value (Paczyński, 1970a; Sugimoto, 1971). In this phase of mixing, the hydrogen-burning shell remains inactive and the effect of superadiabaticity is small. Thus, Sugimoto's result is quantitatively consistent with Paczyński's.

As a result of the penetration of convection and the progress of the helium-burning shell, the hydrogen-burning shell comes very close to the hot helium-burning shell. Then, the hydrogen shell is ignited again. After this, the luminosity of the star increases and the effect of superadiabaticity becomes essential. Assuming $\alpha_p = 1.0$, Paczyński (1970a, b) computed evolutionary sequence, where the convective envelope did not penetrate into the core beyond the active hydrogen-burning shell. As a result of simultaneous hydrogen and helium burning, the core mass grows close to the Chandrasekhar limit and explosive carbon burning takes place.

Assuming adiabatic convection, Sugimoto (1971) obtained a different picture of evolution, where the convective envelope penetrates into the core beyond the active hydrogen-burning shell.

Uus obtained some models by fitting detailed envelopes to isothermal cores. The convective envelope does not penetrate into the core for the case of $\alpha_p = 1.0$ (Uus, 1970a), while the penetration is marginal for the star of mass 5 $M_\odot$ with $\alpha_p = 1.5$ (Uus, 1972c).

Since the penetration depends both on the assumption of α_p and on the stellar mass, it is time consuming to compute many cases for these parameters. Instead of this, Sugimoto and Nomoto (1973) computed growth of the core assuming the boundary condition similar to Equation (5). Such a boundary condition is found to be a good approximation as far as the bottom of convection is very close to the hydrogen-burning shell in its temperature. Assuming no penetration of convection, they obtained the value of entropy at the core edge $\sigma_{1e}^{(core)}(M_1)$ and the stellar luminosity $L(M_1)$ as functions of the core mass. As shown in Figures 5a and 5b, our $L(M_1)$ is close to Uus' result for $\alpha_p = 1.5$, but somewhat different from Paczyński's and Uus' results for $\alpha_p = 1.0$. The reason lies in the fact that the radiative zone is rather wide for the case of $\alpha_p = 1.0$, which makes our boundary condition inappropriate.

When the hydrogen-burning shell is active, the temperature thereof is almost constant. Then the condition (3) against the penetration of convection is conveniently rewritten as

$$L \lesssim L_{\text{mix}}(\alpha_p, M, M_1, T_1) < L_{\text{cr}}(M_1). \qquad (7)$$

Though some rough extrapolations were made, values of L_{mix} were obtained using the work by Nomoto and Sugimoto (1972), which are shown in Figures 5a and 5b for different values of α_p and M. Inspection of these figures leads to the following conclusion: The penetration of convection does not take place in the case of

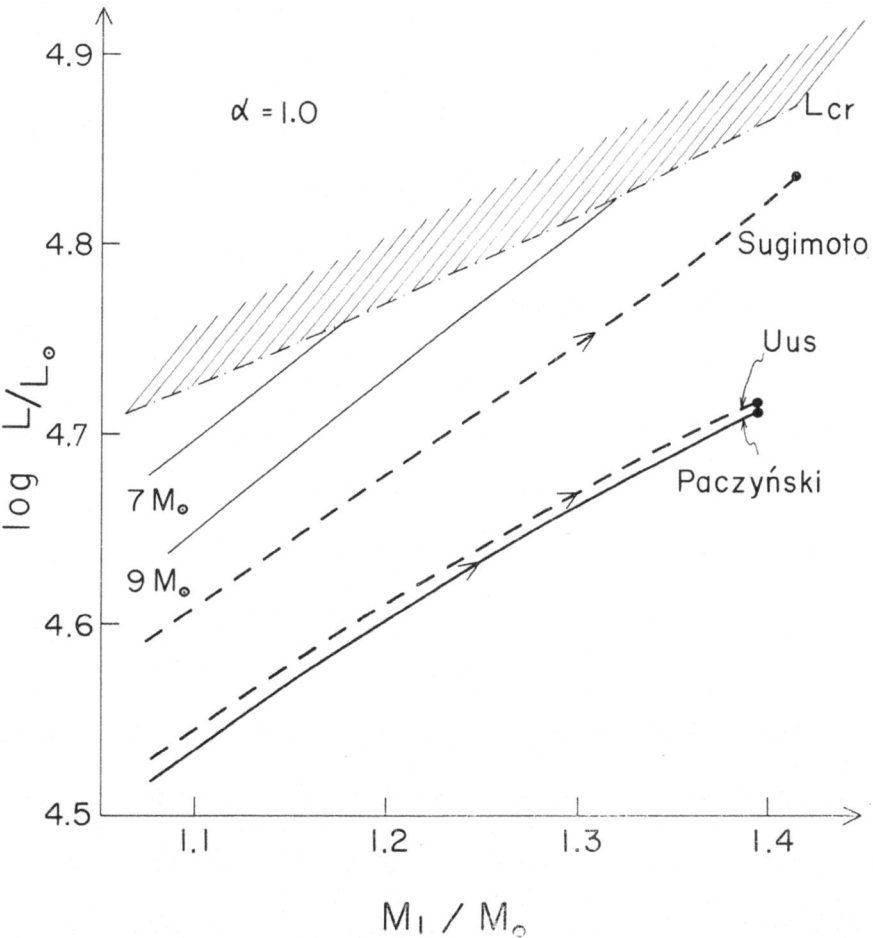

Fig. 5a. Evolutionary changes of stellar luminosity $L(M_1)$ (thick lines), the limiting luminosities against the mixing $L_{\text{mix}}(M_1, M)$ (thin lines labeled with stellar masses), local critical luminosity L_{cr}-(M_1) at the core edge (dash-dot) are plotted against the mass of the growing electron-degenerate carbon-oxygen core. The case of $\alpha_p = 1.0$. Sugimoto's $L(M_1)$ is also included for comparison. No static solution is permitted in the shaded region. Convective envelope does not penetrate into the core when $L \leqslant L_{\text{mix}}$.

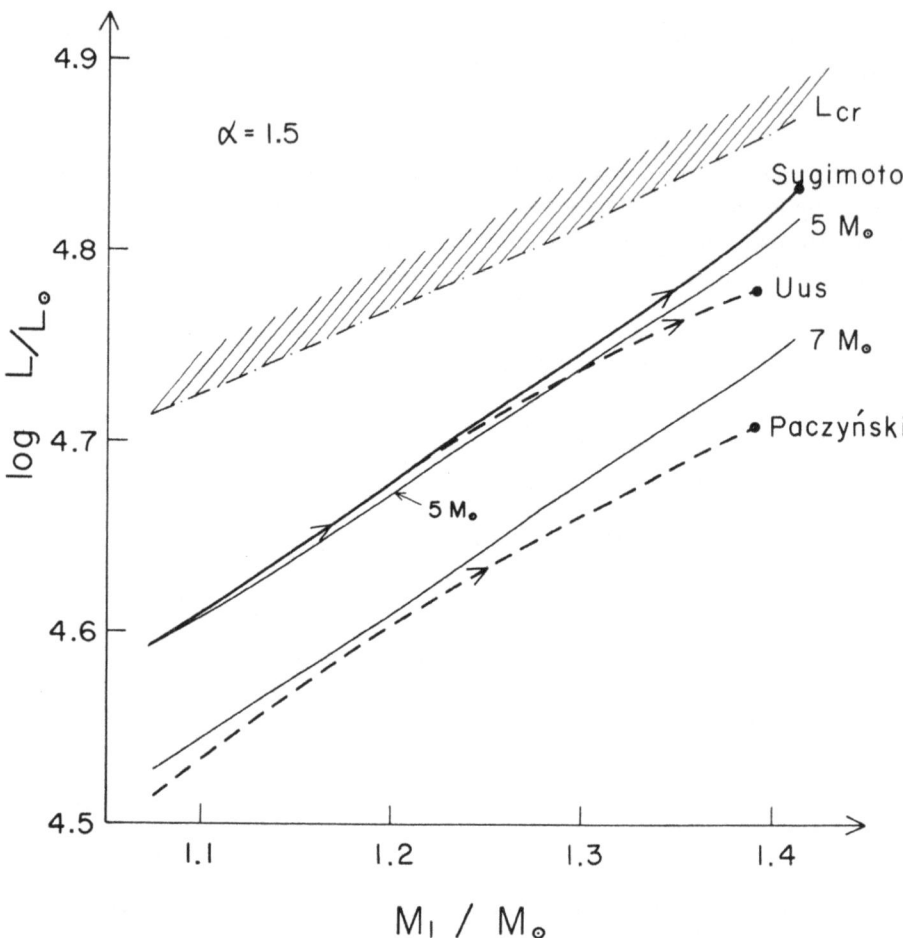

Fig. 5b.   The same as Figure 5a, but for $\alpha_p = 1.5$. Paczyński's $L\,(M_1)$ is added for comparison.

$\alpha_p = 1.0$ (Figure 5a), but it does take place for the star of mass greater than $5\,M_\odot$ in the case of $\alpha_p = 1.5$ (Figure 5b).

When the convection is penetrating into the core, material processed by the hydrogen-shell burning will be distributed throughout the envelope, and the hydrogen in the envelope is depleted uniformly. The lifetime of this phase of hydrogen depletion in envelope is 7×10^6 yr, up to the formation of the star having helium envelope (Sugimoto, 1971). Of this lifetime, the star spends about $\tau_p \simeq 5 \times 10^6$ yr as a peculiar star, which has overabundance of carbon and oxygen. We shall assume that such a process happens for stars in the mass range of Δm near $8\,M_\odot$, where m denotes stellar mass in units of $M_\odot$.

Stars of smaller mass evolve into the carbon detonation supernovae without the mixing. Using computations by Meyer-Hofmeister (1969), their lifetimes as red giants are approximately expressed by

$$\tau_r(m) = 2.2 \times 10^7 \, (m/5)^{-3.6} \, \text{yr}. \tag{8}$$

Taking into account the mass function proportional to $m^{-2.3}$ (Reddish, 1966), the ratio of the peculiar stars to normal red giants is expressed as

$$8^{-2.3} \tau_p \Delta m \Big/ \int_3^8 \tau_r(m) \, m^{-2.3} \, dm = 6 \times 10^{-3} \, \Delta m. \tag{9}$$

If we assume Δm to be of the order of unity, the number of peculiar stars will be of the order of one percent, which is consistent with star counts by Iwanowska (1966).

5.2. EFFECT OF THERMAL PULSES IN THE HELIUM-BURNING SHELL

In all computations discussed above, the thermal instability of the helium-burning shell was artificially suppressed by assuming appropriate ratio of the energy generations by the helium- and the hydrogen-shell burnings. However, Uus (1970b) showed that the helium-burning shell is probably unstable.

Taking two models with $M_1 = 1.07$ and $1.39 \, M_\odot$ from the computation discussed in the preceding subsection, Sugimoto and Nomoto (1974) computed the development of thermal pulses. The case of $M_1 = 1.39 \, M_\odot$ is shown in Figure 6, which is qualitatively similar to those for the stars of $5 \, M_\odot$ (Weigert 1966) and of $1.0 \, M_\odot$ (Sweigart, 1971), both of which have the core of $M_1 \simeq 0.8 \, M_\odot$. Convection appears in the helium zone and the helium is redistributed as shown in Figure 7. After the peak of the helium burning, the helium zone expands and the hydrogen-burning shell is extinguished. As shown in Figure 6, entropy at the edge of the core increases to a very large value, until the condition against mixing (3) is violated.

Though the progress of mixing has not been computed, it will develop as follows. As the mixing proceeds, the hydrogen-burning shell will move inward in mass. Then the temperature will rise and the hydrogen-shell burning will be ignited again. At the same time, the entropy at the core edge will decrease and the penetration of convection will stop.

As seen in Figure 7 the products of helium burning are distributed almost all over the helium zone. The width of the zone consisting of pure helium is relatively smaller for the pulses with a larger core mass. When the core mass is larger than a certain value, the products of helium burning may be brought to the stellar surface by the penetrating convection. This implies that the stars with luminosity higher than a certain value will be peculiar. Unfortunately, the proportion of the peculiar stars cannot be predicted, until detailed models with penetrating convection are computed.

6. Small Mass Stars

The stars of mass around $1 \, M_\odot$ have a phase of electron-degenerate helium core. When such a star reaches the red giant branch of globular clusters, the surface

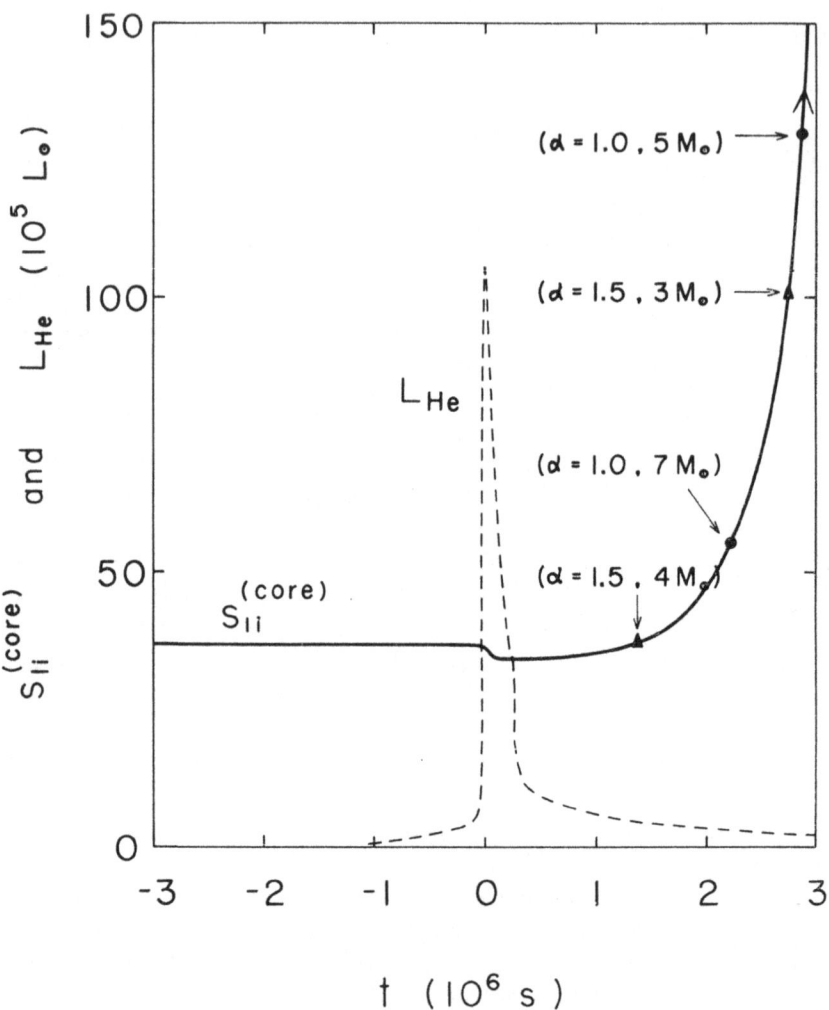

Fig. 6.   Thermal pulse of the helium-burning shell for the carbon-oxygen core of mass 1.39 $M_\odot$.
Change of entropy at the core edge and the energy-generation rate of the helium burning L_{He} are
shown. Critical values of entropy are shown by dots and triangles for different cases. For the entropy
above the critical value, the convective envelope should have penetrated into the core.

convection zone becomes deeper. The convection penetrates into the core and the
helium content of the envelope is increased by about 2% (Thomas, 1967; Demarque
and Mengel, 1971). After the thin hydrogen-burning shell is formed, the convection
does not penetrate any more.

Just after the helium flash, Thomas found a partial mixing between the core and
envelope, but Demarque and Mengel did not. This difference came from the fact that
Thomas used the neutrino loss rate too large by a factor of 4, and that the helium
flash took place in a relatively outer shell.

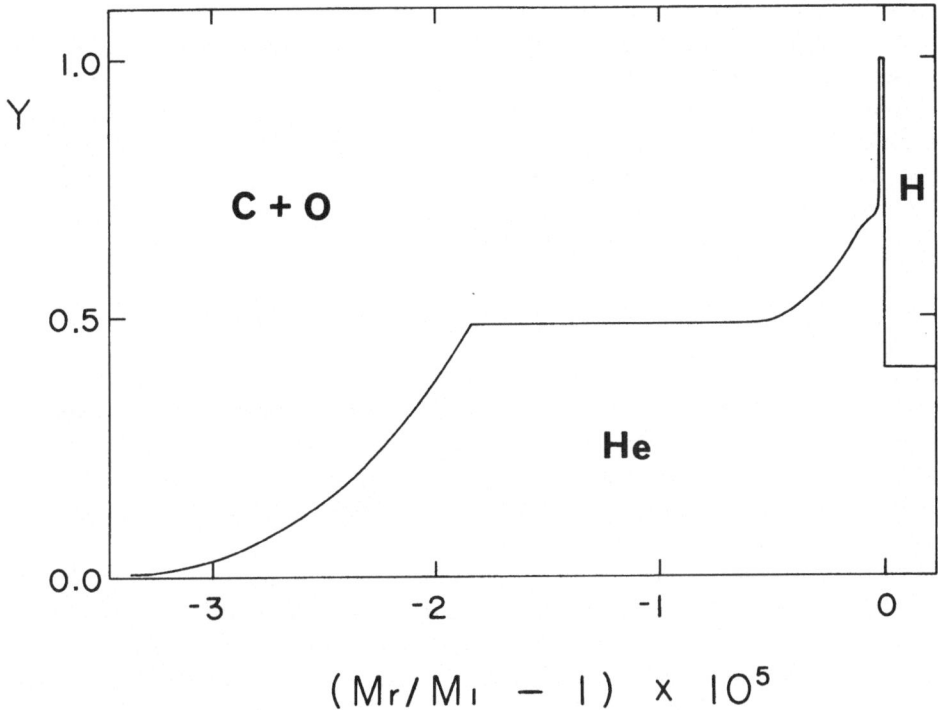

Fig. 7. Distribution of helium concentration Y in the thin helium zone of the star for a stage after the peak of the thermal pulse.

After the core helium burning, the thermal instability in the helium-shell burning takes place as discussed in Section 5.2. Schwarzschild and Härm (1967) found that the convection appearing in the pulsing helium shell touches the tail of the hydrogen distribution. In this case the hydrogen is brought into the helium zone, but nothing is brought into the surface convection zone, because the surface convection zone is too shallow. Sweigart (1971) obtained a similar result for a later phase of the same star. Unfortunately, their results cannot be compared with ours in Section 5.2, because they did not include the effect of radiation pressure and entropy of radiation, which play important roles for the problem of mixing especially in later phases of evolution.

For the star of mass between 4 and 1 $M_\odot$, the total energy of the red giants' envelope becomes positive and the envelope becomes pulsationally unstable, if the luminosity of the red giant exceeds a certain critical value L_0. Paczyński and Ziołkowski (1968) proposed this to be a cause of Mira type variables, mass loss from the red giants and formation of planetary nebulae. Such an effect takes place, if the mass of the star is smaller than a critical mass M_0. Assuming $\alpha_p = 1.0$, they determined approximate value of M_0 to be 4 $M_\odot$.

We have repeated such computation as done by them, using different values of α_p. For the range of $\alpha_p = 1.5$–0.7, M_0 lies in the range of 3–4 $M_\odot$, though a somewhat larger L_0 corresponds to a larger α_p (Fujimoto *et al.*, 1973). The critical mass is rather insensitive to α_p, because almost all mass of the envelope is in the ionization zone irrespective of the value of α_p.

Apart from the mass loss, the penetration of convection into the core is less likely, because the effect of superadiabaticity is larger for the stars of smaller mass (Nomoto and Sugimoto, 1972). When mass loss takes place, the core mass and the stellar luminosity hardly change, while the stellar mass becomes smaller. Then, the convection is more difficult to penetrate into the core.

7. Concluding Remarks

Though the general properties of the deepening convective envelope were discussed, uncertainty is left in the parameter of the mixing length theory. Moreover, quantitative results depend also on details of stellar models, especially on the distribution of hydrogen both in the intermediate zone and within the thin hydrogen-burning shell. Mechanism of penetration of convection beyond the barrier of the change in the mean molecular weight awaits for detailed investigation.

However, the effect of the parameter α_p is relatively small, when the stellar luminosity is relatively small compared with the critical luminosity at the bottom of the envelope, and when the temperature at the bottom of the envelope is relatively low and the hydrogen-shell burning is inactive. In such cases and in some extreme cases, we can draw conclusions with some confidence.

For the star of mass 60 $M_\odot$, the convection penetrates into the carbon-oxygen zone and the star becomes a peculiar star. For the star of mass in the range of 8–30 $M_\odot$, the convection penetrates, at the deepest, only to the helium zone. However, there exist carbon and S-type stars more luminous than $1 \times 10^5 L_\odot$ (Gordon, 1968; Motteran, 1971). They seem to exceed the maximum luminosity that can be attained by the star evolving toward the carbon detonation supernova. These carbon and S-type stars should be attributed to stars of mass larger than 8 $M_\odot$ or stars devoid of hydrogen. We must seek other mechanisms to interpret these highly luminous peculiar stars. If the neutrino loss were completely negligible, on the other hand, there would be too many peculiar stars, as far as the α_p is not much smaller than 1.0.

For the star of mass in the range of 8–3 $M_\odot$, there are two possibilities of the origin of peculiar stars. One is the hydrogen depletion in the envelope. This type of origin is interesting, because it may be connected with the origin of the star having helium envelope and the hydrogen deficient carbon stars such as R CrB. The other possibility is the mixing which takes place just after each cycle of the thermal pulses (see also Scalo and Ulrich, 1973). However, the number of peculiar stars cannot be predicted at present, for the above two possibilities.

There exist carbon stars of luminosities fainter than $10^4 L_\odot$ (Gordon, 1968). Such stars seem to be attributed to the stars of relatively small mass. Effect of thermal pulses for such small mass stars are still an open question.

References

Arnett, W. D.: 1969, *Astrophys. Space Sci.* **5**, 180.
Böhm-Vitense, E.: 1958, *Z. Astrophys.* **46**, 108.
Bruenn, S. W.: 1971, *Astrophys. J.* **168**, 203.
Chandrasekhar, S.: 1939, *An Introduction to the Study of Stellar Structure*, The University of Chicago Press, Chicago; 1957, Dover Publ., New York.
Demarque, P. and Mengel, J. G.: 1971, *Astrophys. J.* **164**, 317.
Eggleton, P. P.: 1967, *Monthly Notices Roy. Astron. Soc.* **135**, 243.
Fujimoto, M., Nomoto, K., and Sugimoto, D.: 1973, private communication.
Gordon, C. P.: 1968, *Publ. Astron. Soc. Pacific* **80**, 597.
Gurr, H. S., Reines, F., and Sobel, W.: 1972, *Phys. Rev. Letters* **28**, 1406.
Hayashi, C. and Hōshi, R.: 1961, *Publ. Astron. Soc. Japan* **13**, 442.
Hayashi, C., Hōshi, R., and Sugimoto, D.: 1962, *Prog. Theor. Phys. Kyoto Suppl.*, No. 22, 1.
Hoyle, F. and Schwarzschild, M.: 1955, *Astrophys. J. Suppl.* **2**, 1.
Iwanowska, W.: 1966, in M. Hack (ed.), *Colloquium on Late-Type Stars*, p. 398.
Meyer-Hofmeister, E.: 1969. *Astron. Astrophys.* **2**, 143.
Motteran, M.: 1971, in M. Hack (ed.), *Colloquium on Supergiant Stars*, p. 292.
Nomoto, K.: 1974, to be published in *Prog. Theor. Phys. Kyoto* **52**.
Nomoto, K. and Sugimoto, D.: 1972, *Prog. Theor. Phys. Kyoto* **48**, 46.
Paczyński, B. and Ziolkowski, J.: 1968, *Acta Astron.* **18**, 255.
Paczyński, B.: 1969, *Acta Astron.* **19**, 1.
Paczyński, B.: 1970a, *Acta Astron.* **20**, 47.
Paczyński, B.: 1970b, *Acta Astron.* **20**, 287.
Reddish, V. C.: 1966, *Vistas in Astronomy* **7**, 173.
Scalo, J. M. and Ulrich, R. K.: 1973, *Astrophys. J*, **183**, 151.
Schwarzschild, M. and Härm, R.: *Astrophys. J.* **150**, 961.
Stothers, R. and Chin, C.-W.: 1969, *Astrophys. J.* **158**, 1039.
Stothers, R.: 1972, *Astrophys. J.* **175**, 717.
Sugimoto, D.: 1970a, *Prog. Theor. Phys. Kyoto* **44**, 375.
Sugimoto, D.: 1970b, *Prog. Theor. Phys. Kyoto* **44**, 599.
Sugimoto, D.: 1971, *Prog. Theor. Phys. Kyoto* **45**, 761.
Sugimoto, D. and Nomoto, K.: 1974, in preparation.
Sweigart, A. V.: 1971, *Astrophys. J.* **168**, 79.
Thomas, H. C.: 1967, *Z. Astrophys.* **67**, 420.
Uus, U.: 1970a, *Nauch. Inform. Akad. Nauk SSSR* **17**, 3.
Uus, U.: 1970b, *Nauch. Inform. Akad. Nauk SSSR* **17**, 48.
Uus, U.: 1971, *Nauch. Inform. Akad. Nauk SSSR* **20**, 64.
Uus, U.: 1972a, *Nauch. Inform. Akad. Nauk SSSR* **21**, 46.
Uus, U.: 1972b, *Nauch. Inform. Akad. Nauk SSSR* **21**, 51.
Uus, U.: 1972c, *Nauch. Inform. Acad. Nauk SSSR* **23**, 85.
Weigert, A.: 1966, *Z. Astrophys.* **64**, 395.
Ziolkowski, J.: 1972, *Acta Astron.* **22**, 327.

EVOLUTIONARY ABUNDANCE CHANGES IN THE ENVELOPES OF MODERATE MASS RED SUPERGIANTS

(Abstract)

U. UUS

W. Struve Tartu Astrophysical Observatory, Estonia, U.S.S.R.

In stars with about 3 to 8 $M_\odot$, undergoing the stage of growth of degenerate carbon-oxygen core, the bottom of the convective envelope may descend into nuclear burning region. The more effective the convective heat transport, and the more massive the star, and the lower the heavy element content of the envelope, the more successfully the convection invades into burning shell.

Convection, if penetrating into burning shell, embraces only its upper part. The lower part of burning zone remains always radiative because of the small value of energy flux there. So the mass of the stellar core continues to increase.

When the bottom of convective envelope descends into the region of nuclear transmutations, the chemical composition of envelope begins to change. Soon after the penetration of convection into hydrogen burning shell, the CN-cycle equilibrium ratios are established in the envelope. If convection penetrates deep into burning shell, the carbon to oxygen ratio increases considerably after a preceding steep drop, and these may even be achieved a carbon prevalence over the oxygen. As for the hydrogen content in the envelope, there is not enough time for it to decrease considerably during the C–O core growing stage.

SURFACE ABUNDANCES RESULTING FROM
DEEP MIXING*

(Abstract)

I.-JULIANA CHRISTY-SACKMANN, RICHARD L. SMITH**,

and

KEITH H. DESPAIN

California Institute of Technology, Pasadena, Calif., U.S.A.

Dr Sugimoto has just presented a cause for deep mixing. We wish to report here the kind of surface abundances that result if one *assumes* deep mixing.

The stage of evolution was that near the red giant tip. The interior structure of the star was that during the helium-shell flashes. We assumed that convection could exist from the surface right down to the center of the helium-burning shell. Elements from envelope could be swept down and elements synthesized in the interior could be transported to the surface. In our models, for simplicity, the temperature-density grids in the star were not considered varying with time as we followed 17 reactions involving the 11 elements H, ^{3}He, ^{4}He, ^{7}Be, ^{7}Li, ^{12}C, ^{13}C, ^{14}N, ^{15}N, ^{16}O, and ^{17}O. There were three free parameters: namely, (i) T_{base}, the temperature at the base of the convective region (for which values $\sim 2 \times 10^8$ K, $\sim 1 \times 10^8$ K and $\sim 5 \times 10^7$ K) were used; (ii) Δt, the duration of the deep mixing phase, and (iii) N, the number of deep mixing flashes.

We found that we could easily bring a large amount of ^{12}C to the surface. It appeared possible to form carbon stars both from stars of low ($\sim 1 \, M_\odot$) and intermediate masses ($\sim 5 \, M_\odot$) and both population types. A single helium-shell flash followed by deep mixing sufficed for the former case but on the order of 100 such flashes were required for the latter. No more than the order of 100 flashes are possible for the former and 1000 for the latter.

We found a great amount of ^{7}Li can be formed. The Cameron-Fowler mechanism for the production of high *lithium* abundances in late-type stars was shown to work satisfactorily, producing ^{7}Li/H up to about 10^{-7} in a single flash. This value is nearly independent of mass and population type.

For high values of T_{base}, namely, ~ 1 or 2×10^8 K, *eruptive stars* result as the convective envelopes reach into helium-burning layers. Large amounts of energy can be liberated in a very short time, e.g., the binding energy of the envelope, 10^{48} erg, can be exceeded in $\sim 10^6$ s.

Many of the gross features of the *R Coronae Borealis* stars can be understood by

---

* This paper was presented by I.-J. Christy-Sackmann and was supported in part by the National Science Foundation (GP-36687X, GP-28027).
** Present address: Rensselaer Polytechnic Institute, Troy, New York, U.S.A.

erupting carbon star models. For example, observed surface abundances of H, He, ^{7}Li, ^{12}C/^{16}O, ^{14}N/^{12}C, and ^{12}C/^{13}C can be accounted for.

For the lower base temperature case $(T_{\text{base}} \sim 5 \times 10^7 \text{ K})$ there were the cases of unrepeated deep mixing $(N = 1)$ or repeated deep mixing $(N > 1)$. For an *unrepeated* deep mixing, an upper limit could be placed on the mass of a carbon star, roughly $\sim 2\,M_\odot$. For short duration mixing, ^{7}Li/H reaches its peak value $(^7\text{Li/H} \simeq 10^{-7})$, ^{12}C/^{16}O > 1 or < 1, depending on the mass, $10^{-3} < {}^{14}\text{N}/{}^{12}\text{C} < 1$ and $5 < {}^{12}\text{C}/{}^{13}\text{C} < 100$; for intermediate duration mixing, ^{7}Li/H $< 10^{-8}$, ^{12}C/^{16}O < 1, $1 < {}^{14}\text{N}/{}^{12}\text{C} < 40$, and $2.4 < {}^{12}\text{C}/{}^{13}\text{C} < 5$; for long duration mixing, ^{7}Li/H $= 0$, ^{12}C/^{16}O > 1, ^{14}N/^{12}C $= 40$, and ^{12}C/^{13}C $= 3.2$. These abundance ratios agree quite well with the observations in the oxygen and carbon stars, S, BaII, C, and CH, ordering these spectral types in a sequence determined by the duration of deep mixing.

Repeated deep mixing can explain the observations known for S, BaII, and C stars when they are considered to be intermediate mass objects. S stars would be interpreted as those with few flashes, BaII stars as those with an intermediate number of flashes. However, repeated deep mixing phases cannot explain the observations of ^{7}Li and ^{13}C seen in CH stars, when the latter are considered to be low mass objects.

MIXING IN STARS

(Abstract)

PETER P. EGGLETON

Institute of Astronomy, Cambridge, U.K.

Mixing of composition in a star may be due not only to convection and semiconvection but also to circulation, as Prof. Kippenhahn has reminded us, and perhaps also to instability, for instance as Dilke and Gough (1972) have suggested for the Sun. I have tried to show in the past that convective and semiconvective mixing can be described by a diffusion equation

$$\frac{\partial}{\partial m}\,\sigma\,\frac{\partial X}{\partial m} = \frac{DX}{Dt} + XR,$$ (1)

where σ is a suitable mixing rate, e.g.

$$\sigma \propto (\nabla_r - \nabla_a)^2$$ (2A)

or

$$\sigma \propto \left(\nabla_r - \nabla_a - \frac{\beta}{4 - 3\beta}\frac{d\log\mu}{d\log p}\right)^2$$ (2B)

in an unstable region. Of course $\sigma = 0$ in a stable region. Provided equation (1) is solved *simultaneously* with the structure equations condition (2A) readily leads to convection and to Schwarzschild's semiconvection. I have not used (2B), but I suspect that in at least some cases it may not lead to a unique answer.

Recently Dr J. Perdang and I have tried using Equation (1) to describe a hypothetical mixing that might arise in and near a shell source in a red giant. We suppose there may be an instability, perhaps of a non-radial gravitational mode which grows on a thermal time scale, whose overall effect on the evolution of the star may be described qualitatively as a diffusion of composition on a thermal timescale. This suggests that we take a mixing rate

$$\sigma \sim (4\pi r^2 \varrho)^2 \, (l^2/\tau)$$ (3)

with l being a typical length scale for the unstable mode and τ being a thermal time scale. With such a mixing we find that the luminosity of a red giant increases rapidly.

A. J. C. Bolton and I have used an equivalent analysis to re-examine the stability of thin burning shells to spherical perturbations on a thermal time scale. Although our equilibrium model of such shells is rather crude, and not applicable to all kinds of burning shells that may actually occur, the perturbation analysis of the model was carried out exactly. Our conclusion is broadly that the stability of a shell, as well as

Tayler (ed.), Late Stages of Stellar Evolution, 125–126. All Rights Reserved.
Copyright © 1974 by the IAU.

PETER P. EGGLETON

its structure, is mainly determined by η, where

$$\eta \equiv \left(\frac{\partial \log \varepsilon_{\text{nuc}}}{\partial \log T} \right)_{X, \, p}. \tag{4}$$

For hydrogen shells instability (with a complex eigenvalue) sets in when $\eta \geqslant 14$; for helium burning, when $\eta \geqslant 24$. This suggests that even hydrogen burning shells, particularly in their early stages, may be unstable by a narrow margin.

Gough and Dilke (1972) have shown that the Sun, which is generally found to be stable to radial thermal instabilities, may be unstable to certain dynamical modes (g-modes) which grow on a thermal timescale. Their work may be applicable to all stars where burning takes place in a radiative region, e.g. stars with thin shells, and I think there is a strong possibility that all such shells are unstable on a thermal time scale to such non-radial dynamical modes.

In the late stages of stellar evolution the thermal and nuclear timescales in the central regions may be not very different, and the perturbation of the composition profile may not be negligible. Equation (1), when solved along with the conventional structure equations, should allow this effect to be easily included.

Reference

Dilke, F. W. W. and Gough, D. O.: 1972, *Nature* **240**, 262.

INTERPRETATION OF CARBON STARS*

(Abstract)

C. BARBARO and N. DALLAPORTA

Istituto di Astronomia, Universita di Padova, Italia

This communication aims at presenting the results of a research, whose purposes are to gain some insight into the evolutionary state of C stars, when the assumption is made that the main features of their spectra are due to mixing of the envelope with thermonuclear processed layers. Approximate indications concerning their population type and their mass range have been derived by:
  – the study of the galactic distribution of several phenomenological groups of C stars;
  – mass evaluation for some C stars belonging to binary systems or to clusters.
From the whole analysis three groups have been identified:
  (i) a Population I group in the mass range $(5–15 \ M_\odot)$ mostly constituted of constant, irregular or semiregular N stars;
  (ii) an old disk Population group with masses between one and two $M_\odot$, formed by R stars and long period variables;
  (iii) a group of halo stars.
The data concerning chemical abundances for these stars have been collected: they are rather scanty except for the determination of the C_{13}/C_{12} ratio. From them, it is tentatively surmised that abundances for the large mass star group are compatible with the assumption that mixing at the surface has occurred with CNO cycle processed layers; while, for the low mass star group, data are insufficient to allow one to decide whether mixing was due to CNO or 3α processed layers.
The theoretical analysis based on the evolutionary tracks of different groups shows that for the large mass group, mixing with CNO processed layers may have occurred at the reaching of the Hayashi border line. However, the amount of material brought to the surface would be insufficient to obtain the appearance of a C spectrum unless mass loss of the envelope has occurred during the He-burning phase. It is shown that mass loss rates of the order of 10^{-6} to $10^{-5} \ M_\odot \ \mathrm{yr}^{-1}$, quite compatible with the experimental observation are sufficient to reach the desired effect. However, it may be suspected that only for the larger mass stars $> 10 \ M_\odot$, the total mass loss would not be too large to allow them to preserve the red supergiant configuration.
For the low mass star group, the analysis of the available tracks leads in all cases to the interpretation that mixing may have occurred with the 3α processed material during the flash either of the He core or of the He shell.
These predictions could be checked by the C/N and C/H abundance determinations both for the large mass and low mass groups, which should be rather different in the two cases.

* This paper was presented by C. Barbaro.

Tayler (ed.), Late Stages of Stellar Evolution, 127. All Rights Reserved.
Copyright © 1974 by the IAU.

DISCUSSION AFTER PAPER BY SUGIMOTO

Massevitch to Sugimoto: Could you please give an estimate of how your results for late stages (for 12 $M_\odot$ and 30 $M_\odot$) will change, if the initial chemical composition is changed?

Sugimoto: I do not think that the results depend very much on composition.

Massevitch: But the depth of the convective envelope would be changed – in what direction?

Sugimoto: I do not think that the change is very great.

Fricke to Sackmann: Did you use a self-consistent method to calculate the joining up of the inner and outer convection zones?

Sackmann: No, we looked at the problem from the other way around. For this project, we have simply *assumed* that such a link-up of the convection zones does occur and then asked what kind of surface abundances would result. It turned out that we obtained all of the peculiar abundances of the light elements observed in the late-type stars of spectral types, S, C, BaII and CH. We could, in fact, even account for many abundance peculiarities of R Coronae Borealis stars. Such argument strongly supports our initial assumptions.

However, independent of the project discussed above, we have investigated plausibility arguments for this assumption. We have found (Smith and Sackmann, 1973, submitted to *Astrophys. J.*) that, from the point of view of stellar evolution, such a link-up would not be impossible.

Schwarzschild to Sugimoto: I very much admire the thorough and beautifully systematic results which you, Dr Sugimoto, have just shown us. I feel quite persuaded that the sequence of evolutionary events as a function of stellar mass which you have presented may qualitatively turn out to be right. But I am somewhat worried that quantitatively these results may still be beset by a rather greater uncertainty than might appear because the range of possible convective efficiency which you have used ($\alpha = 1.0$ to 1.5) seems to me substantially smaller than the range of uncertainty in our knowledge regarding convection.

Sugimoto: Of course you are right but we can say something in extreme cases. In some situations no penetration of convection occurs even if infinite mixing length is used.

Arnett to Sugimoto: I am impressed by Sugimoto's results, and am particularly intrigued by his results for massive stars with variable effectiveness of the lepton neutrino losses. If the neutrino losses are too low, mixing of the carbon-oxygen zone will overproduce ^{14}N. I estimate that on the average (20 $M_\odot$ star?) one should mix less than 6% of the C, O region. Can Dr Sugimoto tell me what value of the Universal Fermi interaction that implies?

Tayler (ed.), Late Stages of Stellar Evolution, 128–129. *All Rights Reserved.*
Copyright © *1974 by the IAU.*

Sugimoto: I am not certain of the exact figure but it should be more than 10% of the standard value.

Ergma to Sugimoto: I should like to point out that the structure and depth of the convection zones in stars with deep convective envelopes depends not only on the assumed value of α but also, as was shown by Henyey, Vardya and Bodenheimer (*Astrophys. J.* **142**, 841, 1965) and more recently by Ergma and Massevitch (*Astron. Nachr.* **293**, 145, 1971) is very sensitive to the values of coefficients α_1, γ and ν which are used for description of convective transfer in the mixing length theory. Taking, for example, $\alpha = 1.0$ but varying α_1, γ and ν, one can obtain a large variety of depths for convective envelopes. Therefore one has to be very careful in using these results for interpretation of physical processes in convective envelopes.

PRESUPERNOVAE*

V. S. IMSHENNIK and D. K. NADYOZHIN

Institute of Applied Mathematics, USSR Academy of Sciences, Moscow, U.S.S.R.

Abstract. A survey is given of the present state of theories concerned with the mechanism of supernova outbursts. Special attention is paid to the state of the star immediately before the supernova explosion (presupernova).

The following topics are discussed.
(1) The thermonuclear model of supernova explosions.
(2) The gravitational collapse model of supernova explosions.
(3) The characteristics of presupernova models resulting from the comparison of theoretical light curves with observational ones.

1. General Survey

The problem of supernovae is an extremely interesting one from very different points of view both theoretical and observational. It is closely connected with a number of various astrophysical topics from the origin of cosmic rays and chemical elements to the origin of pulsars and black holes.

The internal structure of the star immediately before a supernova outburst is very important for elucidation of mechanisms of supernova explosions. Thus, a problem of presupernovae arises. At the present time it is indisputable that the stars explode as supernovae at the late evolutionary stages when main stores of nuclear fuel in stellar core have exhausted due to thermonuclear fusion. However, all intermediate products up to primordial hydrogen and helium at the outermost layers of a star may be in stellar envelope, provided they have not been ejected because of mass loss, or mixed with interiors.

As the theory of late stages of stellar evolution is going on, overcoming great difficulties connected with consideration of convection, mass loss, kinetics of thermonuclear reactions, multishell heterogeneous stellar structure, neutrino energy losses, and so on, we obtain a still more clear idea of presupernova structure. For the present, however, it is difficult to deduce a definite point of view. Therefore, the theory of supernovae began its own independent development based on very simple and crude presupernova models which are usually chosen from general considerations. Such a method is fully justified as it gives a chance to obtain some additional information about presupernovae by means of comparing theoretical results with observational data. It seems for us that the double approach to the problem of presupernovae, both from the side of stellar evolution and of hydrodynamical explosion theory, will finally permit us to solve the problem and, hence, to construct the full picture of star life from its birth to death. The problem of presupernovae will be discussed in this report mainly from the hydrodynamic point of view.

* This paper was presented by D. K. Nadyozhin.

Tayler (ed.), Late Stages of Stellar Evolution, 130–148 All Rights Reserved.
Copyright © 1974 by the IAU.

Roughly speaking there is a rather trivial alternative for the present. In the first place a star may pass over all late stages of its evolution just quietly without disruption including possible stages of slow mass loss. In that case the central regions of the star burn nuclear energy supplies up to full exhaustion and the matter transmutes to the iron group elements. In the second place a star may experience the thermonuclear explosion due to carbon, carbon-oxygen or another flash. This phenomenon was first discovered as a helium flash. It is one of the greatest difficulties in the theory of stellar evolution up to now.

In the first case dynamical instability leading to gravitational collapse develops at centre of the star. The gravitational collapse is followed by powerful neutrino radiation and thermonuclear explosion of nuclear fuel remaining in the star's envelope. Various supernova models of that type have been constructed for a wide range of presupernova masses (Fowler and Hoyle, 1965; Colgate, White, 1966; Arnett, 1966, 1967; Ivanova et al., 1967, 1969; Rakavy and Shaviv, 1967; Fraley, 1968; Finzi and Wolf, 1967; Hansen and Wheeler, 1969). These models differ one from another by the mechanisms of dynamic instability and by nature of the physical conditions involved. The investigation of dynamic instability and a survey of theoretical works on such types of supernovae can be found in the book by Zel'dovich and Novikov (1971). In these works, largely using numerical solutions of hydrodynamic equations, there is a difficulty which consists in the low efficiency of adopted physical mechanisms of ejection of a star's envelope such as thermonuclear explosion of remaining nuclear fuel and neutrino energy deposition. However, the difficulty doesn't concern presupernovae with very large masses $(M \geqslant 30 \, M_\odot)$ and sufficiently low ones $(M \approx \approx M_{Chandr} = 1.4 \, M_\odot)$, (Fraley, 1968; Hansen and Wheeler, 1969). These presupernovae probably don't reach the evolutionary stage with an iron core. The very massive stars are dynamically unstable due to electron-positron pair creation followed by thermonuclear explosion of a carbon-oxygen core (Fraley, 1968; Rakavy and Shaviv, 1967). Dynamic instability due to neutronization developing on the time-scale of beta-processes (Bisnovatyi-Kogan and Kazhdan, 1966; Bisnovatyi-Kogan and Seidov, 1969, 1970; Imshennik and Chechetkin, 1970) is the cause of explosion of the stars with masses near Chandrasekhar's limit. The iron core must be formed as result of evolution of the intermediate-mass stars. The main mechanism of dynamic instability in this case is disintegration of the iron group elements to α-particles and free nucleons (Fowler and Hoyle, 1965). As gravitational collapse proceeds, the central regions of the star become opaque first to the electron neutrino and then to muon neutrino (Domogatsky, 1969). At the moment of ignition of the nuclear fuel remaining in the stellar envelope, the central regions of the star form a small-size core opaque to neutrino radiation characterized by the following parameters: $T_c = 100\text{--}300 \times 10^9$ K, $\varrho_c = 10^{12}\text{--}10^{14}$ g cm^{-3}, $R \approx 10^7$ cm. The interior of the star having dimensions from 10^7 to 10^9 cm absorbs diluted radiation of the neutrino photosphere and simultaneously loses the energy because of volume radiation. That model seems the most plausible one for the problem of heating external layers by neutrino radiation (deposition). In the first works involving the effect of deposition, the deposition power apparently was strongly overestimated (Colgate and White, 1966; Arnett, 1966). Ivanova *et al.* (1967) found more moderate estimation of deposition power and in addition took into consideration the kinetics of oxygen burning. The total energy of a supernova explosion found in that work proved to be of the order of 3×10^{50} erg. It is quite enough to explain observational data (Poveda and Woltjer, 1968). To answer a question about the role of deposition it is necessary to solve a problem characterized by consideration of energy transport by the way of neutrino heat conductivity in a stellar core opaque to neutrino radiation (Imshennik and Nadyozhin, 1972), and in a time allowing for kinetics of β-processes in the field of diluted neutrino radiation in outer stellar layers.

It should be noted that a star's rotation and magnetic field favouring ejection of the envelope have not yet been taken into account appropriately. The investigations in that direction involving great difficulties have been started quite recently (Bisnovatyi-Kogan, 1970; LeBlanc and Wilson, 1970).

In the alternative discussed above, there was mentioned the second possibility of supernova outburst as a result of catastrophic development of carbon or, in evolutionary sense, more late flash. The idea of a thermonuclear explosion in degenerate matter was known long ago. However, only in 1969 was the intensive theoretical elaboration of that version of supernova started (Arnett, 1969). At present time a great number of works are devoted to the problem (Colgate, 1971; Barkat *et al.*, 1970, 1971; Buchler *et al.*, 1971; Bruenn, 1971, 1972). In the first work by Arnett (1969) a dense $(\varrho_c \gtrsim 2 \times \times 10^9 \text{ g cm}^{-3})$ degenerate carbon core has been shown to suffer first a thermal instability and then to explode in regime of detonation. In contrast with previously adopted presupernovae for which initial models had been generally constructed rather arbitrarily (without necessary connection with the late stages of stellar evolution), Arnett (1969) has pointed out appropriate evolutionary models $(3.5 M_\odot < < M < 8 M_\odot)$ characterized by a natural origin of a degenerate carbon-oxygen core. The models have in more detail been calculated also by Rose (1969), Paczyński (1970), and Uus (1970). It is impossible, however, to say that a full fitting of the calculations of stellar evolution with the initial state of stellar core before explosion is now available. Therefore, there is rather great arbitrariness of the choice of the initial presupernova models connected with the mass exchange in binary stars (mass loss for a single star), the neutrino energy losses, the nuclear reaction rate etc. (Bruenn, 1972). The most complete hydrodynamic calculations of the thermonuclear explosion of the stellar core have been made by Bruenn (1971, 1972). These works, as distinct from previous analysis by Arnett (1969), involve the following complex physical processes which relax the detonation: (1) the decomposition of iron group elements at high temperatures $(T \gtrsim 10^{10} \text{ K})$ produced both inside and behind the detonation front, (2) various β-processes such as URCA-process followed by the powerful neutrino radiation and neutronization. As compared with Arnett's work in which full expansion of star with total energy 2×10^{51} erg had been obtained, consideration of the relaxation effects mentioned above has resulted in the possibility of formation of a gravitationally bound remnant and even a neutron star.

Now is the time to say that by contrast with dynamically unstable presupernovae (without regard to stars with extreme masses), it is very difficult to find a gravitationally bound remnant in the case of thermonuclear presupernovae. In this respect the theory of supernovae perhaps fatally contradicts the observations confidently indicative of the relation of pulsars or rotating neutron stars with the remnants of supernova explosions (Zel'dovich and Novikov, 1971). Bruenn's calculations resulted in the formation of a gravitationally bound remnant for central densities $\varrho_c \gtrsim 1.5 \times$ $\times 10^{10} \text{ g cm}^{-3}$. From the standpoint of stellar evolution theory $(T_c \sim \varrho_c^{1/3})$, in that case the central region of the star most likely has already passed over the stage of carbon burning and is composed of its products: O, Ne, Mg, ... The analysis of self-consistency of the detonation (with allowance for nonideality of ion component) has led to doubtful results namely for a such mixture (Bruenn, 1972). Therefore, in spite of a gravitationally bound remnant having been obtained in hydrodynamic calculations, we have to recognize that it was formed under very trying conditions. In short, the neutron stars indeed hardly form in the supernova theory based on a flash of nuclear fuel in a dense stellar core. Certain new possibilities in this respect seem to arise from the really existing pulsational regime of thermonuclear explosion which has been found as result of nuclear kinetics being taken into account in hydrodynamic calculations (Ivanova et al., 1973). The previous idea of the way of detonation development proved to be inexact. The central regions of the star burn without forming a detonation front and the star as a whole suffers radial pulsations. At the contraction phases the portions of matter are burnt out with gradually increasing pulsational amplitude due to resulting energy release. This is the way to obtain mass ejection at one of expansion phases followed by burning extinction and the subsequent formation of a gravitationally bound remnant. Decomposition of iron group elements and β-processes, unfortunately not allowed for the calculations, favour the conclusion. The above calculations resulted in full destruction of the star after the detonation front had been formed in the third burning cycle (the third contraction) at $m/M \approx 0.5$. A remaining half nuclear fuel burnt out in the regime of detonation. From the analysis of results it follows, however, that precise values of the initial data (especially temperature distribution) are very significant.

Finally, another difficulty should be mentioned. There is a doubt as to the fact of thermonuclear explosion due to carbon burning. Paczyński (1972) assumed that the neutrino losses by URCA-processes proceeding in a convective core may stabilize carbon burning. This is a generalization of URCA-shell mechanism taking place in the case of white dwarf vibrations (Tsuruta and Cameron, 1970). Although the first calculations of selfconsistent carbon-oxygen models allowing for URCA-losses have not led to sufficiently large burning relaxation (Ergma and Paczyński, 1973), the effect does change our notion of the conditions of the thermonuclear explosion. In this respect it is necessary to point to the calculations of a carbon-oxygen star's evolution for a large range of stellar masses $1.5 \leqslant M/M_\odot \leqslant 30$ (Ikeuchi et al., 1971, 1972). The calculations resulted in passing through the stages of carbon and later fuels burning well, without appreciable development of thermal instabilities. The dynamically unstable presupernovae with different masses of iron core (in dependence on initial stellar mass) have been obtained in the calculations by the evolutionary method.

When investigating supernova dynamics and also late stages of stellar evolution one often considers only a stellar core neglecting the extensive envelope. It is, of course, reasonable since in the absence of mixing a rarefied envelope has essentially no influence on processes proceeding in the core. The theory, however, must answer a basic question: what will happen to a star with a mass M_{ms} on the main sequence? In this connection it is useful to give Table I (all masses are given in solar units).

TABLE I

M_{ms}	M_{conv}	M_{CO}	M_{Femin}	M_{Fei}
64	48	26	2.8	26
32	19	10	2	10
16	6.8	3	1.5	3
8	2.4	~ 1.4	~ 1	~ 1.4
4	0.8	~ 1.4	~ 1	~ 1.4

Here M_{ms} denotes the mass of the star at main sequence; M_{conv}, the mass of the convective core at main sequence; M_{CO}, the mass of carbon-oxygen core; M_{Femin}, the mass of iron core in absence of mixing due to circulation currents; $M_{Fei} = M_{CO}$, the mass of iron core in the case of carbon-oxygen core evolving with full mixing. The table was made up in accordance with Varshavsky and Tutukov (1972, 1973), Ikeuchi *et al.* (1971, 1972), and Paczyński (1971). As appears from the table the iron cores of massive stars $(M_{ms} \gtrsim 4\ M_\odot)$ finishing their evolution are of the order of 1–3 $M_\odot$. One must have in view, however, that for the present time the role of circulation (and partially convective) mixing is not understood in the theory of advanced stellar evolution. With a sufficiently effective mixing the masses of iron cores may be as high as M_{Fei}.

At this time there are two basic ideas of the presupernova structure. Firstly, there is a star having fully exhausted nuclear fuel in its central region and being just on the boundary of dynamical instability. Secondly, there is a star with the degenerate dense core consisting of carbon or other later nuclear fuels immediately before ignition. Thermal and vibrational instabilities are of importance here.

A lot of information about presupernovae can be obtained by comparing theoretical calculations with observational data on the light curves, the velocities of ejected matter etc. A number of works (Imshennik and Nadyozhin, 1964, 1967; Grassberg and Nadyozhin, 1969a, b; Grassberg *et al.*, 1971) deal with the ejection of supernova envelope by shock waves and the construction of theoretical light curves. The fundamental idea of these works consists in a supernova light curve being a result of illumination of thermal and recombination energy kept in outer layers of the star which have been heated by shock and (or) thermal wave. If a supernova has a compact (homogeneous) structure, almost all shock energy is converted into kinetic energy of envelope expansion and only a small part of the order of 10^{-3}–10^{-4} is lost in the form of radiation. If the dense core of a presupernova is surrounded by a rarefied extended envelope, the process proceeds in a quite different way: the powerful thermal wave is generated in front of the shock propagating in stellar envelope. The time scale of thermal relaxation of the external layers proves to be less than a hydrodynamic one. Therefore, the part of energy lost in form of radiation increases considerably up to 0.02–0.05 for the presupernovae with the dimensions of envelope of the order of $10^4 \, R_\odot$.

The presupernovae models with the extended envelopes explain the basic observational data of light curves fairly well. The compact models are in a good agreement with anomalous light curves of such supernovae as in NGC 5457. Thus there is a principal possibility to find out from observational data whether a presupernova was compact or whether it had an extensive envelope. A peculiar cooling regime of expanding supernova envelopes in the form of cooling and recombination wave was found out in the calculations. In extended models this regime takes place in a few weeks after light curve maximum and may be responsible for the 'hump' observed on the light curves of supernova of type II. In compact models the cooling wave explains almost the whole of observed light curve except a sharp peak of radiation resulting from the emergence of the shock wave. For an explanation of the light curves of supernovae of type II it is necessary to use the presupernova models with very extended envelopes $(\sim 10^4 \, R_\odot)$. In that case it is sufficient that the envelope mass should be only 0.1–$1 \, M_\odot$. It must be emphasized that the envelope may not at all be in hydrostatic equilibrium. For example, it may be in the state of ejection with velocities of the order of a few hundreds km s^{-1}. The above thermal theory of light curves refers basically to supernovae of type II.

In the case of supernovae of type I nonequilibrium processes and also some additional energy sources such as β-decay may be of great importance. The problem was thoroughly discussed by Colgate and McKee (1969). They found that the postmaximum part of the light curve may be explained by two-step decay of radioactive nuclei ^{56}Ni(^{56}Ni $\rightarrow$ ^{56}Co $\rightarrow$ ^{56}Fe). It is necessary that about $0.3 \, M_\odot$ of ^{56}Ni should be in stellar envelope. Radioactive theory is in a good agreement with the exponential part of the light curve and with a lack of ultraviolet radiation (near the maximum of light the theoretical effective temperature equals ~ 6500 K). Though the fluorescence effects may be of some importance for supernovae of type I, an attempt to explain light curves by fluorescence only (Morrison and Sartori, 1969) is connected with great difficulties (Grassberg et al., 1971).

The lack of equilibrium between radiation and matter may result from the dominating contribution of Compton scattering to opacity because of energy exchange between photons and electrons being very small. In that case the propagation of radiation in a moving medium should be considered on the basis of equation of radiative transfer with the motion of matter taken into account (Imshennik and Morosov, 1969).

A lot of information can be in principle derived from the spectra of supernovae. The data on composition of supernova envelopes are particularly important. For example, an almost complete lack of hydrogen in external presupernova layers follows from the interpretation of supernova spectra of type I suggested by Mustel (1971). This fact may result either from the star having lost the hydrogen-helium envelope during its previous evolution or evolving with full mixing since some time.

2. The Computations of Supernova Outbursts and Gravitational Collapse

In this section of the report we shall discuss the results of the calculations of the instabilities developing in presupernovae. It is quite clear, that the analysis of the results and its comparison with observational data give us another opportunity for choosing more real presupernova models, besides its deriving by direct computation from the late evolutionary stages. We shall give below some details of our calculations concerning the problems discussed above: (1) the catastrophic development of a carbon flash followed by thermonuclear explosion in a dense carbon core, (2) the gravitational collapse of a completely evolved stellar core followed by explosion of nuclear fuel remaining in the envelope, (3) propagation of thermal and shock waves through a presupernova envelope, the dynamics of envelope expansion, and the supernova light curve.

2.1. Thermonuclear explosion of dense carbon core

In contrast with previous works we have included the kinetics of thermonuclear reaction. The hydrostatic equilibrium of an initial configuration with mass 1.4 $M_\odot$ was obtained by the relaxation method. The adopted equation of state is characterized by allowing both for degeneracy and relativity of electrons and positrons and also by taking ions in the approximation of perfect gas. For the sake of simplicity it was assumed that stellar core was composed of pure carbon. The rate of the nuclear reaction $^{12}C + ^{12}C \rightarrow ^{24}Mg + \gamma$ was taken into consideration. The initial data for hydrodynamic calculations are given in Figure 1. The initial temperature distribution is approximately indicated by solid line ($T_9 = 0.8$ quite near the centre). The distribution imitates the thermal stage of a carbon flash which is very long in comparison with hydrodynamic time scale ($\sim 2 \times 10^5$ s against ~ 2 s respectively). The equations of hydrodynamics in Lagrangian form

$$\frac{\partial r}{\partial t} = u \tag{1}$$

$$\frac{\partial u}{\partial t} = -4\pi r^2 \frac{\partial P}{\partial m} - \frac{Gm}{r^2} \tag{2}$$

$$\frac{\partial E}{\partial t} + P \frac{\partial}{\partial t}\left(\frac{1}{\varrho}\right) = \varepsilon_{nuc} - \varepsilon_\nu \tag{3}$$

$$\frac{\partial r^3}{\partial m} = \frac{3}{4\pi\varrho} \tag{4}$$

were solved numerically together with the equation of carbon burning

$$\frac{\partial X_c}{\partial t} = -f(\varrho, T, X_c). \tag{5}$$

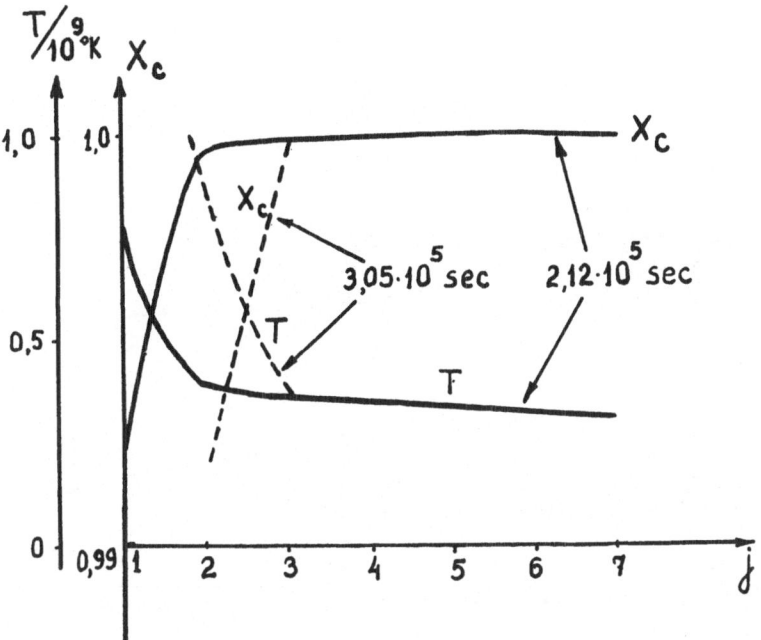

Fig. 1. Temperature and concentration of carbon before the beginning of calculations as the functions of Lagrangian coordinate j (the mass scale is indicated in Figure 2).

In (1)–(5) r denotes the radius; ϱ, the density; T, the temperature; P, the pressure; u, the velocity; E, the specific energy; and X_c, the concentration of carbon by weight. In the approximation of instantaneous conversion of ^{24}Mg to ^{56}Ni one may derive the following relation for nuclear energy generation

$$\varepsilon_{\text{nuc}} = 7.64 \times 10^{17} f\,(\varrho, T, X_c)\,\text{erg g}^{-1}\,\text{s}^{-1}. \tag{6}$$

The function $f\,(\varrho, T, X_c)$ with corrections due to screening and ^{24}Mg burning up to ^{56}Ni was taken according Fowler and Hoyle (1965):

$$\log f\,(\varrho, T, X_c) = 27.00 + \log\,(\varrho X_c^2) - \tfrac{2}{3}\log T_9 +$$
$$- \frac{36.57}{T_9^{1/3}}\,(1 + 0.08\,T_9)^{1/3} + 1.65\,\varrho_9^{1/3}/T_9. \tag{7}$$

Only URCA-proceses were not included in the neutrino losses ε_ν. Figures 2–6 show the results of the calculations. The distributions of T, ϱ, P, and u (Figures 2–5) are presented for a set of times given in Table II.
Figure 6 shows the variation of integral energies:

$$\mathscr{E}_{\text{in}} = \int_0^M E\,\mathrm{d}m, \qquad \mathscr{E}_g = -\int_0^M \frac{Gm}{r}\,\mathrm{d}m, \qquad \mathscr{E}_k = \int_0^M \frac{u^2}{2}\,\mathrm{d}m \tag{8}$$

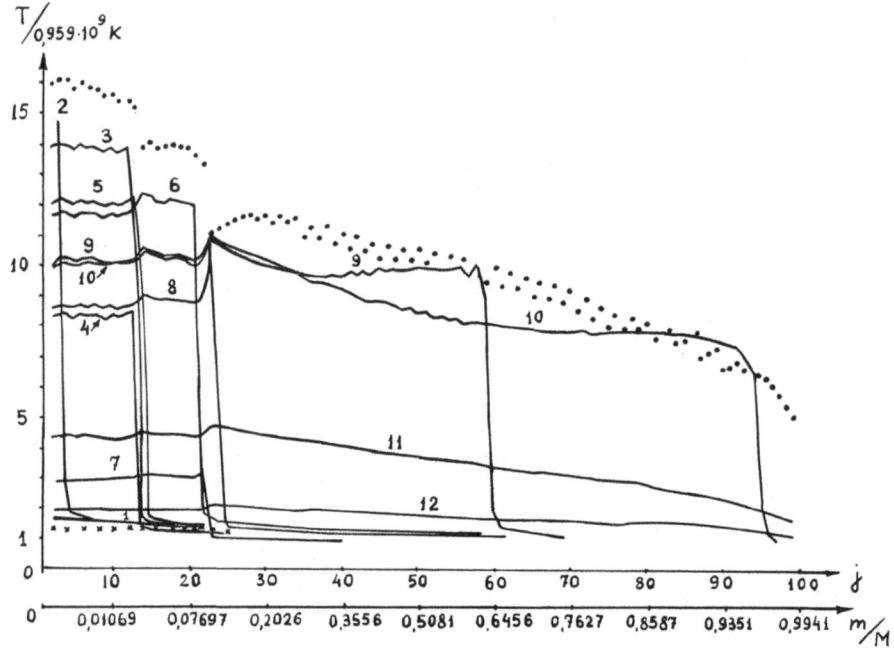

Fig. 2. Temperature as function of m and j for various moments of time. The circles indicate the absolute temperature maxima for every zone. The crosses correspond to temperature distribution in hydrostatically equilibrated configuration.

with time. The intervals of time corresponding to carbon burning are underlined on the axis of abscissas. The basic result of the calculations consists in finding the pulsational regime of nuclear burning. The burning first begins at the centre and gives rise

TABLE II

The numbers in the figures	The moments of time (s)	Comments
1	10.14	The end of relaxation. The configuration in hydrostatic equilibrium is found.
2	10.18	The beginning of the first burning phase.
3	10.40	The end of the first burning phase.
4	11.65	The maximum of the first expansion.
5	12.57	The beginning of the second burning phase.
6	12.78	The end of the second burning phase.
7	15.42	The maximum of the second expansion.
8	17.90	The beginning of the third burning phase.
9	17.96	The beginning of the detonation. The detonation front is in point $j = 59$.
10	18.01	The detonation front is in point $j = 94$.
11	18.34	The beginning of total expansion.
12	18.98	The star is in the state of total expansion.

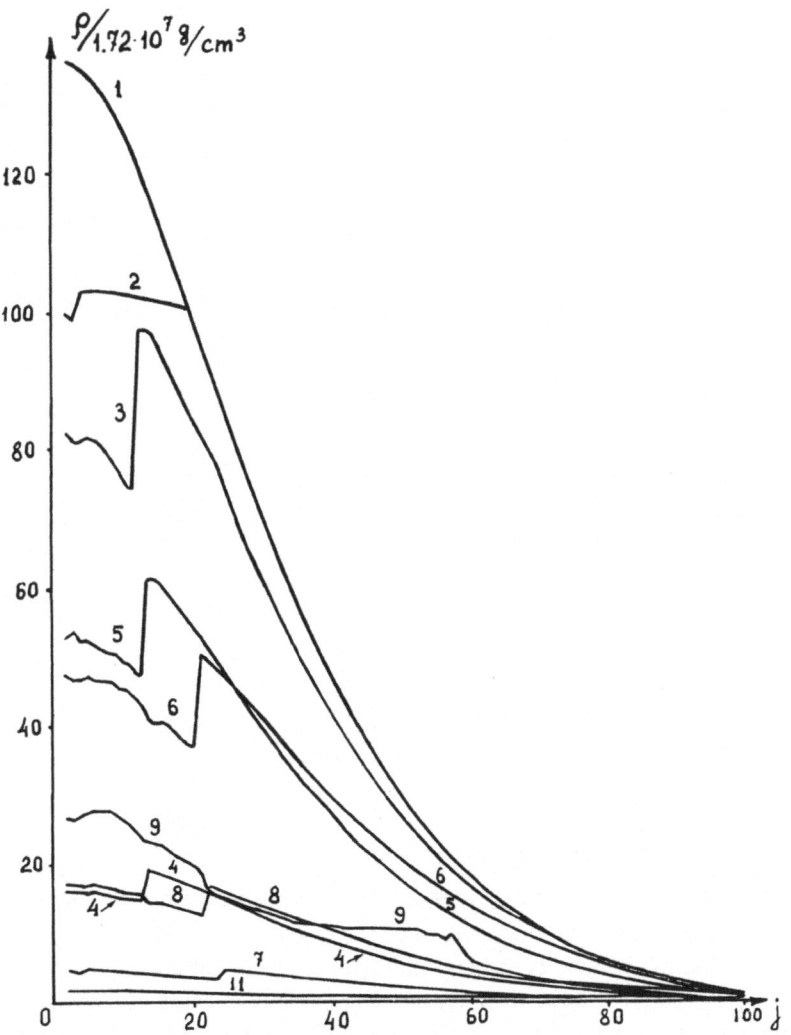

Fig. 3. Density as function of j for various moments of time. The moment t_{10} is excluded for technical reasons. For t_{12} the density practically goes together with the axis of abscissas.

to expansion of stellar matter which in turn extinguishes the burning itself. As distinct from detonation we obviously deal with tearing the motion of matter away the zone of burning. The burning recommences in the phases of contraction of the star and then extinguishes again with following expansions. Three stages of burning were found in the calculations. The detonative burning appeared only in the third contraction on a level with relative mass $m/M \approx 0.5$. As a result the total expansion of the star occurred with complete carbon burning. However, from analysis of the calculations one may conclude that the pulsational regime of thermonuclear explosion may result

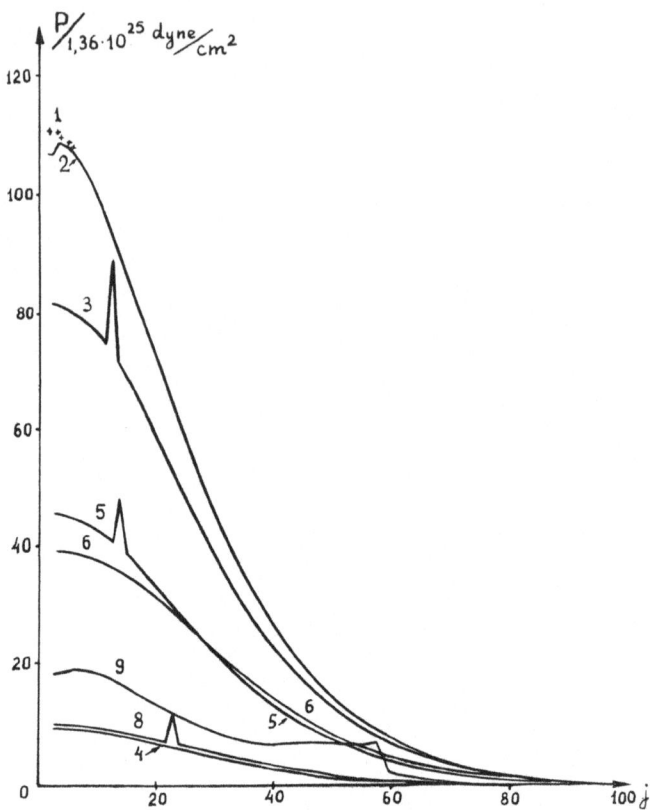

Fig. 4.   Pressure for the same moments as in Figures 2 and 3.

in principally new inferences. In particular, it may facilitate the formation of a gravitationally bound remnant. Another version of the calculations with δ-shape initial temperature distribution which fits thermal theory of the flash better (Figure 1) resulted in
numerous pulsations without any appreciable carbon burning. In that case there
would be needed a lot of pulsations in order that carbon should burn completely.
Apart from great sensitivity to initial conditions in the case of pulsations, the calculations point to the importance of some physical processes such as: (1) decomposition
of iron group elements to nucleons and more light nuclei, (2) β-processes, neutronization of matter, and URCA-losses with due regard of pulsational intensification of the
latter. The next series of our calculations devoted to consideration of the above effects
is under way.

2.2. Gravitational collapse

At initial stages the gravitational collapse of a stellar core devoid of nuclear fuel may
be considered without recourse to GTR-effects. The Equations (1)–(6) should be used
in the case. In the stellar core one must put $\varepsilon_{nuc} = 0$, but in the envelope one should

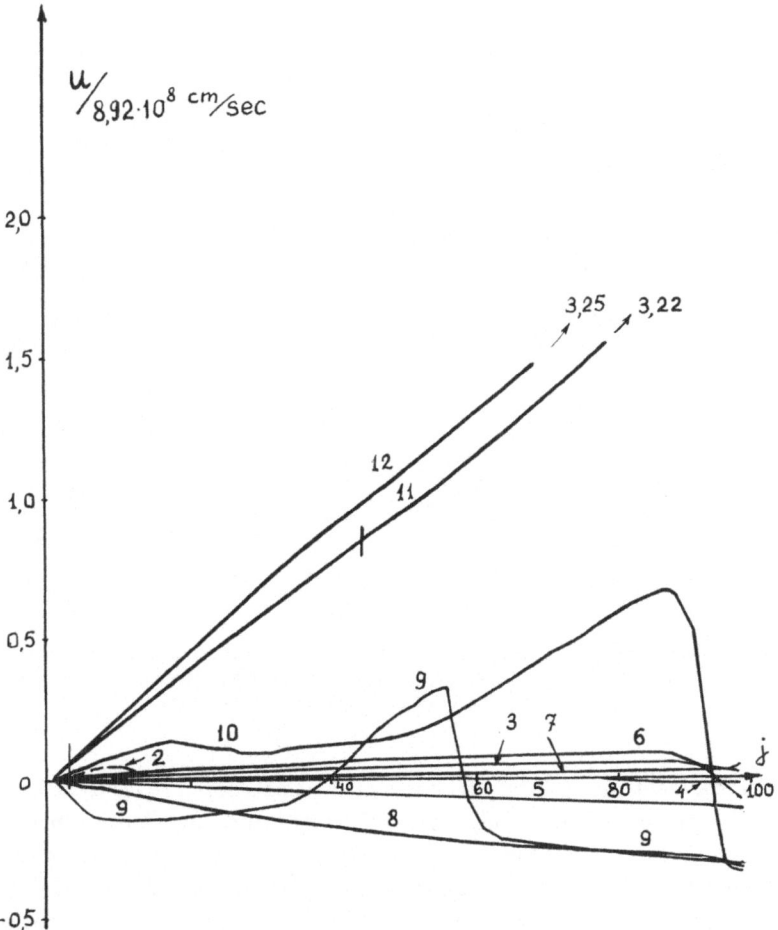

Fig. 5. Velocity for various moments of time. In moment t_1 the velocity equals to zero.

use ε_{nuc} corresponding to unburned matter, which is most likely composed of α-nuclei: ^{12}C, ^{16}O, ^{24}Mg, ... It is necessary to use the complex equation of state taking account of the decomposition of iron group elements to free nucleons (Imshennik and Nadyozhin, 1965; Chechetkin, 1969; Imshennik and Chechetkin, 1970). The violent reduction of the adiabatic index due to decomposition of iron is the reason of dynamic instability transiting to gravitational collapse.

Ivanova et al. (1967, 1969) have pointed out that, immediately after decomposition of iron, neutrino emission was due mainly to an URCA-process on free nucleons, while at the moment of stability loss neutrino emission was due largely to processes of electron-neutrino interaction (pair annihilation, plasmon neutrino, photo-neutrino). In these works the formulae for neutrino emission due to URCA process were also derived.

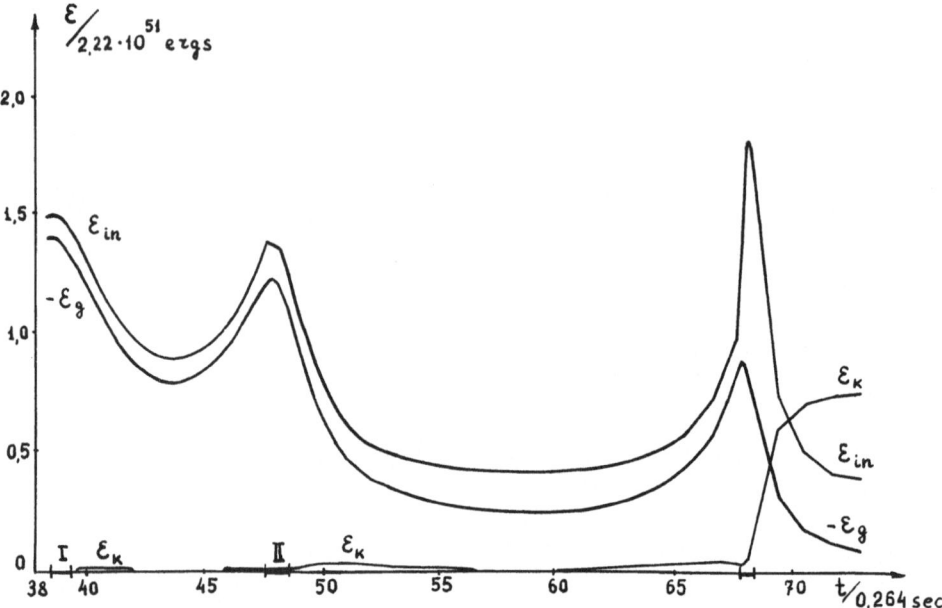

Fig. 6. Dependences of total energies (ε_k – kinetic energy; ε_{in} –- internal energy with electron-positron rest energies included; ε_g – gravitational energy) on time. Roman numerals are used for denoting of carbon-burning at phases of contraction.

The electron neutrino 'optical' thickness of the star's core proves to be of the order of 1 when in process of the collapse central temperature and density increase up to 40×10^9 K and 10^{11}–10^{12} g cm^{-3} respectively. Therefore the processes of electron neutrino absorption are of a great importance at that stage. The absorption of neutrinos may be taken into account by multiplying the rate of volume energy radiation ε_v at every point of the star by some factor K, which in very crude approximation may be used in the following form:

$$K = e^{-\tau_v(r)}, \tag{9}$$

where $\tau_v(r)$ is neutrino optical depth at the distance r from the centre of the star. It is connected with mean free path of neutrinos l_v by the next relation:

$$\tau_v(r) = \int\limits_r^R \frac{dr}{l_v}. \tag{10}$$

Such a method doesn't allow for redistribution of energy transported by neutrino radiation (or deposition) which happens by the way of diffusion in the case of $\tau_v \gtrsim 1$. Somewhat different method was used by Ivanova *et al.* (1969). An analytical formula for K in the star's centre was found from the exact solution of the equation of the neutrino transfer with the aid of quadratures. The formula gives the same order of K

as follows from (9) with $r = 0$. It was then assumed that K is constant all over the star. In the terms of (9) and (10) it was assumed that

$$K = e^{-\tau_\nu(0)}. \tag{11}$$

So the deposition was in a sense taken into consideration. A special calculation of gravitational collapse with K taken from (9) and (10) was carried out to illustrate the above method. The full lines (Figure 7) indicate the dependence of $\tau_\nu(r)$ and $\varepsilon_\nu e^{-\tau_\nu(r)}$ at some moment of collapse. The dotted line was used for dependence $\varepsilon_\nu e^{-\tau_\nu(0)}$ which on the order of magnitude corresponds approximately to the work by Ivanova *et al.* (1969). The deficiency of neutrino radiation presented by shading (Figure 7) may be regarded as some kind of consideration of the deposition.

The results of the work in question point out that the deposition itself doesn't lead to sufficiently intensive energy transfer. However, it favours the detonation of nuclear fuel (oxygen) remaining in the stellar envelope. Energy release due to the detonation is quite sufficient for the explanation of the supernova outburst.

In collapsing supernova models it is very easy, as compared with thermonuclear models, to obtain a gravitationally bound remnant. In the course of subsequent evolution it may change into pulsar or black hole depending on its mass. Figure 8 taken from Ivanova *et al.* (1969) shows the distributions of velocity in Lagrangian coordinates for

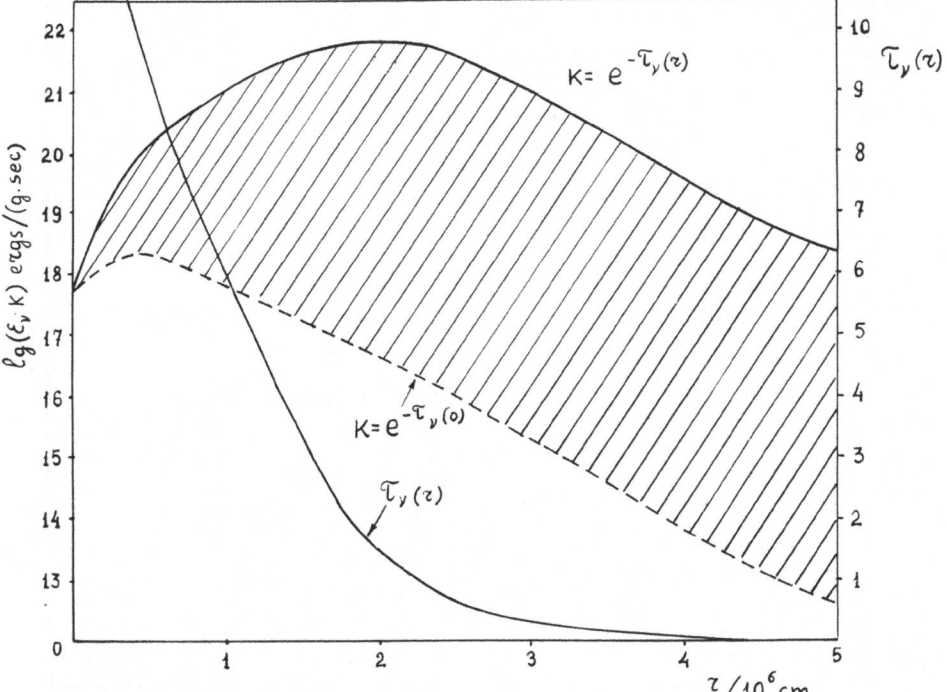

Fig. 7. Optical neutrino depth and neutrino energy losses for the various methods of considering neutrino absorption as function of Eulerian coordinate at some moment of gravitational collapse.

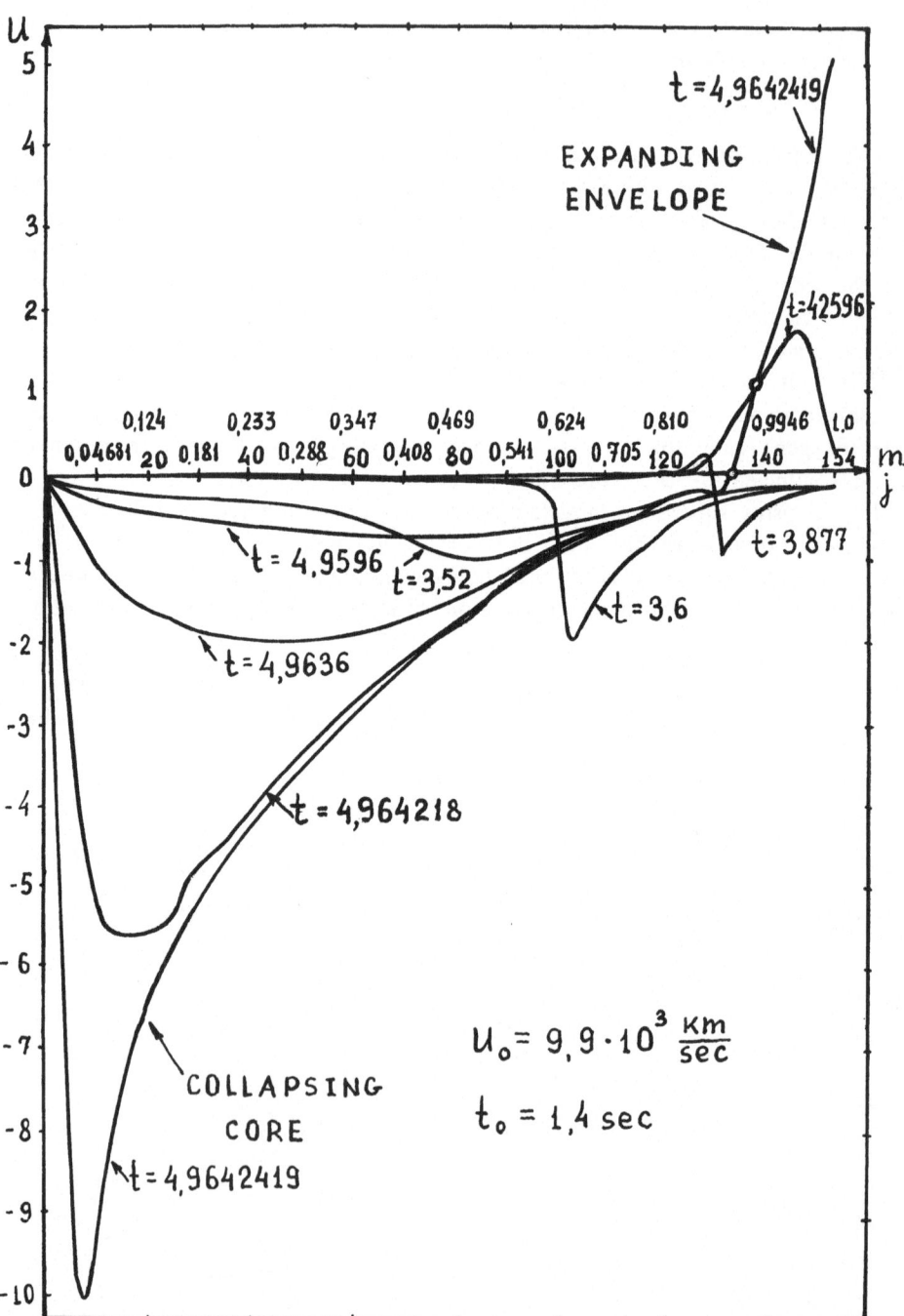

Fig. 8.   Distributions of velocity in the star of 10 $M_\odot$ at various moments of development of dynamic instability. The velocity and time units are $u_0 = 9.9 \times 10^3$ km s^{-1}, $t_0 = 1.4$ s. The collapsing core and expanding envelope may be seen at the same time.

different moments of time. It should be noted that we have a collapsing core and an expanding envelope at the same time.

3. Theoretical Light Curves of Supernovae

Figures 9–10 shows theoretical light curves of the compact and expanded models, and

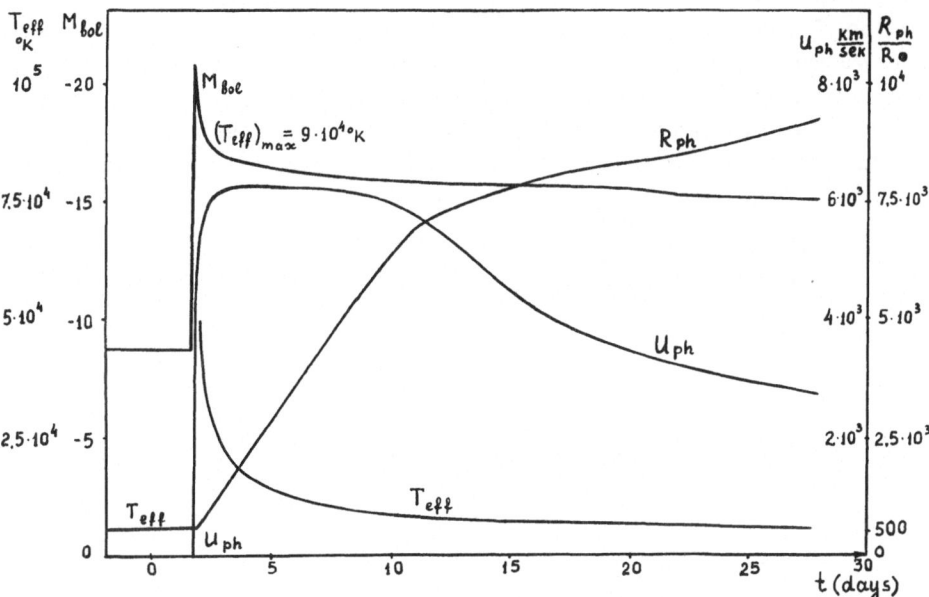

Fig. 9.   Theoretical light curve and characteristics of supernova photosphere for presupernova with $R = 500 \, R_\odot$.

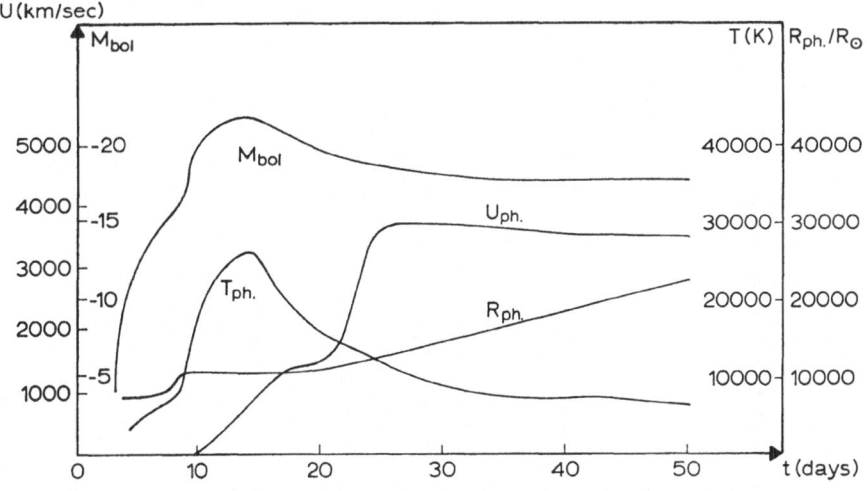

Fig. 10.   The same as in Figure 9 but for presupernova with $R = 10^4 \, R_\odot$.

also characteristics of the photosphere such as radius, effective temperature and the speed of matter emerging through the photosphere. From the comparison of observed light curves with theoretical ones (Figure 11), it follows that theoretical and observed light curves are characterized by the same parameters. However, there is some discrepancy basically involving in the shape of light curves. One should think here of theoretical results having been derived from the very crude presupernova models. For example it was assumed, for models with extended envelopes, that the density of the envelope falls off with the cube of the distance and is broken off at the surface of the star. We believe that the fitting between the theory and observations may be improved by a suitable selection of the law of density variation.

The lack of hydrostatic equilibrium of the extended envelopes has not practically any influence on the theoretical light curves since the speed of ejected matter is con-

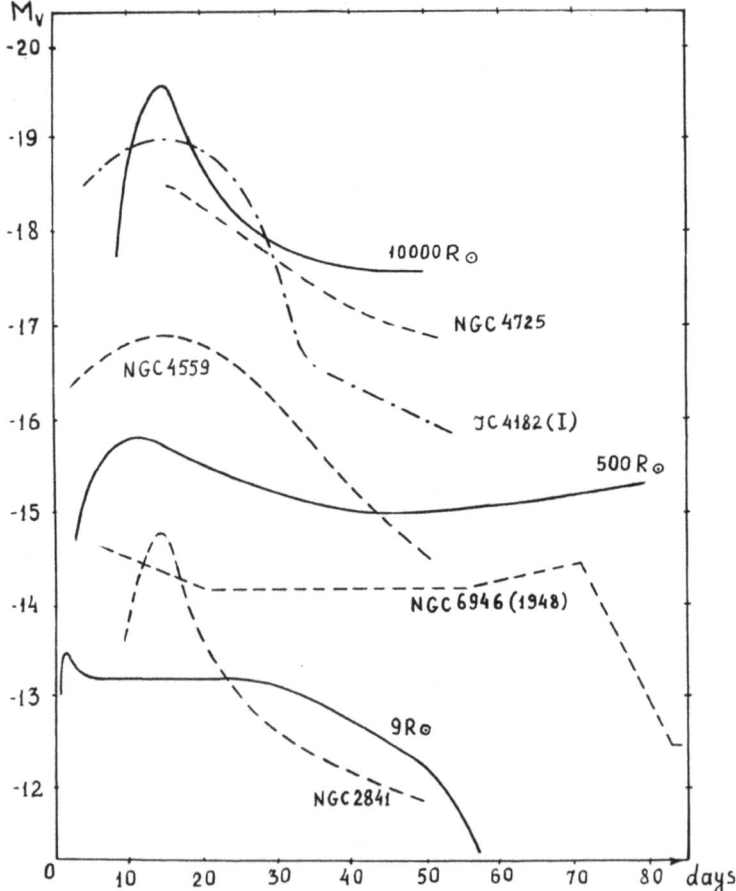

Fig. 11. Comparison of theoretical light curves (continuous curves) with observations (dotted curves).

siderably greater than the speed of sound. The extended envelope may be formed as a result of mass loss in the late stages of stellar evolution. For the stationary mass loss the density at great distance from the star is in inverse square proportion to the distance:

$$\varrho = \frac{|\dot{M}|}{4\pi v} \frac{1}{r^2},$$ (12)

where $\dot{M}$ denotes the rate of mass loss; and v, the velocity of ejection. The position of the photosphere R_{ph} is determined by the relation:

$$\tau(R_{ph}) = \int_{R_{ph}}^{\infty} \varrho \varkappa \, dr = \tfrac{2}{3}.$$ (13)

After substituting (12) to (13) and integrating one finds

$$|\dot{M}| = 10^{-9} \frac{\bar{v}}{\bar{\varkappa}} R_{ph} (M_{\odot} \, yr^{-1}),$$ (14)

where $\bar{\varkappa}$ and $\bar{v}$ are some means of the opacity and the velocity; $\bar{v}$ is in km s^{-1}, and R_{ph}, in $R_{\odot}$. The velocity of ejection is of the order of 10–100 km s^{-1}. According to (14) with $\bar{\varkappa} = 0.4$ cm^2 g^{-1} the rate of mass loss of the order of

$$|\dot{M}| = 2.5 \times (10^{-4} - 10^{-3}) \, M_{\odot} \, yr^{-1}$$

is necessary that the radius of photosphere should be as large as $10^4 \, R_{\odot}$. Thus, a quite moderate loss of mass is needed for producing the extended atmosphere. It should be noted that the latter may be cold and transparent immediately before the supernova outburst. It was just assumed in the above calculations. From the theoretical point of view it is only important that the dimensions of the photosphere should be of the order of $10^4 \, R_{\odot}$ after the sharp increase of opacity due to heating of the matter by the thermal wave.

Hence, presupernova may not surely look like a red (or infrared) supergiant. It may be a star with intermediate effective temperature (of spectral type G or F), embedded in an extended, cold, and transparent envelope.

References

Arnett, W. D.: 1966, *Can. J. Phys.* **44**, 2553.
Arnett, W. D.: 1967, *Can. J. Phys.* **45**, 1621.
Arnett, W. D.: 1969, *Astrophys. Space Sci.* **5**, 180.
Barkat, Z., Buchler, J. R., and Wheeler, J. C.: 1970, *Astrophys. Letters* **6**, 117.
Bisnovatyi-Kogan, G. S. and Kazhdan, Ya. M.: 1966, *Astron. Zh.* **43**, 761.
Bisnovatyi-Kogan, G. S. and Seidov, Z. F.: 1969, *Astrofizika* **5**, 243.
Bisnovatyi-Kogan, G. S. and Seidov, Z. F.: 1970, *Astron. Zh.* **47**, 139.
Bisnovatyi-Kogan, G. S.: 1970, *Astron. Zh.* **47**, 813.
Bruenn, S. W.: 1971, *Astrophys. J.* **168**, 203.
Bruenn, S. W.: 1972, *Astrophys. J. Suppl.* **24**, 283.
Buchler, J. R., Wheeler, J. C., and Barkat, Z.: 1971, *Astrophys. J.* **167**, 465.

Chechetkin, V. M.: 1969, *Astron. Zh.* **46**, 202 and 206.
Colgate, S. A. and White, R.: 1966, *Astrophys. J.* **143**, 626.
Colgate, S. A. and McKee, Ch.: 1969, *Astrophys. J.* **157**, 623.
Colgate, S. A.: 1971, *Astrophys. J.* **163**, 221.
Domogatsky, G. V.: 1969, *Nauch. Inform. Astron. Soviet. Acad. Sci. USSR* **13**, 94.
Ergma, E. and Paczyński, B.: 1973, in preparation.
Finzi, A. and Wolf, R. A.: 1967, *Astrophys. J.* **150**, 115.
Fowler, W. and Hoyle, F.: 1965, *Neutrino Processes and Pair Formation in Massive stars and Super-novae*, University of Chicago Press, Chicago-London.
Fraley, G.: 1968, *Astrophys. Space Sci.* **2**, 96.
Grassberg, E. K. and Nadyozhin, D. K.: 1969a, *Astron. Zh.* **46**, 745.
Grassberg, E. K. and Nadyozhin, D. K.: 1969b, *Nauch. Inform. Astron. Soviet. Acad. Sci. USSR* **13**, 96.
Grassberg, E. K., Imshennik, V. S., and Nadyozhin, D. K.: 1971, *Astrophys. Space Sci.* **10**, 28.
Hansen, C. J. and Wheeler, J. C.: 1969, *Astrophys. Space Sci.* **3**, 464.
Ikeuchi, S., Nakazawa, K., Murai, T., Hoshi, R., and Hayashi, C.: 1971, *Prog. Theor. Phys. Japan* **46**, 1713.
Ikeuchi, S., Nakazawa, K., Murai, T., Hoshi, R., and Hayashi, C.: 1972, *Prog. Theor. Phys. Japan* **48**, 1870.
Imshennik, V. S. and Nadyozhin, D. K.: 1964, *Astron. Zh.* **41**, 829.
Imshennik, V. S. and Nadyozhin, D. K.: 1965, *Astron. Zh.* **42**, 1954.
Imshennik, V. S. and Nadyozhin, D. K.: 1967, in Ya. B. Zel'dovich and I. D. Novikov (eds.), *Relativistic Astrophysics*, Nauka, Moscow.
Imshennik, V. S. and Nadyozhin, D. K.: 1972, *JETP* **63**, 1548.
Imshennik, V. S., Nadyozhin, D. K., and Pinaev, V. S.: 1966, *Astron. Zh.* **43**, 1215.
Imshennik, V. S., Nadyozhin, D. K., and Pinaev, V. S.: 1967, *Astron. Zh.* **44**, 768.
Imshennik, V. S. and Morozov, Yu. I.: 1969, *Astron. Zh.* **46**, 800.
Imshennik, V. S. and Chechetkin, V. M.: 1970, *Astron. Zh.* **47**, 929.
Ivanova, L. N., Imshennik, V. S., and Nadyozhin, D. K.: 1967, preprint IAM Acad. Sci. USSR.
Ivanova, L. N., Imshennik, V. S., and Nadyozhin, D. K.: 1969, *Nauch. Inform. Astron. Soviet. Acad. Sci. USSR* **13**, 3.
Ivanova, L. N., Imshennik, V. S., and Chechetkin, V. M.: 1973, in preparation.
LeBlanc, J. M. and Wilson, J. R.: 1970, *Astrophys. J.* **161**, 541.
Morrison, P. and Sartori, L.: 1969, *Astrophys. J.* **158**, 541.
Mustel, E. R.: 1971, *Astron. Zh.* **48**, 3.
Nadyozhin, D. K. and Chechetkin, V. M.: 1969, *Astron. Zh.* **46**, 270.
Paczyński, B.: 1970, *Acta Astron.* **20**, 47.
Paczyński, B.: 1971, Preprint 1, Warsaw, Polish Acad. Sci.
Paczyński, B.: 1972, *Astrophys. Letters* **11**, 53.
Poveda, A. and Woltjer, L.: 1968, *Astron. J.* **73**, 65.
Rakavy, G. and Shaviv, G.: 1967, *Astrophys. J.* **148**, 803.
Rose, W. K.: 1969, *Astrophys. J.* **155**, 491.
Tsuruta, S. and Cameron, A. G. W.: 1970, *Astrophys. Space Sci.* **7**, 374.
Uus, U.: 1970a, *Nauch. Inform. Astron. Soviet. Acad. Sci. USSR* **17**, 3.
Uus, U.: 1970b, *Nauch. Inform. Astron. Soviet. Acad. Sci. USSR* **17**, 25.
Varshavsky, V. I. and Tutikov, A. V.: 1972, *Nauch. Inform. Astron. Soviet. Acad. Sci. USSR* **23**, 47.
Varshavsky, V. I. and Tutukov, A. V.: 1973, *Nauch. Inform. Astron. Soviet. Acad. Sci. USSR* **26**, 35.
Zel'dovich, Ya. B. and Novikov, I. D.: 1971, K. S. Thorne and W. D. Arnett (eds.), *Relativistic Astrophysics*, Vol. 1: *Stars and Relativity*, The University of Chicago Press, Chicago and London.

DO PULSARS MAKE SUPERNOVAE?[*]

(Abstract)

J. P. OSTRIKER

Princeton University Observatory, U.S.A.

and

P. BODENHEIMER

Lick Observatory, U.S.A.

The hypothesis, that pulsars can produce supernova explosions of type II, is explored with the aid of detailed hydrodynamical calculations. Preliminary calculations performed with Gunn (1971) indicated that the electromagnetic energy radiated by a newly formed rotating, magnetic, neutron star could drive off the remaining envelope of a red giant star for stars having initial masses in the range 3–8 $M_\odot$. Here we present theoretical light curves obtained from hydrodynamic calculations with radiative diffusion and compare them with observations of supernovae. For the case of a star of original mass 4.5 $M_\odot$, light maximum is calculated to occur when the envelope, now in the form of a thin shell, has reached a radius of 9×10^{15} cm, a velocity of 7650 km s^{-1} and an effective temperature of 8000 K, at which point its optical thickness approaches unity. The early part of the light curve can be strongly affected by the presence or absence of a dust-laden circumstellar envelope. The temperature, luminosity, decay rate, and envelope velocity observed in the latter phases of type II supernovae are simulated with fair accuracy by the present model.

Reference

Ostriker, J.P. and Gunn, J. E.: 1971, *Astrophys. J. Letters* **164**, L95.

[*] This paper was presented by J. P. Ostriker.

Tayler (ed.), Late Stages of Stellar Evolution, 149. All Rights Reserved.
Copyright © 1974 by the IAU.

EVOLUTION OF MASSIVE STARS FROM HELIUM BURNING TO THE START OF NEUTRONIZATION

(Abstract)

W. DAVID ARNETT

Dept. of Astronomy, University of Texas, Austin, Tex., U.S.A.

Evolutionary sequences have been constructed beginning at helium burning and ending in the hydrodynamic collapse of the core when neutronization begins. Details of these models during silicon burning as well as at the onset of neutronization are presented. The connection of these results with the calculations of Wilson (1971) and Arnett (1967) is discussed. Calculations by Ostriker and Gunn (1971) and Falk and Arnett (1972) provide the beginning of a detailed understanding of supernova light curves. Taken together a coherent and continuous picture of stellar evolution through supernova explosion and pulsar formation can be obtained. While the theory still needs much more work, the gaps so obvious a few years ago are disappearing.

References

Arnett, W. D.: 1967, *Can. J. Phys.* **45**, 1621.
Falk, S. W. and Arnett, W. D.: 1973, *Astrophys. J.* **180**, L65.
Ostriker, J. P. and Gunn, J. E.: 1971, *Astrophys. J.* **164**, L95.
Wilson, J. R.: 1971, *Astrophys. J.* **163**, 209.

ON THE FINAL FATE OF DIFFERENTIALLY
ROTATING DEGENERATE CONFIGURATIONS

(Abstract)

D. KOESTER

Institute für Theoretische Physik und Sternwarte der Universität Kiel, F.R.G.

Double shell-burning stars of intermediate masses develop a degenerate carbon/oxygen core. Whether this core will ignite carbon depends on several parameters, i.e. mass, neutrino rates and rotation. If angular momentum is conserved, the core may even grow above the Chandrasekhar limit (Sackmann and Weidemann, 1972). In this case mass loss might terminate the nuclear evolution, leaving a remnant star which approximates a differentially rotating Ostriker-Bodenheimer (1968) configuration. The final fate of such configurations on cooling down is not yet clear.

Schwartz and Africk (1970) proposed that evolution leads to uniformly rotating cores with masses above the Chandrasekhar limit which finally end in collapse and supernova explosion. On the other hand Sackmann and Weidemann suggested rotational mass loss resulting in stable configurations below Chandrasekhar limit.

Recently, Durisen (1973) showed that viscosity does not in general lead to uniform rotation above the melting point of the ion lattice. For lower temperatures, however, the viscosity increases by several orders of magnitude such that the crystallized part of the star must certainly be in uniform rotation.

In order to follow the evolution during crystallization we made schematic model calculations, assuming always uniform rotation for the crystalline core and neglecting transport of angular momentum in the non-crystalline part of the model.

The configurations were assumed to be spherical with angular velocity depending only on the distance from the centre. The radial component of the centrifugal force is replaced by its mean value over the sphere. With these simplifications we constructed evolutionary sequences consisting of sequences of equilibrium models with constant angular momentum and mass but with growing crystalline cores. The results show that only if a differentially rotating model starts with very nearly the critical density may it undergo collapse by inverse β-decays (or ignite carbon). In most cases crystallization is terminated by mass loss, leaving as end products uniformly rotating degenerate stars below the Chandrasekhar (or James) limiting mass.

References

Ostriker, J. P. and Bodenheimer, P.: 1968, *Astrophys. J.* **151**, 1089.
Sackmann, I. J. and Weidemann, V.: 1972, *Astrophys. J.* **178**, 427.
Schwartz, R. and Africk, S.: 1970, *Astrophys. Letters* **5**, 141.
Durisen, R. H.: 1973, *Astrophys. J.* **183**, 205, 215.

THE EXPLOSION OF A ROTATING MAGNETIZED STAR
AS A SUPERNOVA

(Abstract)

G. S. BISNOVATYI-KOGAN

Institute of Applied Mathematics, USSR Academy of Sciences, Moscow, U.S.S.R.

It is easy to show that a moderately rotating star, which is contracted to the size of a neutron star, will rotate very rapidly if angular momentum is conserved. In fact, if on the main sequence the equatorial rotation speed is 10 km s^{-1} at $R_{eq} = 10^{11}$ cm, then, after the contraction to $R = 10^6$ cm, $v = 10(R_{eq}/R) > c$; of course the relativistic formula is necessary in order to obtain $v < c$. Thus, when rotation has rather small importance in early stages of evolution, it may play a decisive role in late stages, particularly in a supernova explosion.

The existence of pulsars shows that magnetic fields in neutron stars are very large ($\sim 10^{12}$ G). In Bisnovatyi-Kogan (1970) a mechanism of supernova explosion was proposed in which rotation was the main source of energy and the magnetic field provided the mechanism for its release. After contraction from the rapidy rotating pre-supernova stage, a differentially rotating disc-like configuration is formed, with a rapidly rotating neutron core. The transfer of angular momentum from the core to the envelope due to magnetic connection leads to the formation of a compressional wave in the envelope, which propagating in a medium of decreasing density transforms into a shock wave which appears as a supernova explosion.

A numerical calculation has confirmed this qualitative picture. Although the calculation assumes cylindrical geometry, it seems clear that in the more realistic spherical geometry the magnetorotational mechanism is important. It seems to me that the mechanism of supernova explosion due to pulsar-like emission will, in contrast, not work, because the high density plasma around the neutron star will screen out slow waves.

Reference

Bisnovatyi-Kogan, G. S.: 1970, *Astron. Zh. Akad. Nauk SSSR* **47**, 813.

Tayler (ed.), Late Stages of Stellar Evolution, 152. *All Rights Reserved.*
Copyright © 1974 by the IAU.

IRON CORE COLLAPSE IN 10 TO 30 $M_\odot$ STARS AND THE EFFECT OF μ NEUTRINOS

(Abstract)

JAMES R. WILSON

Radiation Laboratory, Livermore, Calif., U.S.A.

The evolution of 10, 20, 30 $M_\odot$ stars were calculated by Barkat, Rakavy and Reiss up to the point of Fe core collapse. The calculations produced Fe cores of masses 1.49, 1.6 and 2.2 $M_\odot$. The core collapse and the subsequent envelope burning were studied with the Wilson (1971) hydrodynamic neutrino transport computer programme. The results are that the high mass (2.2 $M_\odot$) core bounced with a weak outward shock that did not push back the envelope at all. The low mass core (1.49 $M_\odot$) bounced and sent a fairly strong shock out into the envelope. However, the shock was not strong enough to detonate the envelope. The envelope in this case was also at too low a density to support detonation by itself. The calculations produced no supernova. Lower mass Fe cores are indicated for the production of stronger outward shocks. Shock weakening with increasing core mass is due to neutrino cooling in the shock.

S. Weinberg has suggested that μ-neutrinos may act like electron neutrinos. In the iron core collapse calculations of Wilson (1971), 70 to 90% of the collapse energy was released by μ neutrinos. One therefore might expect large effects by changes of μ neutrino interactions. Two models (1.5 $M_\odot$, 3 $M_\odot$) were calculated with the μ neutrinos treated the same as the electron neutrinos (same opacity). The behaviour was very similar to the standard treatment except that the late cooling was much slower. However, by this time the system had become quiescent. The Weinberg hypothesis was also applied to the burning of a collapsing cold carbon white dwarf model of Craig Wheeler. With the regular μ neutrino model, the fraction of carbon needed to avoid disruptive explosion was less than or equal to 20%. Treating the μ neutrinos like electron neutrinos raised the limit on carbon for the formation of a neutron star to 70%. Changes in the electron neutrino opacity of the order of 10 in the standard treatment produced a smaller effect in the critical carbon composition.

Reference

Wilson, J. R.: 1971, *Astrophys. J.* **163**, 209.

DISCUSSION AFTER PAPER BY NADYOZHIN

Arnett to Nadyozhin: If I correctly understood what was said, Craig Wheeler and I have independently obtained similar results to those for carbon ignition obtained by Nadyozhin. We concluded that these results were produced by inadequate mass zoning. Although it seems clear that the detonation wave will propagate, it is difficult properly to treat the *formation* of the detonation wave under these conditions.

Fowler: As a result of these talks, I am completely bewildered and confused about the scenario for a supernova explosion. Ostriker said that he used a nuclear δ function as the source of the explosion, whereas Arnett said that few nuclear reactions occurred. Arnett said that there was some contribution due to diluted neutrino flux, while Wilson said that there was not. Can these different comments possibly be reconciled?

Ostriker: A time scale of less than 10^3 s is effectively a δ-function. (reply to Bisnovatyi-Kogan). It does not really matter whether the pulsar energy is carried by a low frequency wave or by an Alfvén wave or some other mechanism provided that it is propagated outwards.

Fowler: I find the pulsar luminosity one an attractive one but if this is applicable does any nucleosynthesis occur in the event?

Arnett: The low Z elements are at very low densities and they will not be processed very much. I am uncertain of the effect on elements between silicon and the iron peak.

Fowler: You have been an ardent supporter of explosive nucleosynthesis. Does any nucleosynthesis occur during a supernova explosion? If not, where does explosive nucleosynthesis occur?

Arnett: I cannot be certain; we are learning that our simple-minded models are not realistic.

Bisnovatyi-Kogan: In the model which I described, energy generation occurred in a very short time $\sim 7\text{--}10$ s unlike the very much longer timescale associated with the magnetic explosion.

Nadyozhin: I want to reply to Arnett's remark. The detonation was thoroughly treated in the calculations performed by Ivanova *et al.* (1973). Both the equations of hydrodynamics and of nuclear kinetics were used in contrast with previous works.

Tayler (ed.), Late Stages of Stellar Evolution, 154. *All Rights Reserved.*
Copyright © 1974 *by the IAU.*

THE THEORY OF NOVAE AND NOVA-LIKE SYSTEMS*

JOHN FAULKNER

Lick Observatory, Board of Studies in Astronomy and Astrophysics, University of California Santa Cruz, Calif. 95064, U.S.A.

Abstract. Recent observational and theoretical developments in the study of novae, particularly dwarf novae, are discussed. Mechanisms promoting mass transfer include (i) nuclear evolution or (ii) envelope instability of the red star and (iii) gravitational radiation of orbital angular momentum. Growing observational evidence against (ii) is supported by recent theoretical work on the medium and long term response of stellar radii to mass loss. Mechanisms (i) and (iii) may operate alone or in concert, depending on the circumstances.

1. Introduction

In the limited time at my disposal, it is impossible to review all the relevant recent work that falls within the scope of my title. I hope I may therefore be forgiven for personal bias in selecting points that seem to be of particular interest.

One must first face the problem that counter-examples exist to almost any statement one tries to make about these systems. Furthermore, although one can attempt to contrast and distinguish between the four types: classical novae, recurrent novae, dwarf novae and nova-like variables, the distressing fact is that in many of their properties there is considerable overlap. Nowhere are these points better illustrated than by the virtually ubiquitous property of membership in a short-period binary system with period substantially less than a day. A glance at a compilation of spectroscopic binaries among these variables (Mumford, 1967) reveals that the recurrent nova T CrB stands out with a period of $227^{\text{d}}6$; while the hope that some correlation might exist between type and binary period is dashed by the observation that there is an example of each type among the four shortest periods shown in the same table. One must perforce conclude that apart from the obvious, and defining characteristics (e.g. magnitude and frequency of outbursts – see Table I), these systems are differentiated by some underlying physical phenomena for which we possess inadequate observational handles.

Much of the recent detailed theoretical work has concentrated on studying runaway thermonuclear explanations for outbursts at or near the upper end of the logarithmic range of observed outburst energies, e.g. Rose (1968), Starrfield (1971a, 1971b) Rose and Smith (1972), and Starrfield *et al.* (1972). The latter work in many respects seems to put the theory of classical novae *per se* on a fairly sound theoretical footing, but at the same time raises awkward evolutionary questions. In what follows, I shall in fact concentrate more on the puzzles raised by the dwarf novae in particular, where the situation is even murkier. The term 'novae' will however be used to denote any of the four (or more) types unless a careful distinction needs to be drawn.

---

* *Contributions from the Lick Observatory*, No. 401.

Tayler (ed.), Late Stages of Stellar Evolution, 155–167. All Rights Reserved.

2. The Binary Model

A little over a decade ago, classical work by Kraft (e.g. Kraft, 1963, and references contained therein) established by a variety of means that novae were binary systems: Some are double-lined spectroscopic binaries, some single-lined, while some show regular eclipses. This direct evidence is available for so many of the known novae that one can reasonably draw two inferences:

(1) All novae are binaries, with direct evidence absent in some instances only because of unfavourable orientation of the orbit.

(2) Furthermore, the components must be 'close', that is to say at least one member has dimensions comparable to the separation.

Spectroscopically, a hot blue component is generally present in emission. Excluding T CrB (which, in agreement with point (2) above, at least shows a giant spectrum), the longer period systems are seen to contain red dwarf or subdwarf members. There is a systematic tendency for these to become later and fainter as the periods decrease until, for periods shorter than about six hours, no late-type component can be seen. However, eclipses by relatively dark components can still occur, thereby revealing the presence of what are probably dwarf M's.

In response to these observations, a model has emerged over the last decade. The model (see Figure 1) consists of a late-type, essentially main sequence component filling its Roche lobe and spilling matter towards a white dwarf companion (evidence for the existence of the latter is summarized in the next section). The resulting model has much in common with that first proposed for the nova-like variable UX UMa by Walker and Herbig (1954). In its latest refinements (Smak, 1971; Warner and Nather, 1971), the stream of infalling material creates a ring or disk surrounding the white dwarf. The disk contains a luminous 'hot spot' where fresh infalling material collides with the disk, which is itself the debris from many previous collisions. Thus 'the medium is the wreckage'. With three possible light centres in the system (or at least two, with one of them, the hot spot, variable and permanently displaced from either main mass concentration), it becomes possible to explain many of the otherwise puzzling features of the periodic systems, e.g. eclipse asymmetries, the 'humps' prior to primary eclipse, the disappearance of 'flickering' during eclipse, and so on.

With simple assumptions, the period-spectral type correlation and the disappearance of late-type components at about six hours can be understood, lending further support to the model. Assume that the late components are sufficiently like main-sequence stars that the radius-mass relationship is essentially undisturbed. Assume also that the white dwarf mass is comparable to, or greater than its companion. Then (Faulkner et al., 1972) a $P\sqrt{\varrho}$ relationship exists between the period P and the density, ϱ of the lobe-filling late component which is, to an accuracy of $\sim 3\%$, independent of the white dwarf mass. This relationship,

$$P\sqrt{\varrho} \sim 3.8 \times 10^4 \text{ s}$$

coupled with a crude representation of the lower main sequence (i.e. $R \propto M$) yields

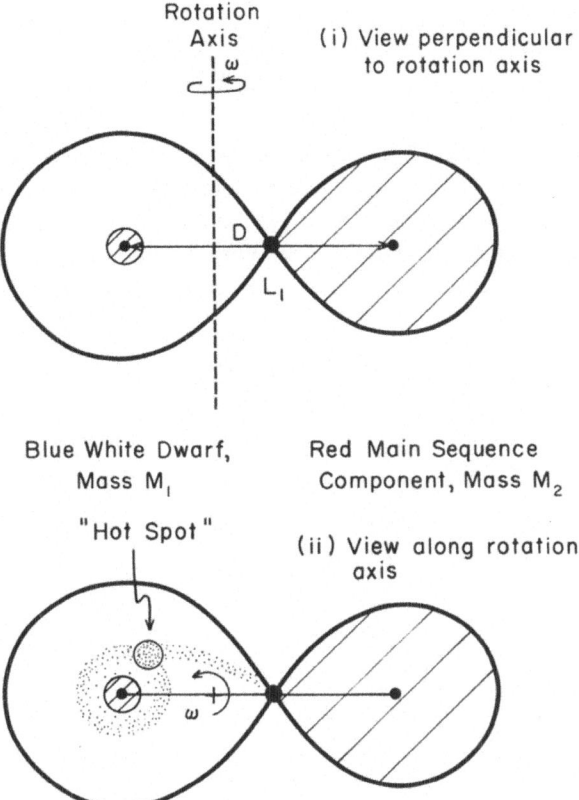

Fig. 1. The typical nova configuration. However, as pointed out in the text, the central parts of the red component may be fairly well evolved. In later stages as suggested for HZ 29 (Faulkner *et al.*, 1972), the main sequence component may be replaced by a degenerate, receding system, mass transfer still proceeding as shown.

$P \propto M$ (Faulkner, 1971) and thence, by appropriate specification of spectral types, the observed correlation and disappearance.

The restriction of the $P \sqrt{\varrho}$ relationship to situations where the lobe-filling mass contains less than half the mass of the system is not, in practice, a serious deficiency. If the lobe filling mass fraction were to exceed a critical value of order 0.5, mass transfer would occur first on extremely short term dynamical time scales followed by medium term thermal time scales before settling down to the ultimate long term time scales which presumably characterize the bulk of these variables. This long term time scale is, of course, intimately related both to the actual mass transfer mechanism and to the precise seat of the quasi-periodic outbursts. In contrast with the general agreement on the model described above, both of these points have remained bones of contention, as has the nature of the progenitors. We shall return to these points later.

We conclude our discussion of the binary model by remarking that the above considerations mean that in practice we may turn the $P \sqrt{\varrho}$ relationship around and use

JOHN FAULKNER

it to deduce candidates, possibly in a variety of evolutionary stages of development, for the lobe-filling component. This is illustrated in Figure 2, where a number of interesting and relevant radius-mass relationships for such lobe-filling stars are crossed by lines indicating the associated orbital period. In the most extreme example to which this has been applied, i.e. HZ 29, it has been shown that the extremely short period of this nova-like variable (~17.5 min) suggests a most natural model in which an unobserved but eclipsing low mass (~0.04 $M_\odot$) lobe-filling degenerate helium star orbits a more massive and observable white dwarf (Warner and Robinson, 1972b; Faulkner

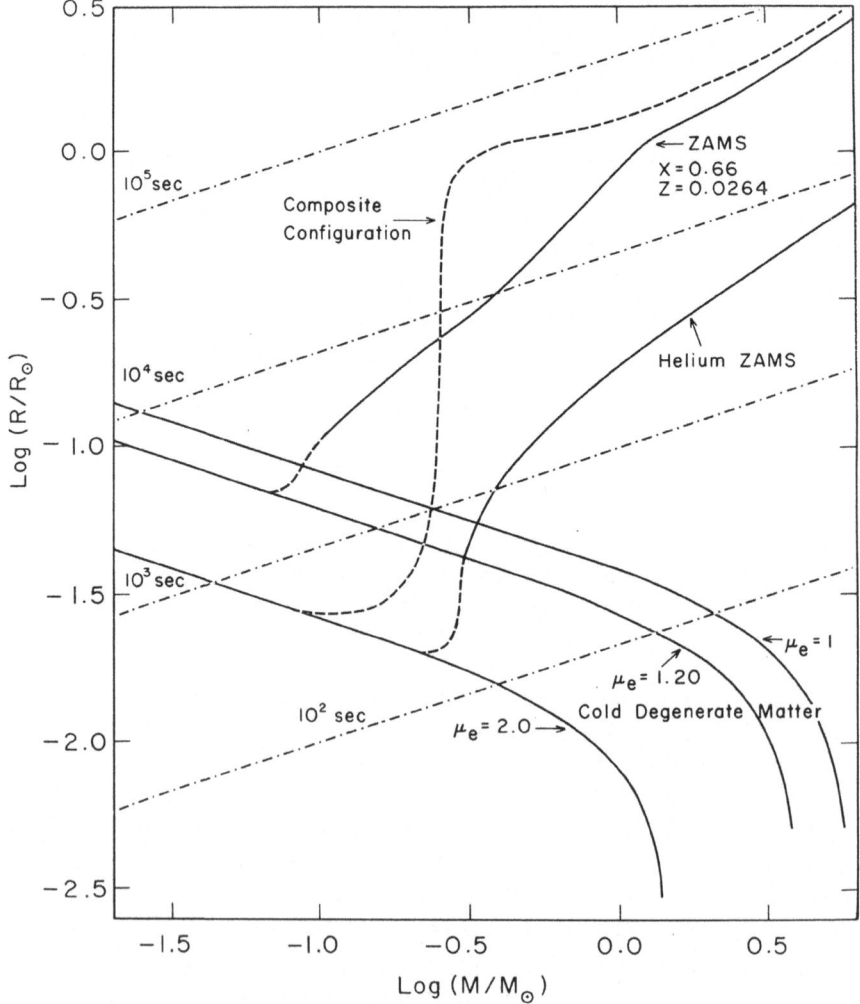

Fig. 2. Lines of given orbital period are shown for lobe-filling secondaries, together with typical hydrogen- or helium-burning zero-age main sequences, and a schematic representation of the possible relationship for a composite configuration. Because of mass loss, evolution proceeds to the left along any given sequence, until a transition ultimately occurs to the appropriate degenerate sequence.

et al., 1972). The small separation of these two degenerate stars ($\lesssim 0.25\ R_\odot$) demands an explanation which must, presumably, arise as a logically possible outcome of nova or nova-like evolution.

3. Evidence for the White Dwarf

Following Kraft, white dwarfs are assumed to be present in all these systems, although until recently WZ Sge was the only certain case because a typical DB spectrum was observed. It was also supposed that the 71s pulsations of DQ Her (Walker, 1956) were associated with a white dwarf. Were these radial pulsations, they would imply a very low mass. However, Warner *et al.* (1972) have shown that a phase shift of $+360°$ which occurs during eclipse may be interpreted as evidence for non-radial modes of pulsation in a more massive white dwarf.

Within the past year, a most dramatic order-of-magnitude increase has occurred in the number of suspected non-radially pulsating white dwarfs. These discoveries have been made mainly, but not exclusively, among the dwarf novae. Iben has reported earlier that Warner discussed many additional examples at the I.A.U. Symposium No. 59. These presumably included Z Cam, CN Ori, UX UMa, AH Her and AM CVn (i.e. HZ 29) with periods near 17, 24, 29, 31 and 115 s respectively (Warner and Robinson, 1972a), SY Cnc and KT Per (25 and 27s; Robinson, 1973c), CD $-42°$ 14462 (29s; Warner, 1973b), Z Cha (28s; Warner, 1974) and VW Hyi (~ 28–34s; Warner and Harwood, 1973; Warner and Brickhill, 1974). With one or two exceptions, the amplitudes are very small (in the range of $\sim 0^{m}_{\cdot}01$ to $0^{m}_{\cdot}001$ or less) and the periods are usually found by combinations of power spectrum and periodogram analysis. In some cases, during lengthy observing runs (duration of hours or days) systematic and possibly quantized monotonic changes of period have been seen, the changes being positive for one object and negative for others. In CN Ori, a 'phase transition' was actually observed. In AH Her (Robinson, 1973b) the pulsations have been shown to be absent at several points on the rising branch of an outburst and during minimum light, in contrast with the situation at and just after maximum light in AH Her, Z Cam and CN Ori (Warner and Robinson, 1972a). The inference has been drawn that one is seeing non-radial pulsations in the white dwarf as a consequence of an outburst in the latter, the amplitudes being small because the outburst has caused the inner parts of the accreting disk to brighten up considerably and dominate the optical output. The lack of pulsations leading up to the outburst is seen (Starrfield *et al.*, 1974) as evidence for the picture in which a major role is assigned to the release of energy in the disk by β^{+} – unstable nuclei following non-equilibrium nuclear burning among possibly enhanced abundances of carbon and oxygen. Starrfield *et al.* claim that the build up of pulsations would be seen in the models of Rose and Smith.

Against the euphoria engendered by these brilliant observational triumphs, one should perhaps set the growing suspicions that one may not actually be seeing white dwarf pulsations at all! Thus Bath *et al.* (1974) have suggested that hot spots from accretion funneling onto a rapidly rotating magnetic white dwarf might be responsible for the DQ Her observations. The sign of their phase shift is however almost certainly negative. While this agrees with a $-360°$ phase shift recently observed in UX UMa pulsations during eclipse (Nather and Robinson, 1974), it appears that one should be looking for an explanation capable of producing $\pm 360°$ phase shifts with equal facility. The positive sign would demand white dwarf rotation in the opposite sense to the orbit, which seems unlikely. In another approach, Bath (1973) proposes that transient hot spots in the orbiting disk close to the white dwarf are responsible for the pulsations seen. This would seem once again to produce only negative phase-shifts. This is perhaps the most serious objection to Bath's proposal, although in all other respects it satisfies four summary criteria for any explanation of the behaviour of UX UMa, as compiled by Nather and Robinson. The latter authors inciden-tally remark that the $l = 2$, $m = 0$ pulsation mode of Warner et al. (1972) fails two criteria, the hot spot funneling of Bath et al. three. Nather and Robinson suggest in-stead $l = 2$, $m = \pm 2$ pulsations (only the m value is really important), but then cast doubt on the whole idea of seeing white dwarf pulsations directly by noting serious discrepancies between several estimates of the blue star radius in UX UMa. Recent pulsation models (Osaki and Hansen, 1973) imply a radius certainly less than $\sim 0.05\ R_\odot$; on the other hand, both the eclipse solutions for UX UMa and the dura-tion of the phase shift seen suggest a radius probably in the range of $\sim 0.3\ R_\odot$ to $\sim 1.0\ R_\odot$. Their tentative way out of this dilemma is that the hot surrounding disk may somehow act as an extension of the white dwarf envelope, creating for some purposes a large apparent photosphere. It becomes imperative to test whether the broad H and He I lines seen in UX UMa are due to rotational Doppler broadening in the disk or to Stark broadening on the white dwarf surface.

The suspicion that the glowing disk may play a far more important and obscuring role than hitherto assigned also occurred apparently independently and simultaneously to Warner (1974) and Starrfield et al. (1974). One may summarise by saying that all the models proposed to explain the pulsations contain at heart a stellar mass at least as compact as a white dwarf, and that on the issue of the proved existence of white dwarfs in these systems, there is room for cautious, if confused, optimism.

4. Proposed Outburst Mechanisms and Evolutionary Scenarios

Variants of at least four basic mechanisms have been proposed for nova outbursts. Briefly, they may be characterized as follows:

(i) Nuclear evolution of the red component causes it to swell, overflowing the sur-rounding Roche lobe (Crawford and Kraft, 1956). That portion of the hydrogen-rich material released which subsequently reaches the white dwarf via the accretion disk ultimately undergoes violent, unstable nuclear burning.

(ii) A thermal instability or relaxation oscillation occurs in the convective envelope of the red component. As a result the thermal energy of the envelope is released, man-ifesting itself as an outburst (Paczyński, 1965; Bath, 1969, 1972). Osaki (1970) pro-duced a variant in which shear turbulence in the surface layers strongly modifies ener-gy transport and induces mild outbursts.

(iii) Gravitational radiation inexorably removes energy and angular momentum from the system so that the Roche lobe continuously encroaches upon the surface of the red component. Mass must necessarily be transferred (Kraft, 1966) and, taking account of stellar structure requirements, the self-consistent rate of transfer can be calculated (Paczyński, 1967; Faulkner, 1971). The outburst follows as in (i) above.

(iv) The X-mechanism (Ostriker, 1973). By analogy with a model produced for self-excited binary X-ray sources (Davidson and Ostriker, 1973), intense radiation from the hot spot and accretion disk induces continued mass loss from the cool red component. The seat of the ultimate outburst would again seem to be the white dwarf. The suggestion is however largely unexplored.

Before discussing how these mechanisms fare against recent observations, it is convenient to draw up a table summarising the energy requirements of nova models (Table I). The table makes no attempt to be complete (indeed, our lack of knowledge precludes it), but it illustrates properties generally attributed to some of the more 'typical' examples of each class.

TABLE 1

Energy requirements of nova models

	Classical novae	Recurrent novae	Dwarf novae
Outburst range (magnitude)	~ 10–12	~ 6–8	~ 2–5
Outburst energy, E(ergs)	$\sim 10^{45}$ or more	$\sim 10^{43}$–10^{45}	$\sim 10^{38}$–10^{39}
Time interval, T(years)	(300–1000??) (great uncertainty)	25–50	~ 0.05–1
Mass ejected, M_{ej} (gms)	$\sim 10^{28}$–10^{29}	$\sim 10^{28}$	?
M_{ej}/T ($M_\odot$ yr^{-1})	??	$\sim 10^{-7}$	$\sim 10^{-9}$?? (one example)
$E/(0.007c^2\,T)$ ($M_\odot$ yr^{-1})	??	$\sim 10^{-10}$	$\sim 10^{-12}$–10^{-11}
*M_{burnt}/M_{ej} (nuclear explanation)	??	$\sim 10^{-3}$	?

*$M_{burnt} = E/(0.007c^2)$, the amount of hydrogen burnt on the nuclear explanation to produce the outburst energy E.

The major question which one might hope to resolve, thereby distinguishing between the above possibilities, is the rate of mass transfer and/or ejection. There are those who favour a canonical figure of $\sim 10^{-7}\,M_\odot$ yr^{-1} on the grounds that (a) such an ejection rate has been observed, for example in the recurrent nova RS Oph (Pottasch, 1967); (b) it might also hold for classical novae which eject ~ 10–30 times as much mass but at intervals suspected to be ~ 10–30 times less frequent (Pottasch, 1959; Boyarchuk* 1970); (c) orbital period changes of order 1 part in 10^7 per annum are taking place for example in the classical nova DQ Her (Nather and Warner, 1969). However, as Smak (1972) has pointed out, the orbital periods are known to both in-

* Note that the shell mass for RS Oph 1958 (Folkart et al., 1964) as quoted in Boyarchuk's Table I is incorrect. It should read $2 \times 10^{-6}\,M_\odot$ (not 2×10^{-7}). This value was however revised upwards by a factor of 3 (Pottasch, 1967).

crease and decrease, sometimes in a cyclical manner. Recent examples include the
~ 29 yr variation of UX UMa (Krzeminski and Walker, 1963; Mandel, 1965, Nather
and Robinson, 1974), the ~ 2 yr variation of HZ 29 (Krzeminski, 1972), and similar
effects in RW Tri (Mandel, 1965) and U Gem. Explanations in terms of third bodies
for all these systems seem a little unlikely, and the effect may occur because disks can
act as temporary reservoirs of mass and angular momentum. In any event, we feel
with Smak, that one should be cautious in interpreting observed rates of orbital period
change as evidence for long term mass transfer at comparable relative rates.

For at least one dwarf nova it appears possible that the mass transfer or ejection
rate may be significantly less than the 'canonical' figure. Robinson has derived values
of $\sim 3 \times 10^{-9} M_\odot$ yr^{-1} (Robinson, 1973a) and $\sim 2 \times 10^{-9} M_\odot$ yr^{-1} (Robinson,
1973d) for Z Cam. While these are admittedly lower limits, it hardly seems likely that
they are underestimates by factors of 30 or 100.

What rate of mass transfer might one expect according to competing theories? The
answers are surprisingly incomplete, or they depend upon a better knowledge of the
circumstances than we possess. If a system had become of the envisaged contact type
relatively recently and if the red component were more massive than the white dwarf
at this stage, as seems likely to be the case, then a period of thermal readjustment involving timescales of $\sim 10^6$ or 10^7 years would indeed be appropriate.

Following this stage, there are two ways in which nuclear evolution timescales
might determine the subsequent events. The white dwarf mass may have been built
up through sufficient retention of added material that its companion, now of comparable mass, is able to evolve in a reasonable length of time. This could well be the case
for Z Cam where the masses are: white dwarf ~ 1.0–$1.3 M_\odot$, red component ~ 0.8–
$1.0 M_\odot$ (Faulkner, 1971; Warner, 1973a). Incidentally, it is unfortunate (to say the
least!) that the author chose to illustrate mechanism (iii) by applying it to Z Cam. The
temptation to apply the theory to a system about which much was known was irresist-
ible. However, as we shall discuss later, it was a borderline case for application, and
it appears from Robinson's observations that Z Cam's red component is indeed currently a frustrated main-sequence leavetaker. One other way in which nuclear timescales could still be important would arise if the red component's core was already
sufficiently far evolved when the system arrived at the standard configuration. In this
case, the previous stripping of the red component's envelope during the thermal
readjustment era might leave it, even though of very low mass, poised for its fruitless
attempt to depart the main sequence. That this, or something like it, must be a possibility is hinted at by the model for HZ 29 (Faulkner *et al.*, 1972), although the critical
point where nuclear evolution ceases to be relevant is not yet known.

Finally we come to a stage with a timescale which is unavoidable if Einstein's
theory of gravity, or anything like it, is correct, i.e. mechanism (iii) above. Where this
takes over from nuclear evolution as the dominant mechanism will depend upon the previous evolutionary history of the red component. If the latter is truly like a pristine
main sequence star, evolution via gravitational radiation could be more important
than nuclear evolution for stars as massive as $\sim 1 M_\odot$ in systems of total mass $\sim 2 M_\odot$

(Faulkner, 1971). The more evolved the interior of the red component, the lower the mass at which gravitational evolution will dominate – but it certainly seems likely that systems with binary periods shorter than ~ 5 hours containing white dwarfs more massive than the red components (themselves $\leqslant 0.7\ M_\odot$) will be so dominated. According to the theory (with its imperfections, based as it is on zero-age main sequence approximations), the rate of mass transfer is very well determined and lies in the range ~ 0.5–$2.0 \times 10^{-10}\ M_\odot\ \mathrm{yr}^{-1}$. These transfer rates lead, in the absence of other mechanisms, to timescales of the order of a few to ten billion years. They have a comfortable margin in hand to satisfy the energy requirements of Table I.

The scenarios presented above have suggested ways in which high mass transfer rates and mechanisms (i) and (iii) might be relevant at different times, even in the same system. We have not discussed mechanism (ii) because we believe it is almost certainly ruled out by recent observational and theoretical developments. It is to these developments that we now turn.

5. The Seat of the Outburst; Observations and Theory

The precise seat of the outburst has been a matter of some controversy during the past decade. Krzeminski (1965) showed that in U Gem, the primary eclipse present at minimum light remained approximately constant in intensity units during outburst. As a result, it was essentially absent (in magnitude) at maximum light. Assuming that the white dwarf was obscured during primary eclipse, Krzeminski concluded that it had nothing to do with the outburst itself. The outburst was therefore associated with the red, or secondary component. It was in response to these observations that Paczyński (1965) developed the theoretical 'secondary hypothesis'.

The secondary hypothesis was challenged by Walker and Chincarini (1968) who, observing SS Cyg during a rise to maximum, found that the outburst was associated with the blue component. However, according to Smak (1969), uncertainties in the elements of the SS Cyg velocity variations permitted a solution in which the contrary conclusion would be valid. Subsequently, Walker and Reagan (1971) checked the orbital period of SS Cyg once again, essentially confirming the value used by Walker and Chincarini. Although some disquieting phase shift seems to have intervened also, the results of Walker and Reagan nevertheless support the contention that the outburst originates in the blue, hot component.

A way of avoiding Krzeminski's conclusion was independently proposed by Warner and Nather (1971) and Smak (1971), namely the 'hot spot' model we have discussed throughout. It is the hot spot which is eclipsed, and it remains of essentially constant brightness during outburst. Smak in particular shows how Krzeminski's observations may be interpreted as evidence for the outburst occurring somewhere near the central parts of the accreting disk. However, this explanation requires us to be in a particularly favourable direction with respect to the orbital plane of U Gem: able to see the white dwarf and central disk regions satisfactorily, but having the hot spot periodically obscured by the secondary. It thus became imperative to search for further evidence to test the model. Robinson (1973a) made observations of Z Cam which indicated that the central disk regions brightened during outburst. However, the matter appears to have been clinched by Warner's observations (Warner, 1974) of the southern eclipsing dwarf nova Z Cha. The eclipse, which is total, shows a disappearance both of the white dwarf primary and the hot spot. During outburst the eclipses become (a) wider and (b) partial, apparently establishing with certainty that the outburst is centred on the primary. At maximum light, the eclipses show that the whole disk around the primary has brightened up, whereas the hot spot remains of constant intensity.

After many years, this observationally disposes of the red star as the seat of the outburst. But Bath (1973), resourceful as ever, maintains that it may still be the red star, undergoing his envelope oscillations, which sends matter towards the white dwarf, the outbursts occurring in direct response to each burst of mass transfer. While we think the behaviour of the bright spot (Warner, 1974) makes this unlikely, other stellar structure considerations cast more doubt on the whole envelope mechanism. This brings us to our final topic.

6. The Response of Stellar Radii to Mass Loss

According to Bath (1972), any stellar model corresponding essentially to stars in or to the red side of the instability strip and its extensions in the HR diagram will be unstable to mass loss if confined by a Roche lobe. This conclusion is, we feel, suspect on at least two grounds: (a) the changing size of the Roche lobe is ignored, and even more importantly, (b) the long term consequences of mass loss for the underlying star are ignored. For, whatever the behaviour of the models on the short timescales studied by Bath, surely the important question if one wishes to understand say dwarf novae, is what happens to the underlying star given mass loss on timescales of 10^7, 10^8, 10^9... years? Should conditions in the underlying star differ significantly from those initially assumed, the calculations will not be relevant to the long term situation.

Accordingly, we and colleagues have investigated the response of stellar radii to mass loss in the surprisingly neglected region of the lower main sequence (Eggleton *et al.*, 1974). Such simple models have been studied because (a) they are thought to be relevant to the short binary period novae, (b) they provide a *reductio ad absurdum* disproof of the contention that the envelope instability hypothesis can work independently of the state of evolution of the underlying star and (c) no one seems to have done it previously!

The models exhibit both transient and long-term effects because of the thermal imbalance which necessarily accompanies long-term mass loss. Models above and below a critical mass ($\sim 0.75\ M_\odot$) initially shrink more or less rapidly than the main-sequence relationship would suggest. At sufficiently high rates of mass loss, the models of lowest mass investigated (0.5 and 0.25 $M_\odot$) do experience transient expansion. However, it is clear that for models appropriate to many of the observed dwarf and other novae, the greater the rate of mass loss, the faster the star shrinks. For the shortest period, lowest secondary mass systems (which must surely be transferring to more massive white dwarfs), the transfer rates would have to be so great to induce expansion which could overcome the increasing size of the swelling Roche lobe, that there would be no chance of observing them. In short, the tail cannot wag the dog following amputation.

We conservatively conclude that the secondary mechanism cannot be a prime theoretical cause, although it may in some fashion modulate the rates implied by some other mode of mass transfer.

7. Summary and Wild Speculations

We have suggested in Section 4 a scenario in which mechanisms (i) and (iii) might play roles of varying importance. Although it is certainly not clear that the various types of novae need be connected in any evolutionary sense, nevertheless we have concluded that if they are, a good, if unconventional case may be made for the dwarf novae to be later and associated with the most massive white dwarfs. This contention indeed receives strong support from Warner's study (Warner, 1973a) of the masses of the components of ten short period cataclysmic variables, predominantly dwarf novae. The white dwarf primaries, with one exception, lie in the narrow range of $\sim 1.2 \pm$ $\pm 0.2\ M_\odot$. The exception, interestingly enough, is U Gem, with a primary of only $\sim 0.65\ M_\odot$ and a more massive secondary, which may explain the relatively short term thermal timescales associated with its period variations mentioned above. Is it possible that dwarf novae explanations are bimodal? Whether this is the case or not, the suggestive proximity of Warner's primary masses to the Chandrasekhar limit may help explain the mild outbursts and short repetition times associated with the dwarf novae. The surface gravitational fields may be so large, and the non-degenerate layers so thin, that an extremely small addition of fresh nuclear fuel can trigger outbursts at very frequent intervals. The thermal history of the white dwarf, completely ignored above, may also be an important factor.

We have throughout avoided the problem of the progenitors. We feel, with Kraft (1967), that W UMa systems are the most likely although others (e.g. Giannone, 1973) are of the opinion that close but not contact systems appropriate to 'case A' of mass exchange may be relevant. Vilhu will speak on some of these points later (Vilhu, 1974). Whichever close systems one favours for the progenitors, one thing seems perfectly clear: the white dwarf masses obtained when a semi-detached system is first formed rarely exceed $\sim 0.3\ M_\odot$, a far cry from Warner's observations, and demanding subsequent build up.

Looking back over the past decade, we have, it seems, come a long way, if not by a direct route. Much work remains to be done, however, before we can confirm Kraft's provocative suggestion that gravitational radiation might ultimately be responsible for the behaviour of some dwarf novae; a suggestion made on a similar occasion (Hamburg, 1964) and published (Kraft, 1966) in proceedings as widely ignored by relativists as these may be.

JOHN FAULKNER

Acknowledgements

I am, as ever, deeply indebted to a large number of astronomers who placed their great knowledge at my disposal, in particular Robert P. Kraft, Edward L. Robinson and Brian Warner. The opprobrium for the use made of that knowledge rests entirely on my shoulders. I am most grateful also to Professor Donald Lynden-Bell for his hospitality at the Institute of Astronomy, Cambridge during the period when the computations described in Section 6 were performed. The author's research in this area is supported by a grant from the National Science Foundation.

References

Bath, G. T.: 1969, *Astrophys. J.* **158**, 571.
Bath, G. T.: 1972, *Astrophys. J.* **173**, 121.
Bath, G. T.: 1973, *Nature Phys. Sci.* **246**, 84.
Bath, G. T., Evans, W. D., and Pringle, J. E.: 1974, *Monthly Notices Roy. Astron. Soc.* **166**, 113.
Boyarchuk, A. A.: 1970, in H. J. Habing (ed.), 'Interstellar Gas Dynamics', *IAU Symp.* **39**, 281–290.
Crawford, J. A. and Kraft, R. P.: 1956, *Astrophys. J.* **123**, 44.
Davidson, K. and Ostriker, J. P.: 1973, *Astrophys. J.* **179**, 585.
Eggleton, P. P., Faulkner, J., and Webbink, R. F.: 1974, in preparation.
Faulkner, J.: 1971, *Astrophys. J. Letters* **170**, L99.
Faulkner, J., Flannery, B P., and Warner, B.: 1972, *Astrophys. J. Letters* **175**, L79.
Folkart, W., Pecker, J.-C., and Pottasch, S. R.: 1964, *Ann. Astrophys.* **27**, 252.
Giannone, P.: 1973, private communication.
Kraft, R. P.: 1963, *Adv. Astron. Astrophys.* **2**, 43.
Kraft, R. P.: 1966, *Trans. IAU* **12B**, 519.
Kraft, R. P.: 1967, *Publ. Astron. Soc. Pacific* **79**, 395.
Krzeminski, W.: 1965, *Astrophys. J.* **142**, 1051.
Krzeminski, W.: 1972, *Acta Astron.* **22**, 387.
Krzeminski, W. and Walker, M. F.: 1963, *Astrophys. J.* **138**, 146.
Mandel, O. E.: 1965, *Peremennye Zvezdy* **15**, 474.
Mumford, G. S.: 1967, *Publ. Astron. Soc. Pacific* **79**, 283.
Nather, R. E. and Warner, B.: 1969, *Monthly Notices Roy. Astron. Soc.* **143**, 145.
Nather, R. E. and Robison, E. L.: 1974, *Astrophys. J.*, in press.
Osaki, Y.: 1970, *Astrophys. J.* **162**, 621.
Osaki, Y. and Hansen, C. J.: 1973, *Astrophys. J.* **185**, 277.
Ostriker, J. P.: 1973, private communication.
Paczyński, B.: 1965, *Acta Astron.* **15**, 89.
Paczyński, B.: 1967, *Acta Astron.* **17**, 287.
Pottasch, S. R.: 1959, *Ann. Astrophys.* **22**, 394.
Pottasch, S. R.: 1967, *Bull. Astron. Inst. Neth.* **19**, 227.
Robinson, E. L.: 1973a, *Astrophys. J.* **180**, 121.
Robinson, E. L.: 1973b, *Astrophys. J.* **181**, 531.
Robinson, E. L.: 1973c, *Astrophys. J.* **183**, 193.
Robinson, E. L.: 1973d, *Astrophys. J.* **186**, 347.
Rose, W. K.: 1968, *Astrophys. J.* **152**, 245.
Rose, W. K. and Smith, R. L.: 1972, *Astrophys. J.* **172**, 699.
Smak, J.: 1969, *Acta Astron.* **19**, 287.
Smak, J.: 1971, *Acta Astron.* **21**, 15.
Smak, J.: 1972, *Acta Astron.* **22**, 1.
Starrfield, S.: 1971a, *Monthly Notices Roy. Astron. Soc.* **152**, 307.
Starrfield, S.: 1971b, *Monthly Notices Roy. Astron. Soc.* **155**, 129.

Starrfield, S., Truran, J. W., Sparks, W. M., and Kutter, G. S.: 1972, *Astrophys. J.* **176**, 169.
Starrfield, S., Sparks, M., and Truran, J. W.: 1974, in press.
Vilhu, O.: 1974, this volume, p. 168.
Walker, M. F.: 1956, *Astrophys. J.* **123**, 68.
Walker, M. F. and Herbig, G. H.: 1954, *Astrophys. J.* **120**, 278.
Walker, M. F. and Chincarini, G.: 1968, *Astrophys. J.* **154**, 157.
Walker, M. F. and Reagan, G. H.: 1971, *Inf. Bull. Var. Stars*, I.A.U. Comm. No. 27, No. 544.
Warner, B.: 1973a, *Monthly Notices. Roy. Astron. Soc.* **162**, 189.
Warner, B.: 1973b, *Monthly Notices Roy. Astron. Soc.* **163**, 25P.
Warner, B.: 1974, *Monthly Notices Roy. Astron. Soc.*, in press.
Warner, B. and Nather, R. E.: 1971, *Monthly Notices. Roy. Astron. Soc.* **152**, 219.
Warner, B. and Robinson, E. L.: 1972a, *Nature Phys. Sci.* **239**, 2.
Warner, B. and Robinson, E. L.: 1972b, *Monthly Notices Roy. Astron. Soc.* **159**, 101.
Warner, B., Peters, W. L., Hubbard, W. B., and Nather, R. E.: 1972, *Monthly Notices Roy. Astron. Soc.* **159**, 321.
Warner, B. and Harwood, J. M.: 1973, *Inf. Bull. Var. Stars.*, I.A.U. Comm. No. 27, No. 756.
Warner, B. and Brickhill, A. J.: 1974, *Monthly Notices Roy. Astron. Soc.* **166**, 673.

POSSIBLE PROGENITORS OF SOME NOVA-LIKE VARIABLES

(Abstract)

OSMI VILHU

Observatory and Astrophysics Laboratory, University of Helsinki, Finland

Evolution of contact binaries was followed with the help of the modified evolutionary codes by Paczyński (Vilhu, 1973), and based on the concepts of a common convective envelope (Lucy, 1968) and energy exchange between the adiabatic parts (Moss and Whelan, 1970). Similarly as was done by Biermann and Thomas (1973), we do not demand equal specific entropies (equal K-values) for the components. The energy losing component should have a smaller K (higher effective temperature). Many observed W UMa-systems can be explained by models with unequal K-values and small degree of contact (superadiabatic energy exchange) (Whelan *et al.*, 1973). For stability reasons, allowing no entropy differences between the components, Hazlehurst and Meyer-Hofmeister (1973) found a qualitatively different evolution for contact systems. To avoid instabilities for models with unequal K-values, some physical mechanism keeping mass and energy exchanges in equilibrium with the 'degree of contact' must exist. Only a qualitative picture of the possible process can be given (see description by Biermann and Thomas, 1971, and also a discussion after the paper by Hazlehurst and Meyer-Hofmeister, 1971).

When studying the hypothesis that U Geminorum stars are descendants of W UMa-systems (Kraft, 1962), we are faced with the problem of how a white dwarf can be formed in a very close system of small mass (small angular momentum). Only case B evolution, which implies large initial separation (large angular momentum) or very large initial mass ratio, can finally produce white dwarfs of considerable mass ($\gtrsim 0.2 \ M_\odot$) (see e.g. Refsdal and Weigert, 1971).

As a rule, case A evolution sooner or later leads to a contact system. Figure 1 illustrates evolution of one possible such a system, starting from two homogeneous main sequence stars ($0.45 \ M_\odot + 1.35 \ M_\odot$, phase A). At phase B the heavier star fills its Roche lobe when hydrogen is nearly exhausted in its core (X nearly zero up to $M_r = 0.02 \ M_\odot$), and a contact system is formed during the rapid phase (phase C). From this point contact evolution is followed, with transfer of mass from the secondary to the primary, and of energy in the opposite direction. Gravitational energy terms were included also in the secondary, showing no essential difference when the secondaries were approximated by thermal equilibrium models. The evolutionary track of the secondary, when energy is pushed into its envelope, turned out to be very sensitive to its hydrogen profile. For the initial parameters used, it moves nearly parallel to its Hayashi-line but, due to its helium-rich envelope, at phase D ($1.635 \ M_\odot + 0.165 \ M_\odot$) it becomes hotter than the primary, and the energy exchange must stop. Numerically the secondary can be made cooler, and the energy transfer continued, if part of the

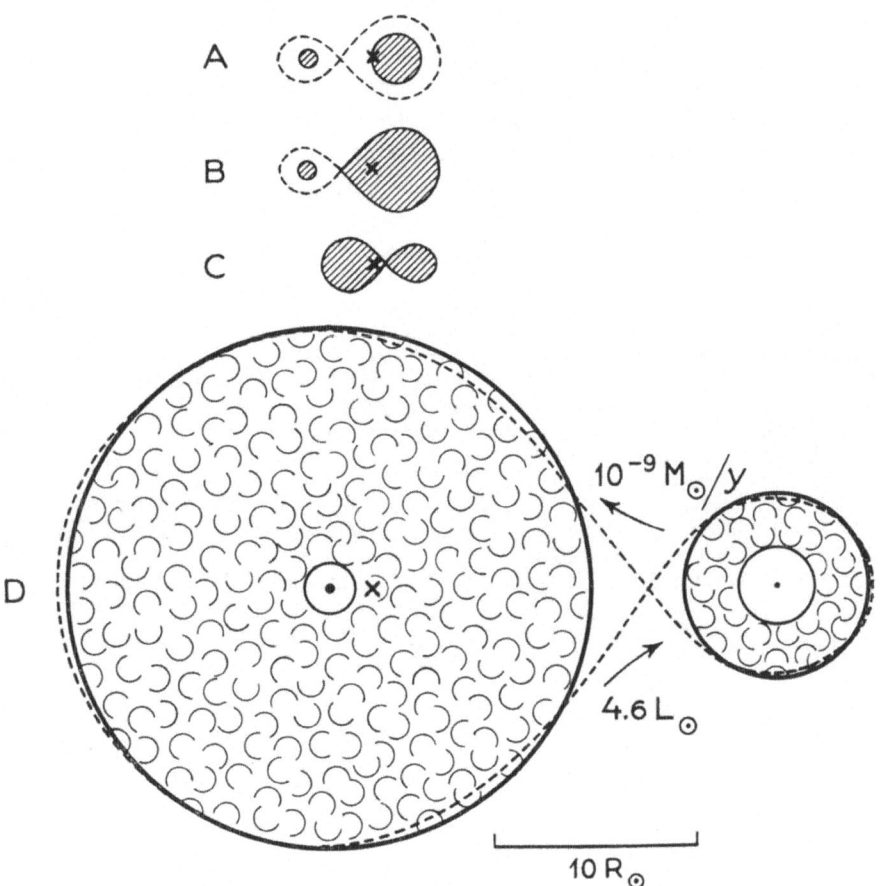

Fig. 1.

energy exchange is assumed to take place at the border of the radiative core and convective envelope, which assumption seems to be quite artificial. At phase D, however, the primary already has a considerably large helium-core of 0.24 $M_\odot$.

Starting from two homogeneous stars in contact (the models by Biermann and Thomas, 1973; the systems 1 and 2 by Vilhu, 1973), a final set-up illustrated as phase C may also be reached, although the time-scale considerations by Biermann and Thomas (1973) do not seem to favour this. This question is important in deciding whether also models for W UMa-stars (nearly homogeneous and contact systems) evolve in the manner illustrated.

After phase D only a qualitative picture of the further evolution can be given. Mass and angular momentum loss seem to be necessary, and finally the system may resemble an U Geminorum star consisting of a white dwarf (0.24 $M_\odot$), and a post main sequence star of nearly the same mass filling up its Roche lobe. Surprisingly, both

W UMa- and U Gem-systems seem to obey the same relation between angular momentum and total mass ($J \propto M_{tot}^{5/3}$), which may be partly due to the definition and partly to the fact that a nearly main sequence star fills its Roche lobe.

The validity of the above evolutionary picture should be critically studied. It remains also to be solved how a still bigger ($\approx 0.5\ M_\odot$) helium-core can be formed, and what initial parameters lead to the evolutionary scheme described.

References

Biermann, P. and Thomas, H.-C.: 1971, '5th Colloquium on Variable Stars', *IAU Colloq.*, No. 15, 285; *Veröffentl. Remeis-Sternwarte Bamberg*, Bd. IX, Nr. 100.
Biermann, P. and Thomas, H.-C.: 1973, *Astron. Astrophys.* **23**, 55.
Hazlehurst, J. and Meyer-Hofmeister, E.: 1971, '5th Colloquium on Variable Stars', *IAU Colloq.*, No. 15, 289; *Veröffentl. Remeis-Sternwarte Bamberg*, Bd. IX, Nr. 100.
Hazlehurst, J. and Meyer-Hofmeister, E.: 1973, *Astron. Astrophys.* **24**, 379.
Kraft, R. P.: 1962, *Astrophys. J.* **135**, 408.
Lucy, L. B.: 1968, *Astrophys. J.* **151**, 1123.
Moss, D. L. and Whelan, J. A. J.: 1970, *Monthly Notices Roy. Astron. Soc.* **149**, 147.
Refsdal, S. and Weigert, A.: 1971, *Astron. Astrophys.* **13**, 367.
Vilhu, O.: 1973, *Astron. Astrophys.* **26**, 267.
Whelan, J. A. J., Worden, S. P., and Mochnacki, S. W.: 1973, *Astrophys. J.* **183**, 133.

DISCUSSION AFTER PAPER BY FAULKNER

Hazlehurst to Vilhu: Have you tested your models for secular stability? Some rough calculations which I have made indicate that they are likely to be secularly unstable.

Vilhu: No, and at the moment I cannot say whether you are right. For models with different K-values for the components there is an unsolved problem of the equilibrium between the energy exchange and the degree of contact. My purpose was just to investigate how a considerably large helium core can be formed starting from a very close system and under the assumptions made about the mass and energy exchange. If it turns out that the stability conditions will put severe restrictions, some other ideas should be investigated.

OBSERVATIONAL DATA ON NOVAE AND SUPERNOVAE

E. R. MUSTEL

Astronomical Council of the Academy of Sciences of the USSR, Moscow, U.S.S.R.

Abstract. The first half of the paper contains a discussion of the chemical composition of the envelopes of supernovae for the period close to light maximum. The principal conclusions are: The abundance of hydrogen in the envelopes of type I supernovae is low, much lower than that of nitrogen; the abundance of oxygen and carbon is also noticeably lower than that of nitrogen; it seems that there is plenty of helium and metals in these envelopes. The information for type II supernovae is more limited. But it is quite certain that the abundance of hydrogen in the envelopes of these stars is much higher than in the envelopes of type I supernovae.

In the second half of the paper the problem of supernova remnants is discussed, the circumstellar shells around supernovae (which according to S. van den Bergh and M. Peimbert are ejected from the star before its explosion) are also included. The discussion of this problem permits to confirm again the idea that there is a very close similarity between supernovae and novae. To be more exact there are reasons to suggest that supernovae as well as novae are double star systems, that they are relatively 'old', that they are quite peculiar objects and that they are not the final stage of evolution of 'normal' stars.

1. Introduction

When describing the data on supernovae and novae we should concentrate our attention on the problems which have direct relation to stellar evolution. These problems are the following:

(a) During the explosion of a supernova we deal with a violent process which is accompanied by the ejection of gases from the relatively *deep* regions of a star. Thus analysing the explosion we have a unique and extremely important opportunity to obtain some information about the chemical composition of these deep regions. Especially interesting are the objects which correspond to the *late* stages of stellar evolution, because in this case we hope to study directly the chemical composition of the internal parts of stars after long series of thermonuclear processes. According to current ideas these objects are type I supernovae, though other types of supernovae are also of great interest.

Here we have two possibilities: (1) The analysis of the chemical composition of a supernova for the moments not very remote from light maximum t_{max}, the moment t_{max} is included. (2) The analysis of the chemical composition of supernova remnants. However there is the following difficulty in this method. We have in mind the fact that these remnants very often sweep up a significant amount of interstellar gases (for example van den Bergh (1971b)) and therefore the observed optical spectra of these remnants are from two non-distinguishable sources. There is also another difficulty, connected with the pre-supernova 'activity', suggested by van den Bergh (1971a), Peimbert and van den Bergh (1971), and van den Bergh (1973); see Section 3 of this paper.

(b) The second problem is that we should try to give an answer to the question,

Tayler (ed.), Late Stages of Stellar Evolution, 172–184. All Rights Reserved.
Copyright © 1974 by the IAU.

whether the supernova phenomenon is inherent to the 'normal' stellar evolution (at a certain stage) or whether supernovae are some peculiar objects in the Universe, similar to novae which constitute quite a distinct group of stars on the HR diagram. Besides we should try to give an answer to the general question why supernovae and novae explode.

2. The Chemical Composition of the Envelopes of Supernovae and Novae Close to Light Maximum

There are different types of supernovae, see Zwicky (1964). But we shall speak mostly about type I and type II supernovae. Our information about types III, IV, V supernovae is very limited.

2.1. TYPE I SUPERNOVAE

A typical light curve of these supernovae is shown in Figure 1, taken from the paper of Zwicky (1964). A peculiar property of this curve is the relatively steep drop in brightness during the first 30–40 days after light maximum. Afterwards the brightness of the supernova decreases much more slowly. It is possible that this drop is due to a rapid decrease of temperature of the supernova after the moment t_{max}. Observations show that this decrease of temperature (manifested for example in the growth of magnitudes $B-V$ and $U-B$) is quite large and takes place approximately during the *same* period of 30–40 days; see papers by Pskowskij (1970), Barbon *et al.* (1973), and Mustel (1974).

As usual the principal source of our information about the chemical composition of stars and their envelopes is the analysis of the spectra of these objects. At present there is a more or less general opinion that the spectra of type I supernovae are mostly *absorption* spectra with very wide absorption lines and with heavy blends of neighbouring absorption lines; see Pskowskij (1968), Mustel (1972), Branch and Patchett (1973), Kirshner *et al.* (1973), Mustel and Chugay (1974), and Mustel (1974). All the absorption lines in the spectrum of a particular supernova are strongly displaced towards shorter wave-lengths and show the *same* Doppler displacement:

$$\varkappa = \frac{\Delta\lambda}{\lambda} = \frac{V}{c} \tag{1}$$

in all the spectral regions.

Velocities V inferred from the analysis of the spectra are different for different type I supernovae and are from 6000 to 15000 km s^{-1}.

It is not excluded that strong absorption lines in the spectra of type I supernovae are accompanied by some emission, the same as in the spectra of P Cygni stars; see Kirshner *et al.* (1973). However all the available data show that these emissions are rather weak; see Mustel and Chugay (1974), Mustel (1974). Thus they can hardly influence significantly the results of the chemical analysis of the envelopes of type I supernovae. Moreover in order to eliminate the influence of emission components

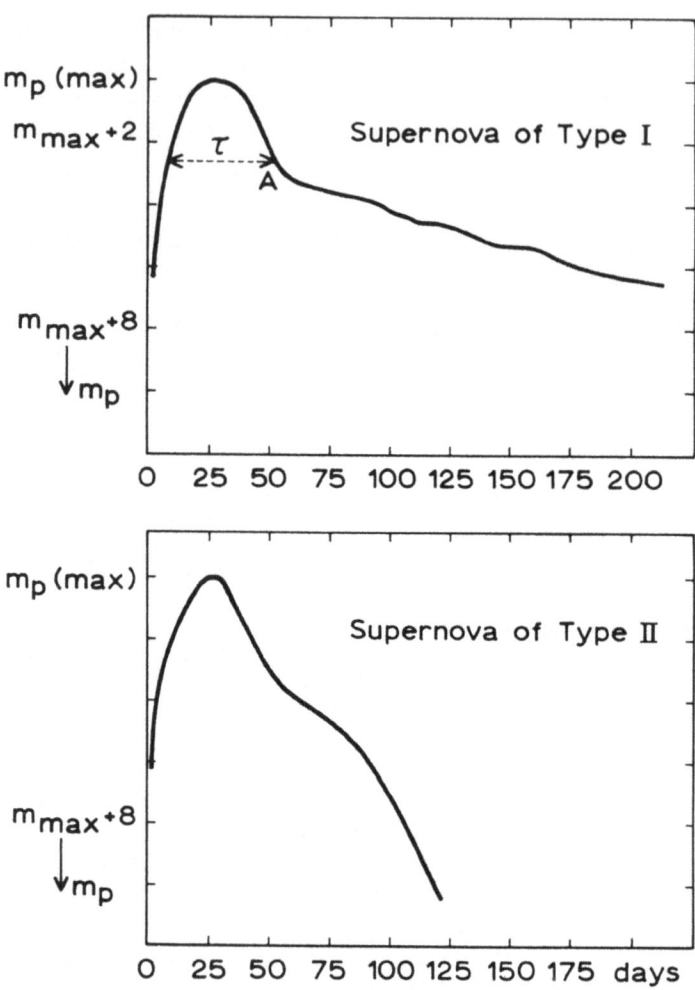

Fig. 1.   Light curves of type I and type II supernovae according to Zwicky (1964).

we may use only the short wave-length halves of the profiles of the absorption lines. And generally speaking it is necessary to point out that at present we cannot require too much from the chemical analysis of the envelopes of type I supernovae. In fact the absorption lines in the spectra of these stars are extremely wide, 30–100 times wider than the same absorption lines in the spectra of common novae. Due to this the effects of blending of the neighbouring spectral lines are extremely strong and as a result of this there are only a few absorption lines in the spectra of typical type I supernovae which are relatively free of the effects of blending. Then, there are all reasons to think that this very large width of the absorption lines is due to the presence of a very large velocity gradient in the envelopes of supernovae. Therefore we must elaborate a special and very complex theory of a model of the envelope with a strong velocity gradient. At last we should mention that the strong blending of absorption lines is accompanied usually by another effect. Namely it is very difficult in this case to draw the line of the continuous spectrum.

Thus at present the chemical analysis of envelopes of type I supernovae may be only a semi-quantitative one. We may identify the spectral lines belonging to different elements, to decide what elements are present or absent in these envelopes and in certain cases to give semi-quantitative estimates for the relative abundances of elements. Correspondingly we shall present the results of such a semi-quantitative analysis for the envelope of the type I supernova 1972e in NGC 5253; see Mustel (1973). This analysis is based on a large number of absolute spectral energy distributions published in the paper by Kirshner *et al.* (1973). These distributions were obtained in the range of the spectrum from 3200 to 11 000 Å. The identification of the absorption lines in the spectral energy distributions is given in Figure 2a and Figure 2b of the present paper. This identification takes into account the principal properties of time-evolution of the spectra of type I supernovae, Mustel (1972). The identifications – all vertical lines in Figures 2a and 2b – are carried out for the Doppler factor $\varkappa = -0.035$, i.e. for $V \simeq 10\,500$ km s^{-1}. The thin vertical lines indicate the calculated positions of individual spectral lines; the dashed vertical lines indicate the outer boundaries of the strongest blends of metals, mostly of Fe II (blends μ, τ, multiplet M73). The dotted vertical lines 'a' indicate the calculated positions of Hα, Hβ, Hγ hydrogen absorption lines; the vertical dotted lines 'e' indicate the normal wave-lengths of Hα, Hβ, Hγ. It is supposed that these normal wave-lengths should correspond to the centres of emissions.

The most important absorptions in Figures 2a and 2b are the following:

2.1.1. *The Absorption Lines of Metals the Lower Atomic Levels of which are ground or metastable*

Generally the effect of metastability of the lower atomic levels plays a very important role in the spectra of type I supernovae. The strongest metallic lines are the lines of Ca II, namely H, K-lines and three lines of the infrared multiplet N2 (their blend) with $\lambda_0 \simeq 8580$ Å. A very important role in these spectra belongs to the absorption lines of Fe II and their blends.

A very prominent absorption in the spectra of type I supernovae at light maximum is due to the blend of two strong lines of Si II, multiplet N2, $\lambda_0 \simeq 6355$ Å. This blend is accompanied by two absorptions of Si II of multiplets NN 3, 4*. After light maximum all these absorption lines of Si II weaken rapidly due to a drop of temperature of the supernova after light maximum and because the lower levels producing these lines are not metastable.

The spectrum of the supernova 1972e contains also absorption lines of Mg II, Sc II, S II. Undoubtedly the spectra of type I supernovae should contain practically all the sufficiently strong absorption lines of metals which are present in the spectra of 'normal' stars, but they are not recognizable due to the strong effects of blending.

---

* The lines of multiplets NN 1, 5 of Si II coincide with the blends of other elements.

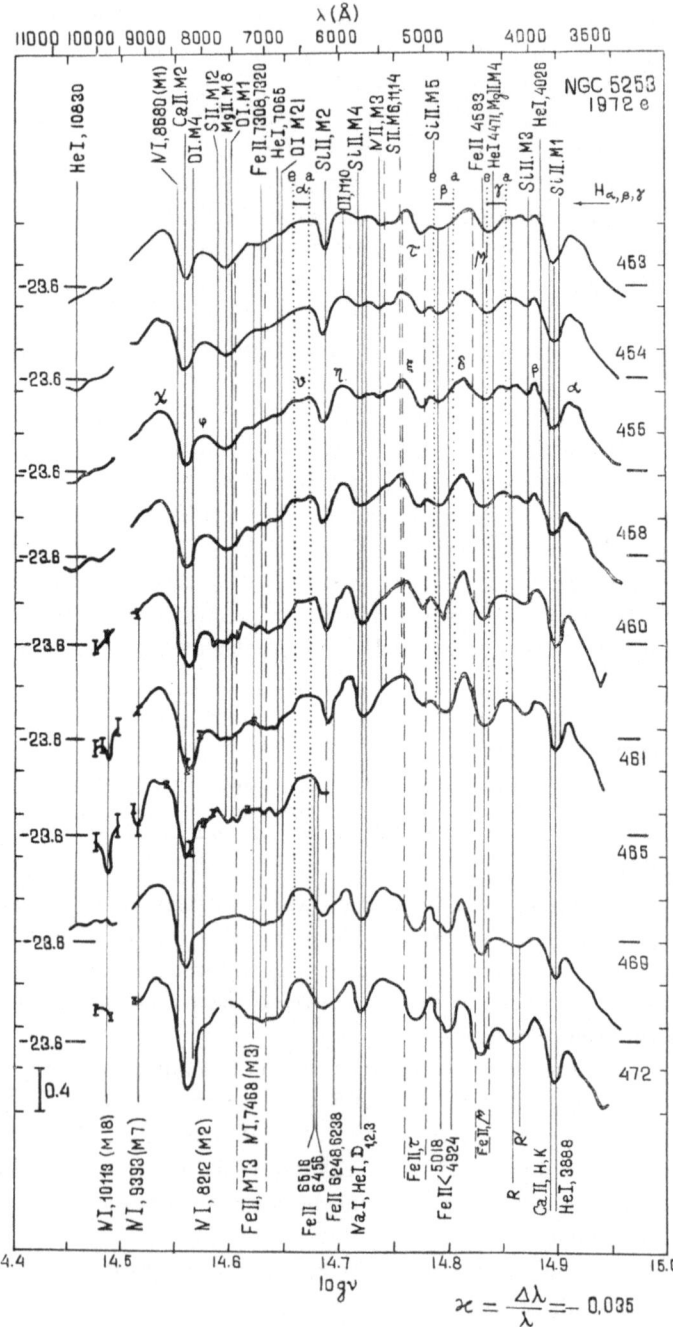

Fig. 2a–b.   Identification of absorptions in the spectra of type I supernova 1972e in NGC 5253 according to Mustel (1973). The absolute spectral energy distributions are taken from the paper of Kirshner *et al.* (1973).

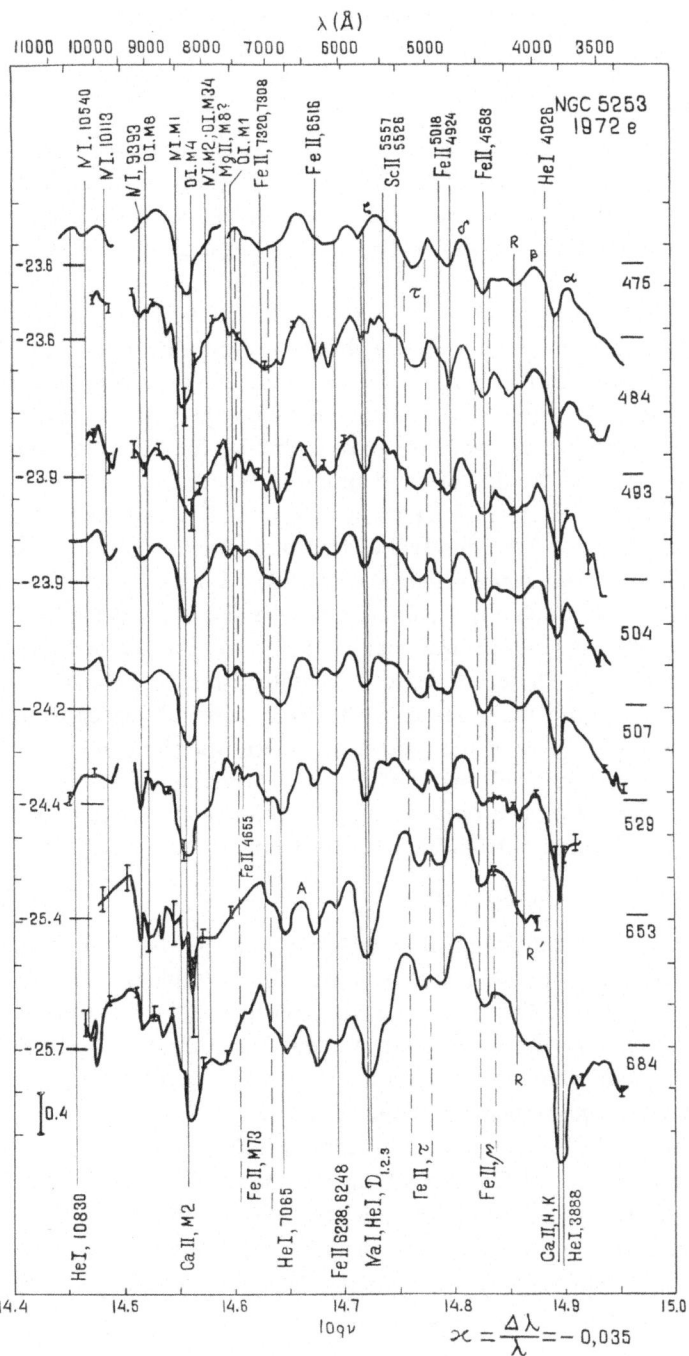

Fig. 2b.

However in the spectrum of supernova 1966j in NGC 3198 with the unusually narrow absorption lines we may record many of these lines, see Mustel (1974).

2.1.2. *Absorption Lines of* He

At first we have to mention an absorption with $\lambda \simeq 5700$ Å which may be attributed to the line of He I D_3, $\lambda_0 = 5875$ Å. This absorption may be attributed partly to the lines $D_{1,2}$ Na I. Nevertheless there are some arguments, which show that the contribution of the helium line D_3 to this absorption is more important, see Mustel (1974).

It seems that an absorption at $\lambda \simeq 6800$ Å is due to the He I line with $\lambda_0 = 7065$ Å. During the first stages of evolution this line was even a little stronger than the D_3-line, though for two last moments on Figure 2b the line D_3 was already stronger than the line $\lambda_0 = 7065$ Å. There are different possible explanations to these alterations in the relative strength of two He I-lines, including for example some uncertainty in the interpolation of the line of the continuous spectrum.

The situation with the strong He I-line with $\lambda_0 = 10830$ Å is a little uncertain because this line is at the end of the observable spectrum. Besides it may be suggested that the weakness of the *absorption* line of He I 10830 Å may be due to some *emission* within the same line. Such a line-emission may be also responsible for the inequality $W_\lambda(D_3) < W_\lambda(7065$ Å$)$ which was mentioned above. At last we may indicate the following possibility. There are reasons to state (Mustel, 1974) that the strongest excitation of He I-lines takes place in the *inner* parts of the expanding envelope. But in the infrared region of the spectra of supernovae there are many overlapping absorption lines of other elements which may originate in the *outer* parts of the envelope. This more or less continuous absorption may weaken considerably (for the observer) the radiation from He I 10830 Å-line.

2.1.3. *Absorption Lines of* C, N, O

The next very important group of elements includes C, N, O. In the optical parts of the spectrum the strongest lines of these elements coincide with the blends of metals and cannot be identified. Only one exclusion is the line of N II, multiplet 3, $\lambda_0 = 5680$ Å. This line was observed during the first moments when the temperature of the supernova was sufficiently high and during the last two moments in Figure 2b when the temperature was *again* sufficiently high, see above.

An analysis* of all the atomic transitions which produce sufficiently strong infrared absorption lines of C I, N I, O I in the spectra of 'normal' stars and a comparison of the results of the analysis with observations (Figures 2a and 2b) permitted us to conclude that nitrogen is a much more abundant element than carbon and oxygen. The strongest lines of N I on 2a and 2b which are sufficiently free from overlapping with other lines and blends are the following: 8212 (M2), 9393 (M7), 10113

---

* The oscillator strengths are taken from Wiese *et al.* (1966) and computations are carried out for the excitation temperature $T_{exc} = 10000°$.

(M18). These lines were prominent also in the spectrum of the type I supernova 1971i in NGC 5055.

On the other hand the identification of those absorption lines of C and O which are usually expected to be strong is difficult. In particular some strong lines of C I, λ_0 9406 M9 and λ_0 8335 M10, may be identified with the absorptions which are ascribed to the strong N I lines. But in this case some other C I lines which are expected to be even stronger than the lines $\lambda_0\lambda_0$9406 and 8335 are practically absent. The same is true for the absorption lines of O I. For example it is expected that the line O I, M10 should be rather strong, but observations do not confirm its existence, see Figure 2a. This conclusion about the very high abundance of N-atoms in comparison with the abundance of C, O-atoms in the envelopes of type I supernovae is in agreement with the computations of Caughlan and Fowler (1962) for the CNO bi-cycle in stars.

2.1.4. *Hydrogen Lines*

Let us consider the spectrum (energy distribution) for the moment JD = 2441465 (Figure 2a). This spectrum shows strong absorption lines of N I but no Hα-absorption line for this moment. At the same time the conditions for the appearance of Hα-line are equally favourable. In fact hydrogen and nitrogen atoms have approximately the same ionization potentials; the excitation potentials for Hα and for the lines of N I (9393 and 10113 Å) are also approximately the same. And the mechanism of widening of absorption lines of N I and H is also the same*; see the article of Mustel and Chugay (1974). Thus we may conclude that the abundance of hydrogen is much lower than the abundance of nitrogen. In addition we may say that Figures 2a and 2b do not show any emission bands of Balmer series.

Now let us summarize our discussion. The presence of sufficiently intense absorption lines of He I and the very high excitation potentials of the corresponding atomic transitions speak in favour of the conclusion that helium is the most abundant element in the envelopes of type I supernovae. It seems that the next element is nitrogen. But further studies are needed to obtain more definite data about the relative chemical composition of envelopes of type I supernovae.

2.2. TYPE II SUPERNOVAE

Type II supernovae are in many respects similar to common Novae. At first we may point out a similarity between the light curve of these supernovae and the light curves of certain novae, especially the 'fast' ones. The spectra of type II supernovae are similar in many respects to the spectra of the 'fast' novae. These supernova spectra contain absorption lines with the accompanying emission at the redward side of absorptions. The evolution of the spectra of these both objects is also similar. At light maximum the emissions are relatively weak but later their intensity steadily increases whereas the absorption lines fade away. The spectra of type II supernovae

---

* The velocity gradient.

differ markedly from the spectra of type I supernovae. For example they contain rather strong absorption and emission lines of *hydrogen.*

According to available information the velocities of the expansion of envelopes of type II supernovae are on the average somewhat lower than the velocities of the expansion of envelopes of type I supernovae.

The problem of time-evolution of the spectra of type II supernovae is very complex. It seems that we should consider separately the spectrum created by the ejected envelope itself and the spectrum emitted by the 'central remnant' of the supernova, see paper of Mustel (1974). Then it is very interesting to note that in the spectra of the type II supernova 19691 in NGC 1058 there was a progressive drift with time toward the red of all the absorption and emission features, see Ciatti *et al.* (1971). After light maximum the mean expansion velocity was decreasing during two months from $V \simeq 9500$ km s^{-1} to 5500 km s^{-1}.

There is no quantitative chemical analysis of the envelopes of type II supernovae. We may mention only the elements which were definitely identified by several investigators: H, He, O, N, Fe; Ca and some other elements. However the relative abundances of all these elements are not known. We may mention also the results of Branch and Greenstein (1971) obtained for a type V supernova 1961 in NGC 1058. These authors find that the chemical composition of ejected material may be similar to that of the Sun, except for a deficiency of hydrogen. However it is not yet clear if the type V supernovae should be classified as supernovae; van den Bergh (1973).

As to the chemical composition of the outer layers of novae we may refer to paper of Antipova (1974). The available data show the following anomaly: in comparison with the 'normal' stars there is a noticeable overabundance of N, C, O in the envelopes of Novae, whereas the relative abundances of metals are practically identical to those in the atmospheres of 'normal' stars. It seems that there is no anomaly in hydrogen.

2.3. The chemical composition of supernova remnants and the activity of supernovae and novae before explosion

When speaking about the supernova remnants we have to include in this conception not only the gaseous envelope ejected during the explosion but also other components. To be more exact we may speak about the following components: (A) The gases which are ejected from the star *during* a relatively short explosion and which constitute the principal envelope of the supernova at light maximum and during some period after it up to the moment when this envelope begins to interact with other components. (B) The circumstellar shells around supernovae which according to the suggestion of van den Bergh (1971a) are ejected from the star before the explosion. (C) The interstellar medium (gases and dust) in the space around supernovae.

Let us consider shortly the first two components. The third component C does not belong to the supernovae themselves and we shall mention its influence only in the cases when this component plays an important role in the physics of the supernova remnants.

2.3.1. *The Envelope Ejected During the Explosion of the Supernova*

This envelope is usually a source of optical and radio emission. As to the optical emission we should like to point out the following circumstances which are important in the interpretation of the optical emissions from component A of type I supernovae. We have in mind the conclusion about the presence of a very large velocity gradient inside the envelopes of these stars; see Section 2 of this article. Due to the velocity gradient the line emission will occupy a very large frequency interval and this will strongly weaken the optical line-emissions. It seems that this is one of the main reasons why the recognition of the optical emissions from the remnants of Tycho's and Kepler's supernovae is so difficult. In particular it may be suggested that Hα-emission from the faint long filaments of Tycho's remnant is due to the presence of component C in the parent envelope of the supernova. In fact according to van den Bergh (1971b) the expanding shell of Tycho's supernova has swept up a mass of *interstellar* gas that is several times greater than its own mass.

Approximately the same situation is true for Kepler's remnant. There is no optical emission from component A in this case. At the same time we know that the radio-emission from the supernova remnants is continuous and therefore the influence of even a very strong velocity gradient here is practically absent. The Kepler's remnant is a fairly thick shell source* with a shell thickness-to-radius ratio of about 0.5, see Herman and Dickel (1973). On the basis of radio observations carried out by Hazard and Sutton (1971), it may be computed that the expansion velocity of the remnant is approximately 12000 km s^{-1}; see van den Bergh (1973).

Thus we may conclude that unfortunately now we do not have any definite information about the chemical composition of component A of both Tycho's and Kepler's type I supernova remnants.

As to the bright radio source Cas A which is connected with a type II supernova the observations show that the fast moving knots of this remnant which certainly belong to component A, exhibit emission lines of [O I], [O III], [S II] and [Ar III]. In these knots oxygen, argon and sulphur are overabundant in respect to hydrogen and nitrogen by at least a factor of 30. This shows that the bright knots in Cas A could not have swept up a significant amount of (hydrogen-rich) interstellar material; see van den Bergh (1973).

We shall not speak here about the Crab Nebula since we do not know what type of supernovae produced this remnant. The information about the abundance of hydrogen in this nebula is somewhat uncertain; see again van den Bergh (1973).

2.3.2. *Circumstellar Shells Around Supernovae*

It is known that in addition to the usual fast moving gases (component A) certain supernova remnants reveal some quasistationary or relatively slow moving gases in

---

* This is probably the influence of the velocity gradient and the absence of a dense interstellar medium in the vicinity of the supernova.

the form of bright knots and filaments. These formations – component B – are observed in the radio source Cas A and in the remnant of Kepler's supernova. The main properties of these knots and filaments are the following: (a) relatively small velocities, much smaller than the average velocity of component A; (b) there is practically no correlation of these knots and filaments with the radio shells and generally with component A. For Kepler's remnant this lack of correlation is pointed out by Herman and Dickel (1973); (c) these knots and filaments have anomalously high intensity of forbidden emissions $\lambda_0\lambda_0 6548$, 6584 Å of N II. Analysing this last property of component B for the radio source Cas A, van den Bergh (1973) writes: "For temperatures $T > 6200$ K Peimbert and van den Bergh (1971) find that the nitrogen-to-oxygen ratio in the quasi-stationary flocculi in Cas A is higher than that prevailing in the Orion nebula. Taken at face value this result implies that the quasi-stationary flocculi cannot represent interstellar gas that was trapped by the expanding supernova shell. The observed overabundance of nitrogen might be understood by assuming that the flocculi were formed by the compression of a pre-existing circumstellar shell. Such a shell might have been enriched in ^{14}N that was produced in the CNO bi-cycle (Fowler and Caughlan, 1962)."

We may notice that this conclusion (which may be applied to the case of Kepler's remnant) is in agreement with the conclusions of Mustel (1973) about the very important role of N-atoms in the *internal* parts of supernovae, see Section 2 of this paper, though the sources of gases in both these cases are different. It seems that the gases which produce B-component are ejected from some subphotospheric levels of the star whereas the component A is produced by the gases which are localized much closer to the center of the supernova. It seems that the important role of N-lines in the spectra of both components shows that the number of N-atoms is enhanced practically at all levels inside the star before its explosion.

All the three points (a), (b), (c) enumerated above do confirm the idea of van den Bergh and Peimbert that there is some ejection of gases from the more or less *deep regions* of supernova before its explosion and that this process may take place during a relatively long period of time. It is difficult to expect that component B is produced by a simple compression of interstellar gases after the supernova explosion. If it would be so then the B-component in Tycho's remnant had to be much more prominent than that in the case of Kepler's remnant. In fact Kepler's remnant is located at a height $Z = 1.4$ kpc above the galactic plane. At the same time we have already mentioned that the interstellar gases played a very important role in the case of Tycho's remnant, see again van den Bergh (1971b).

There are some other very interesting considerations which support the idea that supernovae and novae are the sources of a more or less continuous activity before their explosions. It seems that this activity (at least for novae) is continuing even after the explosion.

In his paper Weaver (1974) presented a model of the envelope around nova V603 Aql 1918. According to this model 'the equatorial' parts ('rings') of the envelope are the result of an interaction of gases ejected during the explosion with the gases ejected before the explosion. This conclusion is based on the results of Kraft (1959) according to which DQ Her even after its explosion possesses some small rather dense emitting disk* (or ring), this disk is moving together with the post nova itself along the orbit of the double system. And it is quite natural to suppose that this disk (or ring) is due to the interactions between the two components in this double system.

Thus it may be supposed that not only after the explosion of the star but also *before* it the nova is surrounded by a 'circumstellar' envelope – component B – according to the previous terminology.

The equatorial belt (or ring) in the envelopes of V603 Aql and DQ Her was discussed in detail by Mustel and Boyarchuk (1970). The principal properties of the envelopes around the post-novae outlined in the above paper were recently confirmed in the papers of Malakpur (1973) and Hutchings (1972). Thus it seems that the presence of these belts (or rings) is an invariable property of the envelopes around *all* novae.

The 'belts' (or 'rings') in the envelopes around post novae produce certain emission components – intensity maxima – in the spectra of these objects. And it is very important to note that the most characteristic property of these 'belts' or 'rings' is the presence of *strong* emissions due to the same forbidden lines $\lambda_0 \lambda_0 6548$, 6584 Å of [N II]! In particular the 'equatorial belt' in the envelope around DQ Her was observed *only* in these lines of [N II], see Mustel and Boyarchuk (1970). A very long persistance of the lines of [N II] in the spectrum of the equatorial 'rings' in the envelope of V603 Aql was pointed by Wyse (1939). Thus we have serious reasons to consider that in the case of novae we have also a continuous ejection of gases from the relatively deep regions of the star which are enriched in ^{14}N.

It seems that this similarity between novae and supernovae is very significant; see Mustel (1970b). It supports strongly the idea that the supernovae are also double star systems. Now there are reasons to think that novae are relatively 'old' objects, see for example Kukarkin (1970, 1973). Therefore a similarity between novae and supernovae may speak in favour of the hypothesis that the supernovae are also relatively 'old' objects. This hypothesis is confirmed by a relatively high abundance of N in the envelopes of type I supernovae and by the absence of hydrogen in these envelopes, see above. Moreover since novae form a somewhat peculiar and isolated group of stars we may suggest that supernovae are also quite peculiar objects and are not the final stage of 'normal' stars at the end of their evolution. All these problems are going to be discussed by the author in *The Astronomical Journal of the U.S.S.R.* in the nearest future.

References

Antipova, L. I.: 1974, in G. Contopoulos (ed.), *Highlights of Astronomy*, Vol. 3, D. Reidel Publ. Co., Dordrecht, p. 501.
Barbon, R., Ciatti, F., and Rosino, L.: 1973, *Astron. Astrophys.* **25**, 241.
van den Bergh, S.: 1971a, *Astrophys. J.* **165**, 457.
van den Bergh, S.: 1971b, *Astrophys. J.* **168**, 37.

* We do not speak here about a very extensive envelope which is distant from the post nova.

van den Bergh, S.: 1973, *Publ. Astron. Soc. Pacific* **75**, 133.
Branch, D. and Greenstein, J. L.: 1971, *Astrophys. J.* **167**, 89.
Branch, D. and Patchett, B.: 1973, *Monthly Notices Roy. Astron. Soc.* **161**, 71.
Ciatti, F., Rosion, L., and Bertola, F.: 1971, *Mem. Soc. Astron. Italiana, Nuova Ser.* **42**, 163.
Fowler, W. A. and Caughlan, G. R.: 1962, *Astrophys. J.* **136**, 453.
Hazard, C. and Sutton, J.: 1971, *Astrophys. Letters* **7**, 179.
Herman, B. R. and Dickel, J. R.: 1973, *Bull. Am. Astron. Soc.* **5**, 284.
Hutchings, J. B.: 1972, *Monthly Notices Roy. Astron. Soc.* **158**, 177.
Kirshner, R. P., Oke, J. B., Penston, M. V., and Searle, L.: 1973, *Astrophys. J.* **185**, 303.
Kraft, R.: 1959, *Astrophys. J.* **130**, 110.
Kukarkin, B. V.: 1970, *Astron. Zh.* **47**, 1211.
Kukarkin, B. V.: 1973, 'Variable Stars in Globular Clusters and Related Systems', *IAU Colloq.* **21**, 8.
Malakpur, L.: 1973, *Astron. Astrophys.* **24**, 125.
Mustel, E. R. and Boyarchuk, A. A.: 1970, *Astrophys. Space Sci.* **6**, 183.
Mustel, E. R.: 1970, *Astrophys. Letters* **6**, 207.
Mustel, E. R.: 1971, *Astron. Zh.* **48**, 665; 1972, *Soviet Astron.* **15**, 527.
Mustel, E. R.: 1972, *Astron. Zh.* **49**, 15; 1972, *Soviet Astron.* **16**, 10.
Mustel, E. R.: 1973, *Astron. Zh.* **50**, 1121.
Mustel, E. R. and Chugay, N. N.: 1974, *Astrophys. Space Sci.*, in press.
Mustel, E. R.: 1974, in G. Contopoulos (ed.), *Highlights of Astronomy*, Vol. 3, D. Reidel Publ. Co., Dordrecht, p. 545.
Peimbert, M. and Bergh, S. van den: 1971, *Astrophys. J.* **167**, 223.
Pskowskij, Yu. P.: 1968, *Astron. Zh.* **45**, 942; 1969, *Soviet Astron.* **12**, 750.
Pskowskij, Yu. P.: 1970, *Astron. Zh.* **47**, 994; 1971, *Soviet Astron.* **14**, 798.
Weaver, H.: 1974, in G. Contopoulos (ed.), *Highlights of Astronomy*, Vol. 3, D. Reidel Publ. Co., Dordrecht, p. 509.
Wiese, W. L., Smith, M. W., and Glennon, B. M.: 1966, NSRDS Natl. Bur. Std. 4, Vol. I.
Wyse, A. B.: 1939, *Lick Obs. Bull.* **14**, Part 3.
Zwicky, F.: 1964, *Ann. Astrophys.* **27**, 300.

EARLY PHOTOGRAPHIC OBSERVATIONS OF RED AND INFRA-RED FEATURES IN THE SPECTRUM OF SN 1972e

(Abstract)

M. F. McCARTHY, S.J.

Vatican Observatory, Vatican City State

When news of Kowal's discovery (1972) of SN 1972e in NGC 5253 was received at Cerro Tololo on 19 May 1972, an infrared plate was obtained with the Curtis Schmidt and its objective prism which revealed the presence of a faint emission feature between 8600 Å and 8700 Å. This was a Kodak IN plate, unhypersensitised and exposed for 25^m without a filter. On 21 May an exposure of 90^m was made on an ammonia-sensitized IN plate through a Wratten filter 89B. The Michigan combination objective prism of 10° with a reciprocal dispersion near 7590 Å of 570 Å mm^{-1} was used, and the prominent emission-like feature centered at 8650 Å was confirmed.

The Schmidt results together with other spectrographic and photometric observations made at Cerro Tololo near maximum phase were reported by Osmer *et al.* (1972). Discussion of the 8650 Å feature and its possible identification with the Ca II triplet at 8498 Å, 8542 Å, 8662 Å, with O I at 8446 Å or with N I 8629 Å and 8683 Å was presented by McCarthy and Araya (1973) and by McCarthy (1973a, b). Further confirmation of this emission-like feature was obtained by image tube observations made later at Asiago by Ciatti (1973). The identification of the 8650 Å feature with the Ca triplet based on a long series of spectrum scanner observations at Palomar has been given by Searle (1973) and later by Kirshner (1974), and here by Mustel (1974). Herbig (1972) first photographed the deep depression near 8250 Å in SN 1972e and this was confirmed in our Cerro Tololo observations. This feature is similar to another deep depression noted in the red at 6175 Å by Ford and Rubin in type I supernovae (1967, 1968) and also observed in SN 1972e by Herbig (1972) and by McCarthy (1973a). Our objective prism spectra show the following features in the red and the near infrared spectra of SN 1972e near maximum phase:

Emission-like feature	Suggested sources	Deep depression feature	Wavelength difference
8650 Å	Ca II, N I, O I	8250 Å	− 400 Å
6550 Å	H, N II	6175 Å	− 375 Å

The association of strong emission-like features with the very deep depression some 400 Å to shortward wavelengths in both red and near infrared regions are the outstanding features present on these objective prism spectra obtained near maxmium phase. The long series of spectra discussed by Searle and Kirschner and by Mustel

Tayler (ed.), Late Stages of Stellar Evolution, 185–186. All Rights Reserved.

will be decisive in establishing at long last the mechanism responsible for the strong and strange features of type I supernovae spectra.

References

Ciatti, F.: 1973, *Astron. Astrophys.* **22**, 465.

Ford, W. K. and Rubin, V. C.: 1968, *Publ. Astron. Soc. Pacific* **80**, 466.

Herbig, G. H.: 1972, *Circular IAU*, No. 2407.

Kirshner, R. P.: 1974, in G. Contopoulos (ed.), *Highlights of Astronomy*, Vol. 3, D. Reidel Publ. Co., Dordrecht, p. 533.

Kowal, C.: 1972, *Circular IAU*, No. 2405.

McCarthy, M. F. and Araya, G.: 1973, *Bull. Am. Astron. Soc.* **5**, 12.

McCarthy, M. F.: 1973a, in C. Batalli-Cosmovici (ed.), *Internatl. Conf. on Supernovae*, University of Lecce (May 1973), in press.

McCarthy, M. F.: 1973b, *Ric. Astron. Spec. Vat.* **8**, No. 21, in press.

Mustel, E.: 1974, this volume, p. 172.

Osmer, P. S., Hesser, J. E., Kunkel, W. E., Lasker, B. M., McCarthy, M. F., and Landolt, A. U.: 1972, *Nature Phys. Sci.* **238**, 21.

Searle, L.: 1973, in C. Batalli-Cosmovici (ed.), *Internatl. Conf. on Supernovae*, University of Lecce (May 1973), in press.

THE RELIABILITY OF PHOTOMETRIC AND SPECTROSCOPIC DATA OF NOVAE

(Abstract)

J. TREMKO

Astronomical Institute of Slovak Academy of Sciences, Skalnaté Pleso, Czechoslovakia

1. Introduction

The photoelectric observations of novae, obtained by different photoelectric photometers, show systematic differences. These systematic differences are the consequence of bright emission lines in the spectrum at the limits of spectral bands of the wideband photometries, and the existing differences of the spectral performance of the instruments used in the corresponding photometric system. The results, which were achieved, do not correspond to the possibilities, which the photoelectric photometry gives us. On the other hand it is very important to know the mutual relations between the changes of the brightness of the central star and the shell of the nova and its spectral characteristics. The systematic differences among various series of the photoelectric observations of novae can be excluded by suitable choice of detectors and mainly by choice of spectral filters. By this means it will be possible to study the behaviour of the central star and the shell separately and will allow a better study of the mutual relations of the physical characteristics of novae.

2. The Method of Obtaining the More Precise Data About the Radiation of the Central Star and that of the Shell of Nova

In present time, the most commonly used photometric system is the UBV one, which is most frequently used for observation of novae, too. In the case of novae the following problem arises: all three spectral bands of the UBV photometric system include both components of the radiation of the nova, the continuum which belongs to the central star and the emission lines of the shell. Further, the bright spectral lines are situated at the borders of spectral bands, which are defined in the UBV photometric system.

We can see, that from the reasons above mentioned, the UBV photometric system is not very suitable for the photoelectric observations of novae. This is the reason why recently I have checked the possibility of using of 18 various photometric systems for the photometry of novae. The results were not satisfactory as no one photometric system fulfils sufficiently the conditions for the photoelectric photometry of novae (Tremko, 1973). The main reason consists in the fact, that in the spectral region from 3600 Å to 6500 Å, in which the electron multiplier phototubes with antimony caesium photocathodes are sensitive, there are many emission lines. Thus it is very difficult

Tayler (ed.), Late Stages of Stellar Evolution, 187–188. All Rights Reserved.

to find a sufficiently wide region with pure continuum. The types of the photomultiplier tubes with other photosensitive surfaces, which are now used in astronomical photometry, have usually lower integral sensitivity.

The recent technical progress in the branch of the development of the photosensitive surfaces and photomultiplier tubes gives us new possibilities for solving this problem. The electron photomultiplier tubes with gallium-arsenide photocathode guarantee high quantum efficiency and spectral sensitivity in a wide spectral region. For example the photomultiplier tube of the type Quantacon has extremely high sensitivity in the region from 2000 Å to 9300 Å, very low dark current and very fast time response. The maximum relative sensitivity is at 8300 Å (*RCA Bulletin*, 1971).

The new photomultipliers tubes can fully substitute in the photographic and visual region the photomultiplier tubes used to this time, and more, they give us a possibility of observation in the near infrared region. Just in the near infrared region there are few regions without emission lines in the spectrum of novae.

In principle the observations in two spectral bands would be sufficient. One of them should include the chosen emission line and the other one the continuum without emission lines. The problem consists in choosing the bands. According to Chalonge *et al.* (1964) the narrow bands of the continuum without emission lines exist near 3900 Å, 3950 Å and 6000 Å. The last mentioned region is bordered by the lines N II 5942 Å and by the forbidden line [Fe x] at 6300 Å. It is necessary to point out, that in the later stages of the development of a nova spectrum, this region can be influenced by the forbidden line [Fe VII] at 6086 Å. The useful band width of about 400 Å is located on the red side of the Hα line. It is bordered by the spectral line He I 6678 Å and He I 7065 Å. The only disturbing factor in the later stages of the nova evolution can be the line of [A XI] at 6919 Å. The widest band is between the spectral lines O I 7772 Å and O I 8216 Å, but this region is disturbed by several very faint emission lines, among them the brightest is the forbidden line [Ni xv] at 8024 Å. The narrower sections, about 200 Å wide, without emission lines are located at 7460 Å and 8340 Å.

The selection of the emission lines is not so difficult and it may be simple to realise by using the narrow band filters.

3. Conclusion

The analysis of the photoelectric observations has shown, that the UBV photometric system is not the most suitable one for the observation of novae. The use of the medium-band and narrow-band filters for the separated measuring of the continuum and emission bands in the near infrared region would be desirable and would bring much more representative results.

References

Bulletin RCA: 1971.

Chalonge, D., Bloch, M., Divan, L., and Fringant, A. M.: 1964, *Ann. Astrophys.* **27**, 255.

Tremko, J.: 1973, *Folia facultatis scientiarum naturalium Universitatis Purkyniane Brunensis*, T. XIV., opus 2, 105.

ON THE FREQUENCIES OF TYPE I AND TYPE II SUPERNOVAE

(Abstract)

N. DALLAPORTA

Istituto di Astronomia, Universita di Padova, Italia

I would like to present some simple considerations concerning the frequency of supernovae of both types and the mass range of their stellar progenitors which I have tried to deduce from the statistics given by Tammann (1970) for Sb and Sc galaxies only, and from some other data of other authors (Bertola and Sussi, 1965; Barbon, 1968) I will just briefly describe the procedure I have followed.

Relying on the data of a fiducial period from 1959 to 1969, Tammann gives the frequency of occurrence of supernovae per unit mass ($10^{10} M_\odot$); in order to separate type I and type II, I have first statistically divided the unidentified cases according to the observed ratio of type I/type II. As it has been observed that type II SN occur only in Sb, Sc and Irregular galaxies, they may be considered as belonging entirely to young Population I. The frequency of type II SN is then obtained by dividing the observed number of events by the number of galaxies considered in the survey, times the average mass young population I content of Sb and Sc galaxies (from 7 to 10%). The mean frequency per unit mass ($10^{10} M_\odot$), per young population I and per 100 years is given by:

Sb	Sc
1.09	1.57

All supernovae appearing in elliptical and So galaxies are of type I. However, many of them are also in spirals; and moreover their location in spirals shows that some of them are located in the central bulk and other in the spiral arms. There is thus some indication that type I SN might be a mixture of two different kinds of events, with similar external phenomenology, but belonging to objects of quite different populations, which one may term as SNIo (old) and SNIy (young). Separating again statistically the observed SNI events according to the observed ratio from the location in Spirals (Bertola and Sussi, 1965) of SNIo/SNIy, we calculate the frequencies of both SNIo and SNIy by dividing the numbers of events thus obtained by the number of galaxies considered and the mean mass amounts of old disk population (76 to 67%) and respectively of young disk Population I (7 to 10%). We thus arrive at the following figures for the mean frequencies per unit mass ($10^{10} M_\odot$) and per 100 years and per population type

	Sb	Sc
SNIy	0.25	1.2
SNIo	0.007	0.058

Tayler (ed.), Late Stages of Stellar Evolution, 189–190. All Rights Reserved.
Copyright © 1974 by the IAU.

The number of SNI observed in elliptical and So galaxies (Barbon, 1968) appears to be compatible with the assumption that they are the same kind of events as the SNIo in spirals.

If now one further assumes the Salpeter birthrate function as being representative also for the young disk Population I of Sb and Sc galaxies, and that death rates equal birthrates, it is then easy to calculate the mass range of SNIy and SNII progenitors yielding the observed frequencies. One finds the following values, for the lower limits of visual, bolometric absolute magnitudes and mass of the progenitors:

	SNII	SNIy
M_v	$-$ 3.7	$-$ 4.0
M_b	$-$ 5.7	$-$ 6.0
Mass	$\sim 13\,M_\odot$	$\sim 15\,M_\odot$

Should the observed frequencies reported by Tammann be somewhat underestimated due to losses, the mass range of the progenitors could perhaps be lowered to some $\sim 10\,M_\odot$. In any way, this value obtained is higher than those generally quoted.

References

Barbon, R.: 1968, *Astron. J.* **73**, 1016.
Bertola, F. and Sussi, M. G.: 1965, *Contrib. Oss. Astr. Univ. Padova*, No. 176.
Tammann, G. A.: 1970, *Astron. Astrophys.* **8**, 458.

CONTRACTING ENVELOPES OF NOVAE AFTER OUTBURSTS

(Abstract)

KYOJI NARIAI

Tokyo Astronomical Observatory, University of Tokyo, Mitaka, Tokyo, Japan

The idea that hydrogen-rich material from one component of a binary falling on to the surface of another component which is a white dwarf causes the nova phenomenon is now currently accepted. The present work is based on this idea, but we study the evolution of the extended envelope which contracts slowly some time after the nuclear event. Of course, there is an expanding envelope outside the contracting envelope.

The luminosity of the contracting envelope is almost constant while the effective temperature rises from about 10^4 K to 10^5 K as the radius decreases until time $\tau_0/2$, where

$$\tau_0 = \frac{\varkappa M_{env}}{4\pi c R_{core}} = 300 \left(\frac{M_{env}}{10^{-4} M_\odot} \right) \left(\frac{0.01 R_\odot}{R_{core}} \right) \text{ days}.$$

Subsequently both the effective temperature and the luminosity decrease. References and more discussion can be found in Nariai (1974).

In constructing models, we assume the functional form of the luminosity distribution and its derivative instead of using the equation for the conservation of energy. This makes the calculation fairly simple because we have ordinary differential equations instead of partial differential equations. Selfconsistent solutions are found by a trial and error method. Details of the method of calculation will be published in Nariai (1973).

References

Nariai, K.: 1973, *Ann. Tokyo Astron. Obs., Ser. II* **XIV**, 1.
Nariai, K.: 1974, *Publ. Astron. Soc. Japan* **26**, No. 1.

ON THE PECULIARITIES OF THE STAR LISTED AS NOVA
AURIGAE 1960–64 (KR AUR)

(Abstract)

MALINA POPOVA

Dept. of Astronomy, Bulgarian Academy of Sciences, Bulgaria

The variability of KR Aur was discovered in 1960 (Popova, 1960, 1961). As a scarce amount of plates was available it could not then be classified. Four years later on the ground of some new observations Hoffmeister supposed that KR Aur is a Nova of the RT Ser type, denoting it as Nova Aurigae 1960–64 (Hoffmeister, 1965, 1970).

We had the opportunity to make use of plate collections in the Sonneberg and Tautenburg Observatories (GDR) and in the Astronomical Institute of Sternberg, Moscow, for a photometric study of this star on about 400 plates taken from 1899 up to this year. It was shown that KR Aur is a blue variable with very unusual brightness variations and most probably is of a new type (Popova, 1965a, b, 1974).

In 75 years 7 rises of the brightness were established.

In Figure 1 observational data for the last 4 cycles are plotted. A length of about 6.5 years for these cycles may be derived. The star is more in maximum than in minimum brightness and according to the light curve it can be called more 'antinova' than 'nova'. Deep relative short minima attaining 18^m or fainter are followed by flat maxima with a mean magnitude of about 13.4 with fluctuations of $0^m.8$.

Noteworthy is the presence of considerable rapid rises and declines of the brightness. For instance in a day (2439442.5–443.5) the brightness declined by $1^m.8$ (13.66–15.50). An outburst of $2^m.4$ at maximum light was also observed in 1963. Five days after the brightness of $13^m.7$ on three plates in a night the star had $11^m.3$,

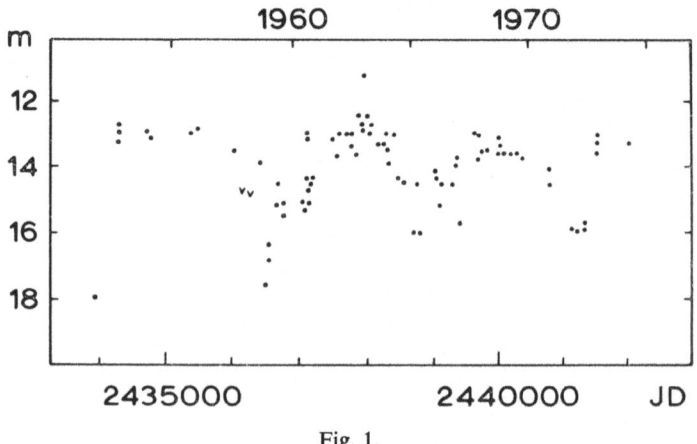

Fig. 1.

Tayler (ed.), Late Stages of Stellar Evolution, 192–193. All Rights Reserved.

and 12 days later during the next observation it was 13.^{m}7 too. On panchromatic plates the variable was of 1.^{m}2 fainter at the outburst and also at normal maximum. Thus the brightness of KR Aurigae varies between 11.3 and 18^m or fainter.

It was worth-while to observe the spectrum of Nova Aur. An attempt was made by Preston, as it is reported in (*Agenda and Draft Reports IAU*, 1967). We took a low dispersion objective-prisma spectrum at maximum light with the 1-m telescope (1:2:1) of the Biurakan Observatory (Arm. SSR) (Popova, 1970). It was established that the star has a considerable UV excess. The absorption hydrogen lines are quite weak and hardly to be seen, but there is an emission of He II 4686. These preliminary results confirm the peculiarity of Nova Aurigae (1960–64).

I would like to emphasize that Nova Aur 1960–64 is now in maximum light and further observations, photometric and spectroscopic ones, with higher dispersion are very desirable.

References

Agenda and Draft Reports IAU: 1967, Prague, 532.
Hoffmeister, C.: 1965, *Inf. Bull. IAU Var. Stars*, No. 93.
Hoffmeister, C.: 1970, *Veränderliche Sterne*, Johann Ambrosius Barth, Leipzig.
Popova, M.: 1960, *Mitt. über veränderliche Sterne* **463**.
Popova, M.: 1961, *Astron. Nachr.* **286**, 81.
Popova, M.: 1965a, *Inf. Bull. IAU Var. Stars*, No. 97.
Popova, M.: 1965b, *Var. Stars USSR* **15**, No. 5 (119) 534.
Popova, M.: 1970, *Bull. Dept. Astron.' Bulg. Acad. Sci.* **4**, 51.
Popova, M.: 1974, *Bull. Dept. Astron., Bulg. Acad. Sci.* **7**, in press.

EVIDENCE FOR CONTINUED EJECTION MODELS OF NOVAE

(Abstract)

M. FRIEDJUNG

Institut d'Astrophysique, Paris, France

New evidence has been obtained by calculating the sizes of regions emitting $H\alpha$ and Fe II. The size of a region emitting a line can be calculated if it is optically thick and the ratio of upper to lower level populations is known. This calculation was performed for $H\alpha$ in the case of Nova Delphini and for Fe II in RR Telescopii. Sizes of the emitting region were obtained which were, especially in the former case, much smaller than would be expected from instantaneous ejection models, indicating that most of the emission comes from a continuously ejecting region near the centre of the envelope. It is very important to continue high dispersion photometric studies of novae and such work on Nova Delphini is being continued in Paris by Malakpur.

A NOTE ON THE POSSIBLE EVOLUTIONARY STATUS
OF THE RECURRENT NOVA T CrB

(Abstract)

PETR HARMANEC

Ondřejov Observatory, Czechoslovakia

I would like to make a comment on the possible interpretation of the recurrent nova T CrB comparing this object with another one, namely with the shell star AX Mon. Basic characteristics of both binaries are:

AX Mon: B2 IV–Ve + K2 II, orbital period $P = 232.5$ days
T CrB: M giant + Be, orbital period $P = 230$ days,

total mass of both systems being probably comparable (depending on the assumed inclinations of orbits).

Some time ago I computed (originally in co-operation with Prof. Plavec) several sequences of models in a case B of mass exchange in an attempt to represent the present stage of AX Mon. I found that for any reasonable combination of initial masses the mass-losing component had to have a deep outer convective zone at the very beginning of mass exchange. It was shown by Paczyński that in such a case mass exchange must proceed on the dynamical time scale of the outer envelope, the radius of the star remaining nearly constant while the absolute dimension of the Roche lobe is shrinking as long as the role of both components is not reversed. Only after that the mass exchange can goes on the Kelvin time scale of the mass-losing star.

Four years ago Bath suggested this instability may be responsible for the nova phenomenon. I think he might be right at least for the case of T CrB. It is noticeable that in T CrB the M giant is *the more massive* component of the system being thus dynamically unstable while AX Mon has the Be-component more massive – being according to my model just at the end of the dynamically unstable phase i.e. only thermally unstable now.

Another fact supporting the principal correctness of Bath's theory is the low limit of the mass of the Be component of T CrB which exceeds the white-dwarf limit.

Thus, I think, a direct evolutionary connection may exist between novae of T CrB type and some shell stars which both are – as I believe – binaries in the stage of rapid mass exchange between components.

References

Horn, J. and Harmanec, P.: 1971, paper presented at the *Czechoslovak Conference on Stellar Astronomy and Astrophysics*, Cikháj, October 12–14, 1971 (see also: *Folia Fac. Sci. Nat. Univ. Purkynianae Brunensis* **14**, ser. Physica, No. 2, p. 71, 1973).
Harmanec, P.: 1974, Submitted for publication in *Bull. Astron. Inst. Czech.* **25**.
Plavec, M., Ulrich, R. K., and Polidan, R S.: 1973, to be published in *Publ. Astron. Soc. Pacific*.

DISCUSSION AFTER PAPER BY MUSTEL

Tayler: I wish to provoke some discussion between observers and theoreticians. Theoreticians require supernovae to produce large amounts of heavy elements and that very large mass loss should occur. However, it is not clear that observers ever see a very high loss of mass or very large amounts of heavy elements. Is this a serious discrepancy?

Mustel: It is very difficult to detect heavy elements from the analysis of the spectra of the envelopes ejected by supernovae. In fact the abundance of all elements of this group is usually very small. Therefore the number of absorbing atoms of any heavy element in the envelope (cm^{-2}) is also small. And due to the very large velocity gradient inside the envelopes the absorption lines of these elements are expected to be very shallow. Thus their detection in the spectra of supernovae with a lot of strong blends is very difficult.

Nevertheless it is interesting to point out that the line of BaII 5854 Å in the spectra of type I supernova 1966j in NGC 3198 was anomalously strong, see Mustel (1974). It is very important to continue the efforts to detect similar lines in the spectra of other supernovae.

Arnett: What is the typical star, which becomes a supernova? Suppose stars more massive than 8 $M_\odot$ do so. The typical one *by number* would be about 15 $M_\odot$, while *by mass* it would be about twice as massive. The nucleosynthesis involves a mass average; observations of supernovae are on averaging by number. *Some* differences in the objects are to be expected.

Another problem is that the observer of a supernova explosion sees only the outer parts of the exploding object, only a study of remnants will tell us about the rest of the matter.

Finally take the 15 $M_\odot$ model which I showed earlier. If it loses its envelope and a 4 $M_\odot$ helium core is left, then an object like the Crab nebula should result: a neutron star and a helium rich nebula of about 2 $M_\odot$.

Audouze: The discrepancy between observations of supernovae, where only light elements are seen, and the theories, where heavy elements are supposed to be produced in large amounts, is only apparent. There is now a rather large consensus that supernovae are the sources of cosmic rays, this is substantiated in particular by energy arguments. In the cosmic rays the heavy elements are observed in larger amounts than in the so-called universal abundances. It seems then that there is experimental evidence that supernovae produce heavy elements in the large amount predicted by theorists.

Tayler: Although the connection between supernovae and cosmic rays is probably valid, this is in reality only assuming one thing which you think plausible in order to prove another.

Tayler (ed.), Late Stages of Stellar Evolution, 196–197. All Rights Reserved.
Copyright © 1974 by the IAU.

McCarthy: Speaking about possible *observational* evidence of heavy elements may we point out that the S stars mentioned by Miss Sackmann as possible forerunners of novae and supernovae show technetium as Paul Merrill and Mrs Sitterly have pointed out. Perhaps studies at higher dispersion of S stars and supernovae at late stages (using image tube techniques) will help to bridge the gap mentioned by Tayler.

Mustel: In a low mass supernova of $\sim 1\ M_\odot$, only about $0.1\ M_\odot$ is ejected and the metals probably stay inside.

Hesser to Arnett and Ostriker: Can you suggest any additional observations that will help in your quest for a satisfactory theoretical model? (This question was asked by Hesser more than once in the Symposium but never produced a reply.)

Schwarzschild: May I ask two questions? First, is the radius of the shell responsible for producing the absorption spectrum which you, Dr Mustel, have discussed in agreement with the extent of the envelope which the theoreticians seem to need to produce the characteristic sharp peak in the observed light curve? Second, could it be that the shell responsible for producing the absorption spectrum is only a small portion of the whole ejected envelope?

Woolf: Dr Mustel's star had a radius at maximum light of 2×10^{15} cm. Dr Ostriker required an envelope of size 5×10^{15} cm. The agreement is therefore excellent.

Arnett: In my model, I used a size of 1×10^{15} cm. These numbers are in agreement to within their uncertainties.

Mustel: I calculated the size of a supernova at light maximum in two different ways. I obtained $40000\ R_\odot$ from the expansion velocity and $20000\ R_\odot$ from the luminosity and effective temperature.

I have one further comment. There are some interesting condensations of nitrogen around supernovae which are visible during the explosion but it seems necessary to assume that they existed *before* the event.

THE CHEMICAL COMPOSITION OF LATE TYPE STARS

A. A. BOYARCHUK

Crimean Astrophysical Observatory, USSR Academy of Sciences, Crimea, U.S.S.R.

Abstract. A discussion is given of the chemical composition of the atmospheres of various types of late type peculiar stars and their compositions are compared with those of normal K and M stars. The groups considered include CN, CH, Ba, S, C, R CrB and HdC stars.

The chemical compositions of the atmospheres of many stars have been investigated at the present time. Although most of stars investigated have a chemical composition of their atmospheres very close to that of the Sun there are many stars which show pecularities of the chemical composition. The pecularities may arise by two reasons:

(i) The differences in the chemical composition of pre-stellar matter.

(ii) Changing of the chemical composition of stellar surfaces during stellar evolution.

Many astronomers have investigated the chemical composition of stars belonging to the different stellar groups which have different ages. It was established that the older stars have the lower abundances of metals in their atmospheres. Most significant variation of the chemical composition took place during the first 2×10^8 years of life of the Galaxy, when the ratio [Fe/H] increased from -2 to 0. (Here and later [Fe/H] means the value $\lg [N(\text{Fe})/N(\text{H})]_{\text{star}} - \lg [N(\text{Fe})/N(\text{H})]_{\text{Sun}}$.) After that variations up to factors of ± 2 from the solar metal abundance occurred (Eggen and Sandage, 1969). The reason of that variation is still unknown (Arp, 1962).

Another important aspect of a problem of chemical compositions of a pre-stellar matter is the relative abundances of individual metals. Pagel (1970) has shown that the s-process elements are less abundant in the stars which have smaller values of [Fe/H]. Especially it is true for the barium.

Thus the general tendency of aging effect of chemical composition is the increasing of the values of [Met/H] and of [heavy met/Fe] during time.

However, from point of view of the present symposium the variations of chemical compositions of stellar surfaces which occurred during stellar evolution are much more interesting. We excluded from consideration the main sequence stars because they are not a topic of this symposium. We concentrate our attention on the late-type giants because the rather detailed observations are known for these stars.

We will consider the different groups of stars which show remarkable peculiarities in their spectra and which are located on the giant branch on the spectrum-luminosity diagram (Figure 1). These groups are following:

(1) Normal K–M giants.

(2) CN-stars, which have strong CN bands in their spectra.

(3) CH-stars, which have strong CH-bands in their spectra.

(4) Barium stars, which show abnormal strong BaII lines.

Tayler (ed.), Late Stages of Stellar Evolution, 198–205. All Rights Reserved.

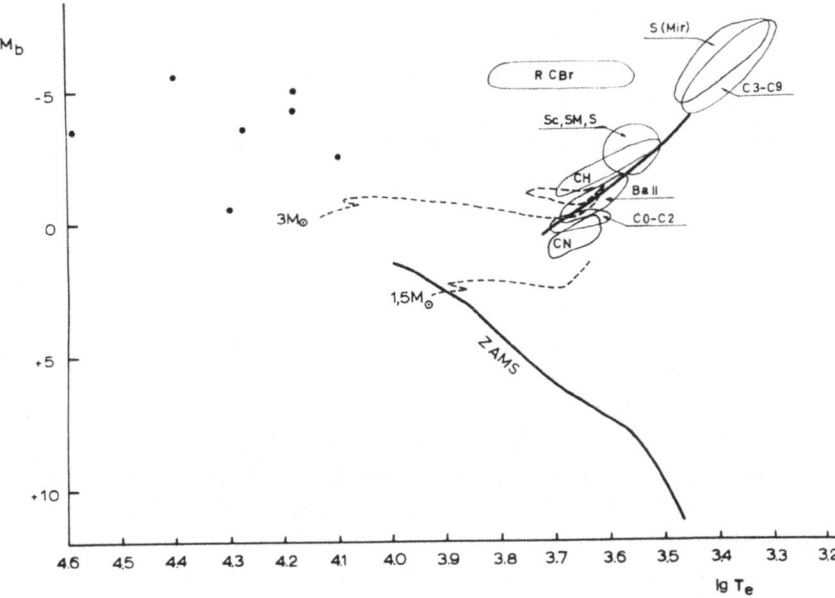

Fig. 1.   The location of the different groups of stars on the HR diagram. Dots are the hot HdC stars.

(5) S-stars. It is possible to say roughly that the S-stars are M-stars with strong ZrO bands.

(6) C-stars. These are carbon stars, their spectra have not TiO-bands, but C_2, CN bands.

It is necessarily to point out that the S-characteristics may be not so strong. In that case we give to the star the class SM or MS. So that there are some intermediate stars between M- and S-type stars. There is a small group of C-stars, which show S-characteristics. These are SC and CS stars. All this means that some groups mentioned above are connected with other groups.

(7) HdC-stars which are hydrogen deficient carbon stars. These stars have rather weak hydrogen lines and remarkable features of carbon both atomic and molecular. This is an inhomogenous group, which contains stars with different characteristics. In particular their temperatures vary in a wide range. There is a subgroup of variable stars. It is so-called R CrB stars, which show suddenly great decreases of their brightness.

There are a few stars, which are difficult to classify into some groups mentioned above. We will not discuss them here. In this report we will concentrate our attention on the typical peculiarities of chemical composition of stars.

Let us start by discussing the normal K–M giants. We called them 'normal' for two reasons. First, the number of these stars is above 80% of yellow and red giants. Second, the abundance pecularities of these stars are smaller than that of other group.

We will concentrate our attention on the following abundance ratios:

(1) The relative abundances of the light elements.

(2) The ratio of carbon isotopes $^{12}C/^{13}C$.

(3) The relative abundances of the metals of the iron group in comparison with hydrogen, M/H.

(4) The relative abundances of the elements, which were produced by s-processes. The ratio [Ti/Zr] characterizes that.

(5) The presence of short-living elements in the atmospheres such as technetium.

(6) The abundances of lithium.

For M-stars we have now not so well established data, but there are estimates of abundances of some elements.

The relative abundance of different elements in the atmospheres of normal K-stars do not differ significantly from that in the solar atmosphere (Conti *et al.*, 1967). There is a strong difference of the carbon isotope ratio only. The K-stars have the ratio $^{12}C/^{13}C$ less than 10, while the solar value is about 90 (Greene, 1969).

The chemical compositions of M stars is not known in detail yet. Spinrad and Vardya (1966) have shown that the ratio $O/C \approx 1.05$ is less than the solar value 1.7. The nitrogen abundance is not the same for all stars. α Her and o Cet have the normal ratio N/H, but α Ori and R Leo have this ratio ten times more. The ratio $^{12}C/^{13}C$ is less than 8 for the M stars investigated (Gaball *et al.*, 1972).

Peery (1971) has found technetium lines in the spectra of two M-type miras o Cet and R Hya. According to Merchant (1967) M stars have a lithium abundance much less than the Sun.

The next group is the stars with strong cyanogen molecular bands. Greene (1969) has shown that the carbon and the nitrogen have little enhancement. The abundance of nitrogen for CN stars seems to be very near to that for normal stars (Schmitt, 1969). The ratio $^{12}C/^{13}C$ is equal 10 for α Ser. The metals show a little deficiency about -0.2 dex.

Then we consider a group of BaII-stars. Here we will follow mainly Warner (1965). On the Figure 2 we can see the differences between stellar abundance ratios and that of the Sun. The main features are following (a) large abundance of lithium, (b) large enhancement of elements heavier than Sr, (c) small overabundance of Eu and Yb; that is r-process elements.

In general, metals are over-abundant by a factor 1.5 in comparison with hydrogen. The $^{12}C/^{13}C$ ratio in the BaII stars is probably more than 20. No Tc-lines were observed in any BaII stars. CH-stars are very similar to BaII stars. These stars have strong CH bands; Figure 3 shows us the differential abundance of two CH stars according to results of analysis by Wallerstein and Greenstein (1964).

We may see that the peculiarities of the abundance ratio log(M/Fe) are very similar to those which we observed in Ba-stars. The heavy elements, which were produced by the s-process are overabundant. The carbon is overabundant also. The metals are deficient in respect to hydrogen by a factor 10. The $^{12}C/^{13}C$ ratio is close to 50. No technetium was observed in CH-stars. The strength of the CH band in comparison with atomic lines is explained by the high abundance of hydrogen.

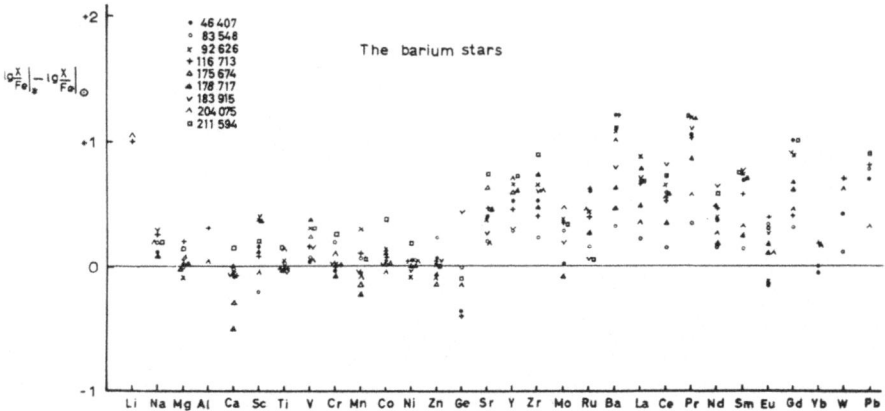

Fig. 2. The relative abundances in the atmospheres of some Ba-stars.

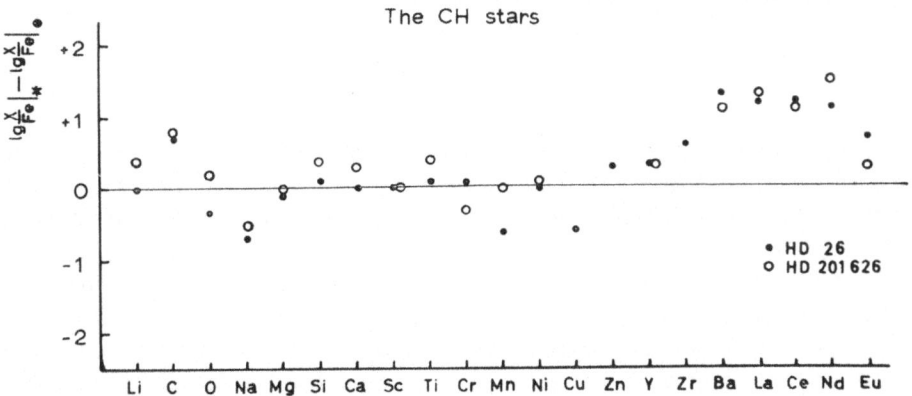

Fig. 3. The relative abundances in the atmospheres of two CH-stars.

The CH stars have large space velocities. Then we can consider them as Ba-stars of the Population II.

Let us consider now S stars. First of all it is necessary to point out that the accuracy of the determination of chemical composition of these stars is much less than that of K-type stars or earlier.

But nevertheless we can draw some conclusions about their chemical composition. The Figure 4 shows the chemical compositions of S and C stars according to Tsuji (1962, 1971) and Hirai (1969). This figure shows that the heavy elements are over-abundant here more than in Ba-stars.

Then we can see that C stars contain more heavy elements than S stars. There are some differences of abundances of r- and s-process elements: the s-process elements such as barium are over-abundant much more than r-process elements as europium. The relative abundance of oxygen group elements in the S stars do not differ from that

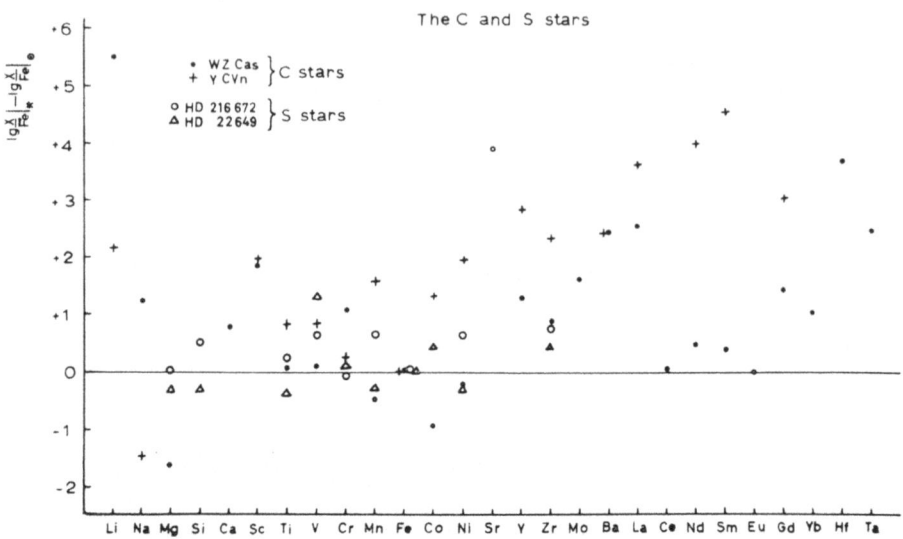

Fig. 4.   The relative abundances in the atmospheres of C and S stars.

of the Sun by more than a factor 2. R And has $^{12}C/^{13}C = 6$ (Tsuju, 1971). According to Peery (1971) the lines of Tc are present in spectra of variable S-type stars and absent in spectra of nonvariable stars. This is true for all stars which have been checked for Tc.

Davis (1971) has suspected the presence of the lines of prometheum in the spectra of two S-stars V Cnc and T Sgr. The abundance of lithium varies in a wide range.

The main feature of C stars is a high abundance of carbon. The ratio O/C is less than unity. The carbon and oxygen form the tightly bound molecule CO. Then the excess of carbon forms such molecules as CH, CN and C_2. But at the same time there is not enough oxygen in order to form oxides such as TiO and ZrO, which are typical for M-type stars. The ratio O/C in C stars is 20 times less than that in the Sun. In most cases the ratio $^{12}C/^{13}C$ is near to the equilibrium value in the CN cycle (4.6). But a few stars have the ratio $^{12}C/^{13}C$ near 20 (Querci and Querci, 1970). 6 stars from 8 investigated have Tc lines (Peery, 1971).

Lithium abundances vary in a very wide range, over 6 orders of magnitude (Torres-Peimbert and Wallerstein, 1966). Some stars have lithium abundances 4 orders of magnitude more than that of the Sun. Usually these stars have also a high abundance of ^{13}C. The ratio $^{12}C/^{13}C \approx 3$ (Gordon, 1971).

One star has been investigated by Catchpole and Feast (1971) for lithium isotopes. It was found that the whole lithium is ^7Li.

Let us go to the left from the red giant branch and discuss R CrB stars. Only two stars were analysed. They are R CrB itself (Searle, 1961) and RY Sgr (Danziger, 1965). The results are shown in Figure 5. We can see the large deficiency of hydrogen and oxygen. The helium, lithium, carbon and nitrogen are overabundant. Approximately the same position on the HR diagram is occupied by the non-variable stars,

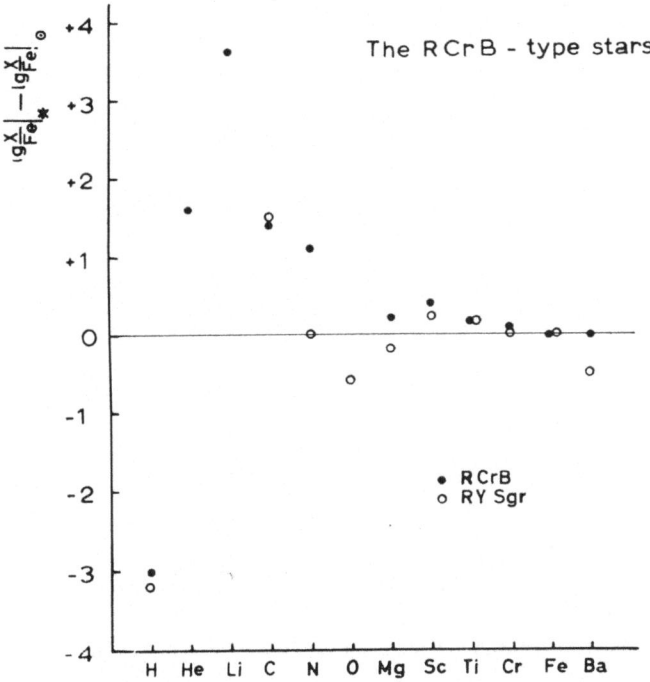

Fig. 5.   The relative abundances in the atmospheres of R CrB type stars.

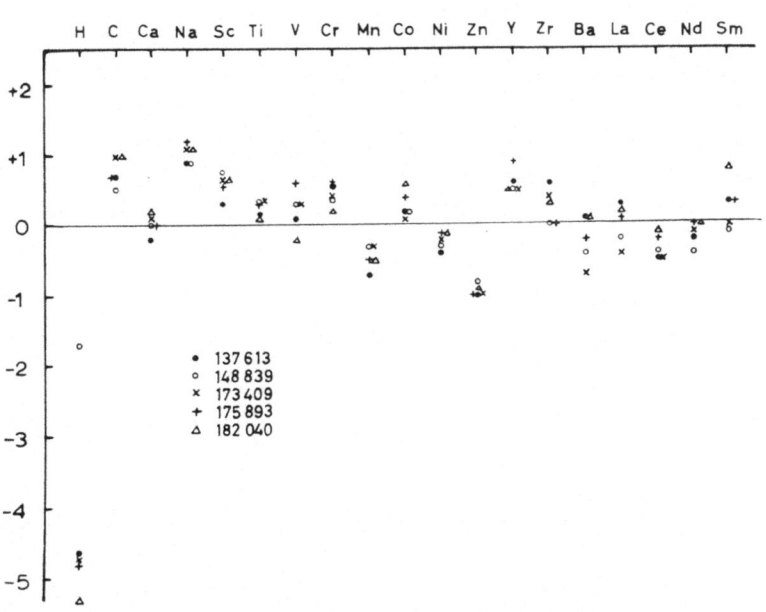

Fig. 6.   The relative abundances in the atmospheres of cool HdC stars.

The hot HdC stars

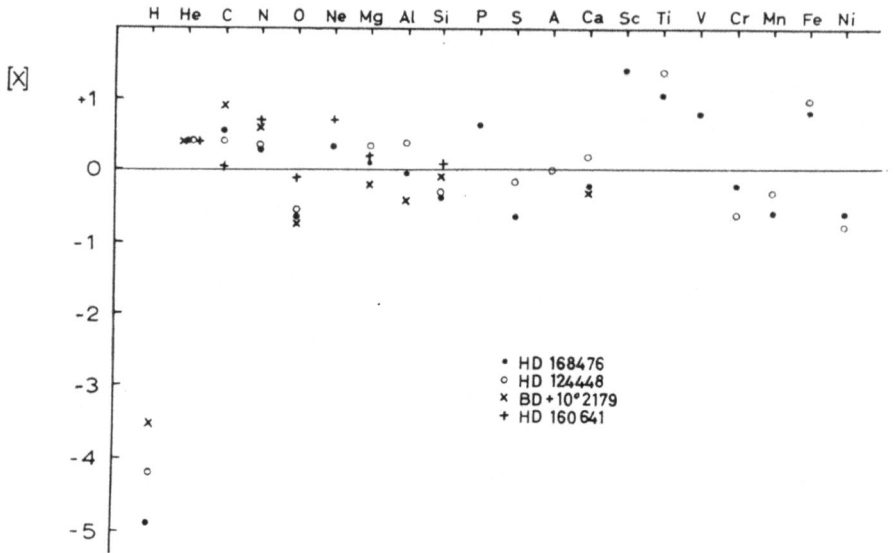

Fig. 7.   The relative abundances in the atmospheres of hot HdC stars.

which have chemical composition similar to R CrB-type stars. These are so-called HdC-stars. Figure 6 represent the results of an analysis by Warner (1967). The main features of the abundances are the deficiency of hydrogen and the excess of carbon.

There are a few stars, which are located to the left of the cool HdC-stars and have the same situation with abundances of hydrogen and carbon. The Figure 7 shows the chemical compositions of some of these stars (Hill, 1965).

We see, that the hydrogen is strongly underabundant. The He, C, N, is little over-abundant and O little underabundant. We have a large scatter for the elements of iron group. This perhaps is a result of weakness of lines of those elements.

Table I gives a summary of abundance differences of some groups of late type.

TABLE I

	[C/H]	[N/C]	[O/C]	$^{12}C/^{13}C$	[M/H]	[Ti/Zr]	Tc	[Li]
Normal M	0	+0.3	−0.2	< 8	0.0	−0.1	yes	−2.0 to +1.0
CN	+0.2	+0.2	0	< 10	−0.2	0.0		
CH	0			50	−1.1	−0.5	no	0
BaII	+0.6			> 20	+0.2	−0.7	no	+1.0
S	0	+0.3	−0.3	6		−0.7	yes	−1.5 to +3.0
C		−0.1	−1.3	4–20		−3.0	yes	−2.5 to +4.3
HdC	+5	0	−1.3			−0.5		
R CBr	+4	−0.3	−2.0	> 50	+3.0			+3.5

References

Arp, H. C.: 1962, *Astrophys. J.* **136**, 66.
Catchpole, R. M. and Feast, M. W.: 1971, *Monthly Notices Roy. Astron. Soc.* **154**, 197.
Conti, P., Greenstein, J. L., Spinrad, H., Wallerstein, G., and Vardya, M. S.: 1967, *Astrophys. J.* **148**, 105.
Danziger, I. J.: 1965, *Monthly Notices Roy. Astron. Soc.* **130**, 199.
Davis, D. N.: 1971, *Astrophys. J.* **167**, 327.
Eggen, O. J. and Sandage, A.: 1969, *Astrophys. J.* **158**, 669.
Gaball, T. R., Wallman, E. R., and Rank, D. M.: 1972, *Astrophys. J.* **177**, L27.
Greene, T. F.: 1969, *Astrophys. J.* **157**, 737.
Gordon, C. P.: 1971, *Publ. Astron. Soc. Pacific* **83**, 667.
Hill, P. W.: 1965, *Monthly Notices Roy. Astron. Soc.* **129**, 137.
Hirai, M.: 1969, *Publ. Astron. Soc. Japan* **21**, 91.
Merchant, A. E.: 1967, *Astrophys. J.* **147**, 587.
Pagel, B. E. J.: 1970, *Quart. J. Roy. Astron. Soc.* **11**, 172.
Peery, B. F. J.: 1971, *Astrophys. J.* **163**, L1.
Querci, M. and Querci, P.: 1970, *Astron. Astrophys.* **9**, 1.
Schmitt, J. L.: 1969, *Publ. Astron. Soc. Pacific* **81**, 657.
Searle, L.: 1961, *Astrophys. J.* **133**, 531.
Spinrad, H. and Vardya, M. S.: 1966, *Astrophys. J.* **146**, 399.
Torres-Peimbert, S. and Wallerstein, G.: 1966, *Astrophys. J.* **146**, 724.
Tsuji, T.: 1962, *Publ. Astron. Soc. Japan* **14**, 222.
Tsuji, T.: 1971, *Publ. Astron. Soc. Japan* **23**, 275.
Wallerstein, G. and Greenstein, J. L.: 1964, *Astrophys. J.* **139**, 1163.
Warner, B.: 1965, *Monthly Notices Roy. Astron. Soc.* **129**, 263.
Warner, B.: 1967, *Monthly Notices Roy. Astron. Soc.* **137**, 119.

ABUNDANCES IN Ap STARS; RESULT OF MAGNETIC BINARY EVOLUTION

(Abstract)

E. M. DROBYSHEVSKI

Institute of Physics and Technic, Leningrad, U.S.S.R.

There is an old theory of Fowler *et al.* (1965) that the peculiar A stars are in the post-giant phase of stellar evolution. This theory tried to explain the anomalous abundances observed in Ap stars but data available today do not confirm it.

Another theory concerns the binary nature of Ap stars. It was suggested by Renson and elaborated by van den Heuvel (1968). The latter believed all the peculiar and metallic-line A stars to be evolved binaries.

The statistical analysis of mass functions for known spectroscopic binaries amongst Am stars developed by myself (Drobyshevski, 1973) shows indeed that about two thirds are evolved binaries.

As for peculiar A stars, the situation is complicated as few are known to have companions. These are the stars having strong magnetic fields and it is natural to suppose that the magnetic field must play a role in the evolution of the binary system.

Resnikov and myself have made approximate calculations of close binary evolution taking into account the influence of the magnetic field on exchange and loss of mass and angular momentum. The main emphasis was on the realization that the field became turbulent through convective motions in the envelope of the primary component – the red giant. The coherent field has a small influence on the process under consideration because of the small value of its flux through any cross-section of the star as compared with the turbulent field flux

$$\phi_{\text{coher}} = \int_s H_{\text{coher}} \, \mathrm{d}S \ll \phi_{\text{turb}} = \int_s |H_{\text{turb}}| \, \mathrm{d}S \,.$$

It follows from the calculations that the magnetic field influences mass and angular momentum exchange in systems with original orbital period $P_0 \gtrsim 100^{\text{d}}$. Then the magnetic field becomes so strong that its pressure in the stream of matter flowing from the primary on to the secondary exceeds the gas pressure

$$\frac{H_{\text{turb}}^2}{8\pi} \gtrsim P_{\text{str}} \quad \text{at} \quad P_0 \gtrsim 100^{\text{d}} \,. \tag{*}$$

The stream parameters were calculated by using the method of Paczyński and Sienkiewicz (1972).

The effect of having a large magnetic field will only be found in wide pairs as the magnetic flux ϕ_{turb} grows when one goes to increasingly wide separations. At the

Tayler (ed.), Late Stages of Stellar Evolution, 206–207. All Rights Reserved.
Copyright © 1974 by the IAU.

same time the relative cross-section of the stream becomes smaller in wide pairs. The details are discussed in an article submitted to *Acta Astronomica*.

When condition (*) is satisfied in the stream, solar-type flares must occur, but their intensity and scale should be much greater. The solar flares are known to be accompanied by the acceleration of heavy nuclei. Brancazio and Cameron (1967) have already shown that the observed abundances of peculiar A stars can be produced by the bombardment of matter with normal abundances by energetic α-particles.

The small observed duplicity of Ap stars is explained because their periods must be long and velocities small. A more detailed analysis, taking account of data concerning cluster membership, kinematics, geometry and magnetic field strength etc., shows that it is the Cr–Eu peculiar A stars which are the result of evolution of wide originally non-evolved binaries with Am components in the manner discussed.

References

Brancazio, P. J. and Cameron, A. G. W.: 1967, *Can. J. Phys.* **45**, 3297.
Drobyshevski, E. M.: 1973, *Astrofizika* **9**, 119.
Fowler, W. A., Burbidge, E. M., Burbidge, G. R., and Hoyle, F.: 1965, *Astrophys. J.* **142**, 423.
Paczyński, B. and Sienkiewicz, R.: 1972, *Acta Astron.* **22**, 73.
Van den Heuvel, E. P. J.: 1968, *Bull Astron. Inst. Neth.* **19**, 326.

HELIUM ABUNDANCE FROM RADIO RECOMBINATION LINES

(Abstract)

E. KRÜGEL

Universitäts Sternwarte, Göttingen, F.R.G.

When stars are losing mass some of the material from which they were formed is being returned to the interstellar medium but, owing to stellar evolution, with a somewhat different chemical composition. This material may again be used in star formation. The chemistry of this new generation of stars can be determined with the help of radio recombination lines arising in the ionized gas, which surrounds the newborn stars of early type. At radio wavelengths the brighter ones of these regions may be traced anywhere in the galactic plane. Assuming a primeval He-abundance observed deviations from this primeval value may be explained by mass loss from stars during their evolved stages. This work has been carried out by Mezger and collaborators at Bonn.

Unfortunately, in order to interpret radio recombination lines in terms of abundances one must know the ratio of the volume occupied by ionized He to that occupied by ionized H. Model calculations which I carried out show that in the absence of dust the relation between this ratio and the effective temperature of the exciting star is a very simple one and independent of the particular density distribution of the gas around the star. But dust will generally be present and it has the unpleasant effect of complicating this relation and making it sensitive to the density distribution.

Additional information may be gained from the IR-data. Dust is believed to absorb stellar radiation and reemit it in the IR. Now one can construct a model of the ionized region which fits the observed radio and IR-data. The theoretical value of the model for the ratio of the ionized He to the ionized H-volume and the then readily derived He-abundance should not be too erroneous.

So a better knowledge of the absorbing properties of the dust and more radio and IR-measurements may help to determine the overall He-abundance.

DISCUSSION AFTER PAPER BY BOYARCHUK

Hesser to Boyarchuk: In your first slide you showed the diagram of Eggen and Sandage relating [Fe/H] to age. This diagram still shows a gap between the Population II clusters (47 Tuc) and the disc objects. Dr David Hartwick and I have investigated 3 previously unstudied southern globular clusters all of which appear to have metal abundances comparable to those of other disc objects (Hartwick, F.D.A. and Hesser, J.E.: *Astrophys. J.* **175**, 77, 1972). This leaves us with a possible dilemma, in that the giant branches of 47 Tuc and NGC 6352 do not overlap those of old open clusters like M67 or of field giant stars (Hartwick and Hesser *Publ. Astron. Soc. Pacific* **84**, 813, 1972), even though the line blanketing, as measured by broadband techniques, is the same for the globular cluster stars as for the disc stars.

Fowler to Krügel: Were you referring to the work of Mezger.

Krugel: Yes, I was asked to do some theoretical calculations on the role of dust.

Fowler: Are you saying that the fact that he observes no helium does not mean that there is none there?

Woolf: For H II regions near the galactic centre, you cannot resolve the infra-red emission of the H II regions from that of nearby molecular regions. However, we can separate these in the Orion nebula, and know from this that most of the infra-red emission comes from the molecular region not the H II region. Therefore in Orion the absorption of UV radiation by internal dust does *not* substantially affect the ionization of the H II region. Therefore, the best assumption which we can make for regions near to the galactic centre is that dust is not important there, in which case helium really is deficient.

Audouze: I think that the abundances deduced from the radio recombination lines must critically depend on the temperature of the regions in front of which the measurements are made. Unfortunately this parameter is not always well known.

Krügel: Temperature uncertainties by a factor of two do not affect the results much.

Fowler: Those who are convinced of a 'universal' He abundance should be warned that there are now sophisticated theories of the pre-baryon stage of the Friedmann universe which predict no production of He in the 'big bang'. R. Wagoner is speaking on these theories at Cracow this week.

FG SAGITTAE: THE s-PROCESS EPISODE*

(Abstract**)

G. E. LANGER,[†] ROBERT P. KRAFT, and KURT S. ANDERSON[††]

Lick Observatory, Board of Studies in Astronomy and Astrophysics, University of California, Santa Cruz, Calif., U.S.A.

The spectrum of the supergiant FG Sge has been studied from a series of high dispersion 120-in. coudé spectrograms obtained during the interval 1969–72, thus continuing the work of Herbig and Boyarchuk (1968). The star, of effective temperature about 6500 K in 1972, is cooling at the rate of 250 K yr^{-1}; it is known to have ejected a still visible planetary nebula some 6000 years ago (Flannery and Herbig, 1973). Abnormally strong absorption lines of Y II, Zr II, Ce II, La II and other s-process species began to appear in the spectrum of the central star some time after 1967 and have progressively strengthened. Present abundances per gram of these elements are about 25 times the solar value. There is little doubt that the atmosphere of the star has been enriched in these elements during the past seven years, but the rate of enrichment now appears to be slowing down.

The present evolutionary state of FG Sge, referred to as the Herbig-Boyarchuk (HB) phase, is discussed qualitatively on the basis of two model scenarios: (1) the HB phase is a post-planetary ejection episode associated with a He shell flash experienced by all stars with masses near 1 $\mathfrak{M}_\odot$; (2) the HB phase is a transient post-helium shell flash phenomenon in a more massive red giant, immediately preceding its transformation into a Ba- or S-type star. Observational and theoretical difficulties with each of these scenarios are discussed.

References

Flannety, B. and Herbig, G. H.: 1973, *Astrophys. J.* **183**, 491.
Herbig, G. H. and Boyarchuk, A. A.: 1968, *Astrophys. J.* **153**, 397.

* *Lick Observatory Bulletin*, No. 653, presented by Robert P. Kraft.
** This paper is only published in abstract as the full text can be found in *Astrophysical Journal*, May 1st 1974.
† Post-doctoral Fellow; on leave from Colorado College, Colorado Springs, Colo.
†† Now at Kitt Peak National Observatory.

DECIPHERING FG SAGITTAE*

(Abstract)

I.-JULIANA CHRISTY-SACKMANN and KEITH H. DESPAIN

California Institute of Technology, Pasadena, Calif., U.S.A.

Dr Kraft has just described a unique astronomical object – FG Sagittae – a new rosetta stone. It is our aim here to attempt to decipher this rosetta stone.

The four major observations on FG Sge, (i) the existence of a planetary nebula, (ii) the apparent brightness increase since the last century, (iii) the extremely fast, continued increase in spectral type, and (iv) the observed surface abundance increases with their respective time scales of ~ 6000, ~ 100, most likely 100, and 1 years, can all be explained in terms of helium-shell flashes. The last flash occurred on the order of 100 years ago, causing the spectral type change but essentially no intrinsic luminosity change. The mass of the carbon-oxygen core is $\sim 0.8\ M_\odot$, that of the helium-rich intershell region of the order of $4 \times 10^{-3}\ M_\odot$, and if the abundances of FG Sge indeed become close to that of the barium star ζ Cap, the envelope mass is less than $\sim 0.5\ M_\odot$, which assigns an upper limit to the present mass of FG Sge of 1.3 $M_\odot$.

The observed abundance increases with the present leveling-off can be interpreted by two types of convective behavior of 'convectively unstable' envelopes deepening into regions which had been enriched in *s*-process products in previous flashes. When the ratio of the masses of the originally unenriched to the enriched part of the mixing envelope is small, the leveling-off can be understood by a deepening of the 'convective' envelope only; when that ratio is large, there must be a link-up of the outer with an inner intershell convective zone. One major difficulty that still remains is which mechanism could produce 'convective' mixing in a supergiant envelope of spectral type as early as F.

Predictions are that lithium lines may show up in the next few years if a few conditions are met. No ^{13}C enhancement should be seen**.

* This paper was presented by I.-J. Christy-Sackmann and was supported in part by the National Science Foundation (GP-36687X, GP-28027).
** There have been revisions in this report since Warsaw.

DISCUSSION AFTER PAPER BY KRAFT

Baschek to Kraft: Is there any evidence of changes in the velocity field (of FG Sge), either shifts or different broadening of lines?

Kraft: There are some changes. The radial velocity is not quite constant. The observations are very difficult as it takes 8 h to get a coudé spectrum but I do not believe that the variations are surprising.

Paczyński to Kraft and Sackmann: I would like to point out that the properties of hydrogen and helium shell sources depend on the core mass and very little on the envelope mass. The same is true for the development of thermal pulses and the interpulse period. The difference between the models of Schwarzschild and Härm and those of Weigert was due to the difference of the core masses, 0.6 $M_\odot$ and 0.8 $M_\odot$ respectively. If you believe that the interpulse period for FG Sge was 6000 years, this indicates a core of 0.8 $M_\odot$, but says nothing about the initial envelope mass. In any case the present envelope mass must be very small otherwise FG Sge would not be able to make an excursion to the hot part of the HR diagram. I understand that the carbon enrichment depends on the intershell mass. This in turn depends on the core mass only.

Kraft: I should like to caution everyone that values such as 6000 yr must be regarded as very uncertain.

Schwarzschild to Kraft: I think we should congratulate Dr Kraft for the most important observations of FG Sge, which he has so perceptively and skillfully carried out. I suspect that it is very dangerous to overinterpret these new data rashly, however tempting. Nevertheless, I feel that we may be reasonably confident of two points. First, the timescale of the observed changes in radius strongly suggest that a shell flash is the basic cause of this remarkable phenomenon and, second, the rapid appearance of the *s*-process elements at the surface indicates that a shell flash can in fact produce these elements, as we had long expected on the basis of detailed flash computations.

Tayler (ed.), Late Stages of Stellar Evolution, 212. *All Rights Reserved.*
Copyright © 1974 by the IAU.

NUCLEI OF PLANETARY NEBULAE

C. R. O'DELL

George C. Marshall Space Flight Center, National Aeronautics and Space Administration, Ala., U.S.A.

Abstract. The nuclei of planetary nebulae are examined both observationally and theoretically. It is seen that the region occupied by these stars in the $\log T - \log L$ diagram is quite wide but consistent with a general progression of stars from high to low luminosity, with a noticeable but not large increase in luminosity during the early phase. The 'evolutionary path' is intrinsically quite wide and may indicate the evolution of stars under different conditions or non-monotonic passage along the mean path. Among the several theoretical approaches to this subject, only the double shell burning models seem to offer enough luminosity and short enough timescales to match the observations.

1. Introduction

The planetary nebulae present a unique opportunity for the study of the late stages of stellar evolution. They obviously represent a very late stage and are probably the immediate precursors of many, if not most, of the white dwarfs. The changes that occur involve most of the star, hence the timescales are not so short as to preclude their study – and, through the presence of the surrounding gas shell, they can be traced through an evolutionary sequence without seeing large changes in any one single star.

This field was given its modern beginning with the work of Shklovsky (1956) with his synthesis of the observational facts and their possible interpretation. Since that time, the observations have been markedly improved and have been subjected to progressive interpretation by many investigators (O'Dell, 1963; Seaton, 1966). At the same time, an even greater effort has been expended on theoretical understanding of the stars and their immediate precursors (Salpeter, 1971).

The presence of the nebular shell around the central stars of planetary nebulae provides the opportunity for their study that far exceeds that of other advanced states of stellar evolution. First, the extended image and its emission line spectrum cause the nebula to be detected with relative ease and, in fact, there are now more than 10^3 nebulae known (Perek and Kohoutek, 1967). Since the nuclei are very hot, they are relatively faint in the visual region; however, the gas absorbs much of the strong ultraviolet radiation from the stars and converts it to visual nebular emission.

Since most of the nebulae are optically thin, observations with slit spectrographs see both sides of the expanding shell, allowing accurate determinations of the expansion velocity and the system radial velocity (Wilson, 1950). The long standing issue of whether the motion was expansion or contraction has finally been resolved in favor of expansion by means of long time interval imaging that has detected the increase in angular size of several nebulae (Liller and Liller, 1968). When this expansion velocity is combined with the size of the nebula, it yields the characteristic time since the shell began expanding. There is no reason to doubt that the larger nebulae are simply more advanced states of the small nebula.

The highest surface brightness nebulae are clearly optically thick to hydrogen

Tayler (ed.), Late Stages of Stellar Evolution, 213–219. All Rights Reserved.
Copyright © 1974 by the IAU.

Lyman continuum radiation. This means that the brightness of the visual hydrogen recombination lines can serve as a count of ultraviolet photons. Combining this with the visual brightness of the central star gives a very wide wavelength base color-index from which the stellar temperature can be determined. Originated by Zanstra (1926), this method has been extended by Harman and Seaton (1966) to include the helium lines. As the nebula expands, the optical depth will decrease, making derived temperatures and luminosities only lower limits. However, if the stellar luminosity decreases rapidly enough, the nebula may again only become partially ionized and hence an accurate temperature indicator.

Since there are no stars close enough to allow trigonometric parallaxes, again the nebula must be used if one is to obtain accurate distances. Although bothersome assumptions must be made, allowing large individual errors, the average luminosities are probably known to a factor of two (O'Dell 1962).

In summary, we can say that it has been the nebula itself that has provided the major information for the study of the stars and it is therefore understandable why their significance did not become apparent until the nebulae themselves were thoroughly studied.

2. The Observed Evolution

The best way of illustrating the current observational picture of the central stars is by way of a plot of bolometric luminosity and temperature. Shown in Figure 1 is a plot used in 1967 at the Tatraska Lomnica Symposium (O'Dell 1968). Although some details have changed, the general picture remains the same. Basically this plot puts together all of the reliable data, using the best temperature method applicable and distances derived astrophysically or by independent means (Magellanic Clouds and M15). The probable error bars reflect the uncertainty in the calibrations, the atmospheric models used (blackbodies) and the probable errors of the parameters used. The gap at about $\log(L/L_\odot) = 3.3$ and $\log T_{star} = 5.1$ is artificial. Certainly stars lie in this region, but there is not a method that gives accurate temperatures since the nebulae are optically thin at this point. The lower limits that are derived for the optically thin nebulae indicate that this region is occupied.

There are systematic differences in the nebulae surrounding stars found in various parts of this diagram. The higher luminosity stars have smaller, higher surface brightness surrounding shells, although the shell velocities are similar. In addition, the spectra of the central stars vary, with the Wolf-Rayet type occurring at the high luminosities along with other strong line spectral types, while the low luminosity stars are hotter and more often continuous in their emission.

The most direct interpretation is that the central stars change in luminosity and temperature on a timescale of about 30000 years, that being about the time required for the nebulae to expand from the small to larger sizes. This is certainly a dramatic change, for it implies a change of stellar radius of a factor of 40 over this same time period. The smallest nebulae at the top are about 0.05 pc radius while the largest are near the bottom at about 0.5 pc.

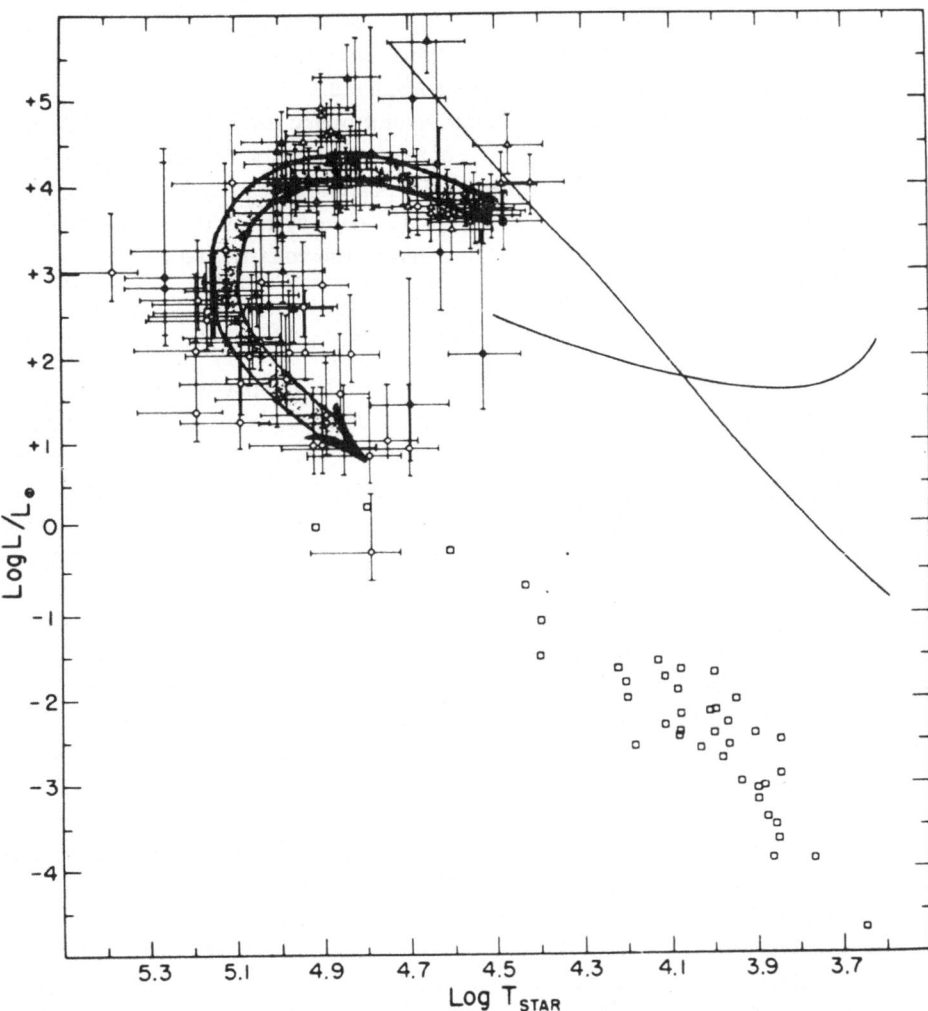

Fig. 1. The composite luminosity-temperature diagram for the nuclei of planetary nebulae and white dwarfs. Open triangles are nuclei in the Magellanic Clouds; filled triangles are optically thin objects with He II Zanstra temperatures; filled circles are optically thick nebulae; open circles are marginally optically thick; and the filled square is K648 in M15. The most probable average evolutionary path is shown superimposed on the basic data. The thin lines represent the initial main sequence and the horizontal branch.

The earliest stage of appearance is probably at about a factor of 3 lower than the peak luminosity and the stars first increase in luminosity while heating up, then monotonically decrease in brightness and then in temperature, finally blending in with the observed white dwarfs. Theoreticians have often mistakenly seized upon the early results of Harman and Seaton, indicating a very large and rapid rise in luminosity. The most complete data show that the average path is flatter at the onset and decreases in temperature at the lower end. An average evolutionary path is superimposed on Figure 1. In the interpretation of this diagram, both the average path and the obvious dispersion should be considered.

Although this smooth variation along an evolutionary track is probably accurate on the average, the issue does have several complications. In particular, it is not certain that only one shell is ejected. There are several planetary nebulae with double shells that cannot be explained otherwise. Perhaps in some systems the phenomenon of shell ejection occurs several times prior to entering the collapse sequence.

3. Theoretical Evolution

There are two families of approaches to the theoretical explanation of the central stars of the planetary nebulae. The first is that the stars are primarily remnant stellar cores, without nuclear fuel burning, going through the gravitational and thermal adjustment necessary for reaching the white dwarf state (Salpeter, 1971; Savedoff et al., 1969; Deinzer, 1967). The second approach is that nuclear burning processes are still important and that helium and hydrogen burning in outer shells occurs (Rose and Smith, 1970; Faulkner, 1968; Paczyński, 1971). Both of these approaches agree in arguing that the stars must be of sufficiently low mass to avoid carbon burning, although not all authors agree on the critical mass for this burning to occur ($1.04–1.3 \, M_\odot$).

Although arguments of the rate of field stars leaving the main sequence imply that the most common original mass for the system is about $1.3 \, M_\odot$, it is unlikely that all of this original mass is still in the star after passing through the horizontal branch stage, since lower masses are required to reconcile the helium burning models with observations. Therefore, a number of original masses can produce the required low mass stars that avoid carbon burning.

Low mass CO stars have been calculated (Beaudet and Salpeter 1969) and are shown in Figure 2 along with the observations. Basically, these are non-burning stars in adjustment. Generally the correct path is followed, but the temperatures are rather high for the required luminosity and the timescales are too long. The real stars must have some light element atmosphere and the theoretical calculations are very sensitive to only a few percent He atmosphere, although it seems unlikely that these models can be brought up to sufficient luminosity and short timescale.

Double shell burning stars have been calculated by Paczyński (1971) with greater success. In this case, he treats in detail the presence of the light element envelope as it is the potential source of most of the energy. The results of his calculations are shown in Figure 3. The timescales are indicated by marks at 10^4 and 10^5 years. The presence of the loops caused by thermal relaxations may be particularly relevant in producing scatter about the average path and to the double shell sources.

In detail, none of the theoretical calculations match the observations; but, the double shell-burning models can probably be matched to them when the problem of correctly handling the atmosphere is resolved. The set of calculations of Paczyński

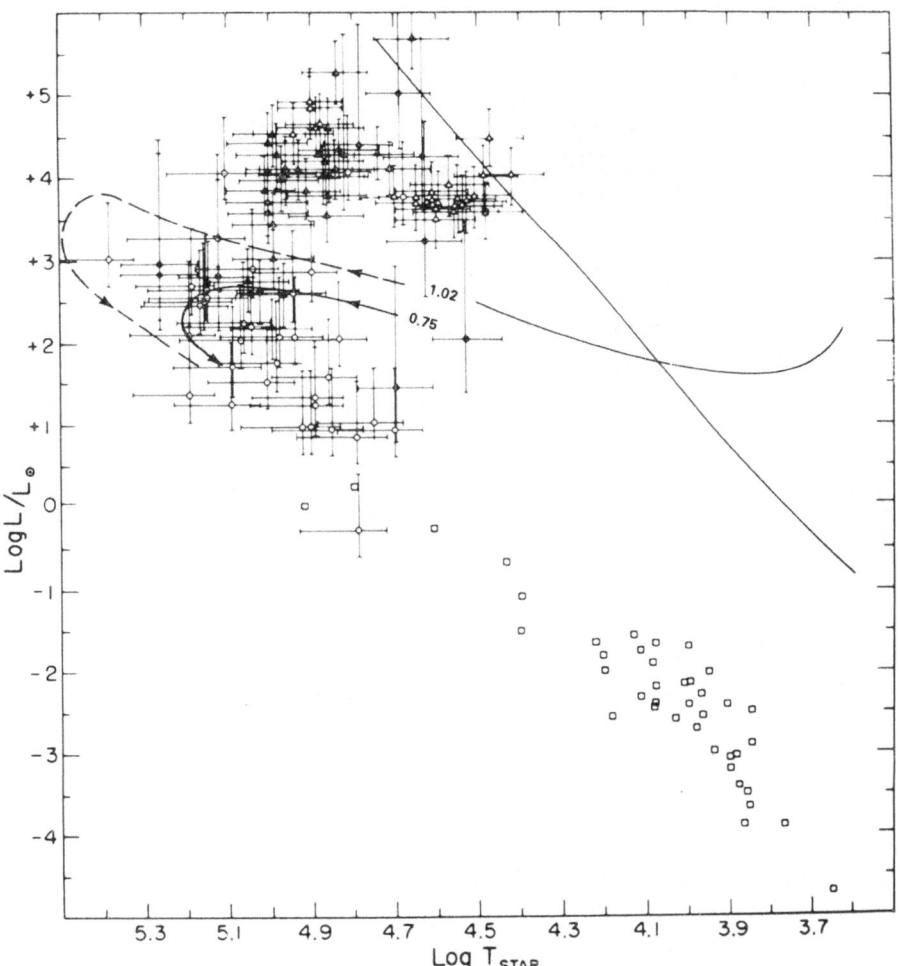

Fig. 2. The same as Figure 1 (without the average evolutionary path) but with the track of 1.02 and 0.75 $M_\odot$ carbon-oxygen stars superimposed. The calculated paths shown do not have light element atmospheres, which can alter the path very significantly.

(1971, Figure 2) of nearly exhausted stars with various fractional envelopes dramatically shows the sensitivity to the amount of atmosphere.

The nature of the star immediately prior and during shell ejection is even more uncertain. There are only a few general considerations that one can use for guidelines.

(1) The original star was probably quite large, approximately 200 $R_\odot$, since ejection from smaller stars would produce a much wider range of kinetic energies than is observed (Abell and Goldreich, 1966).

(2) The ejected shell has not been mixed with the inner star, since the ejected shells have helium abundances comparable to their initial values and the metal poor

C. R. O'DELL

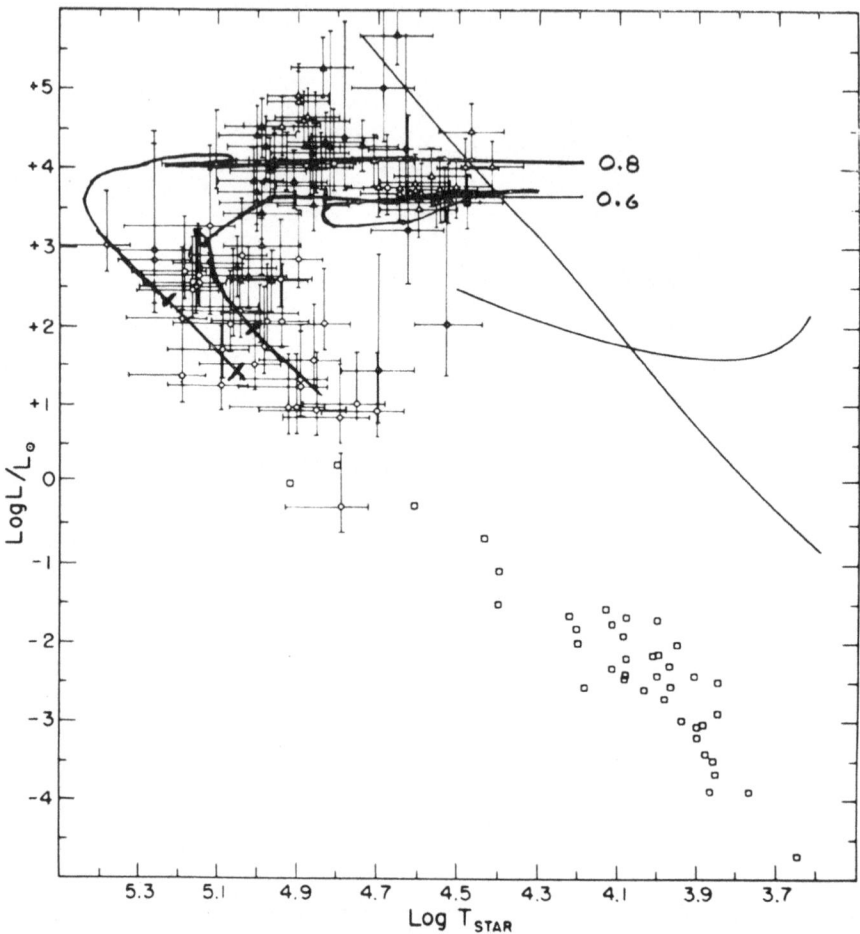

Fig. 3. The same as Figure 2 except that the evolutionary tracks of 0.6 and 0.8 $M_\odot$ shell burning stars as computed by Paczyński (1971) are shown.

globular cluster (M15) planetary has a low metal abundance (O'Dell *et al.*, 1964).

There are currently three approaches to obtaining the ejection: Radiation pressure acting on a star of extended radius (Faulkner, 1970); ionization-recombination equilibrium leading to instabilities; violent thermal relaxations accompanying helium-shell burning flashes (Rose and Smith, 1970). Combinations of these mechanisms may operate together to produce the effect.

In any event, it seems necessary to explain the presence of particulate matter in the ejected shell, which leads to strong thermal infrared emission (Neugebauer *et al.*, 1971). Most likely this is an indication of evolution from a cool-luminous star and requires pushing out the shell prior to the heating up of the star, which would destroy the particles. In addition, there is a growing body of evidence that some of the peculiar late-type stars are precursor planetary nebula systems. Of particular interest along these lines is V1016 Cygni, which has been studied in detail by Baratta *et al.* (1974). They have shown that this star is surrounded by a shell of material ejected earlier and that the central star has only recently ruptured its optically thick atmosphere (producing the late type stellar spectrum) to allow extensive photoionization to occur. We seem to have a similar situation in FG Sge which shows a fossil remnant planetary nebula surrounding a star with an expanding atmosphere possibly representing another shell ejection; but, at a greater time interval than V1016 Cyg. The existence of such a fossil remnant requires that the star did move beyond $T = 30000$ K following the first ejection. Such features together with the observed, resolved double nebulae agree quite well with the loops described by the Paczyński tracks and generally with the multiple thermal relaxations calculated by Rose.

Certainly, more needs to be done both observationally and theoretically. In particular, the burden now seems to be on the observer to study those systems that may be the immediate precursors. In addition, there is the very real possibility of determining the M/R ratio by means of the gravitational redshifts to the central stars by measuring their apparent radial velocities with respect to the nebula. One should not underestimate either the difficulties or advantages of such measurements.

References

Abell, G. O. and Goldreich, P.: 1966, *Publ. Astron. Soc. Pacific* **78**, 232.
Baratta, G. B., Cassatella, A., and Viotti, R.: 1974, *Astrophys. J.* **187**, 651.
Beaudet, G. and Salpeter, E. E.: 1969, *Astrophys. J.* **155**, 203.
Deinzer, W.: 1967, *Z. Astrophys.* **67**, 342.
Faulkner, D. J.: 1968, *Monthly Notices Roy. Astron. Soc.* **140**, 223.
Faulkner, D. J.: 1970, *Astrophys. J.* **162**, 513.
Harman, R. J. and Seaton, M. J.: 1966, *Monthly Notices Roy. Astron. Soc.* **132**, 15.
Liller, M. H. and Liller, W.: 1968, in D. E. Osterbrock and C. R. O'Dell (eds)., 'Planetary Nebulae' *IAU Symp.* **34**, 38.
Neugebauer, G., Becklin, E., and Hyland, A. R.: 1971, *Ann. Rev. Astron. Astrophys.* **9**, 67.
O'Dell, C. R.: 1962, *Astrophys. J.* **135**, 371.
O'Dell, C. R.: 1963, *Astrophys. J.* **138**, 67.
O'Dell, C. R., Peimbert, M., and Kinman, T. D.: 1964, *Astrophys. J.* **140**, 119.
O'Dell, C. R.: 1968, in D. E. Osterbrock and C. R. O'Dell (eds.), 'Planetary Nebulae', *IAU Symp.* **34**, 361.
Paczyński, B.: 1971, *Acta Astron.* **21**, 417.
Perek, L. and Kohoutek, L.: 1967, *Catalogue of Galactic Planetary Nebulae, Academia*, Praha.
Rose, W. K. and Smith, R. L.: 1970, *Astrophys. J.* **159**, 903.
Salpeter, E. E.: 1971, *Ann. Rev. Astron. Astrophys.* **9**, 127.
Savedoff, M. P., Van Horn, H. M., and Vila, S. C.: 1969, *Astrophys. J.* **155**, 221.
Seaton, M. J.: 1966, *Monthly Notices Roy. Astron. Soc.* **132**, 113.
Shklovsky, I. S.: 1956, *Astron. J. Soviet Union* **33**, 315.
Webster, B. L.: 1969, *Monthly Notices Roy. Astron. Soc.* **143**, 113.
Wilson, O. C.: 1950, *Astrophys. J.* **111**, 279.
Zanstra, H.: 1926, *Phys. Rev.* **27**, 644.

DISCUSSION AFTER PAPER BY O'DELL

Ostriker to O'Dell: I do not comprehend how, in the non-burning models for central stars, the addition of a small (non-burning) envelope can alter either the luminosity or the duration significantly. I would have thought that only the radius and hence the temperature would be altered. Thus, discrepancies between the calculated and observed durations may remain a serious problem for the non-burning models. Is my understanding correct?

O'Dell: This is a feature of the work of Salpeter which, I must confess, I do not fully understand.

Paczyński: Was the hydrogen shell ignited?

O'Dell: I think not.

Sugimoto: When we add the hydrogen-rich envelope, the star moves rightwards in the HR diagram, keeping its luminosity constant. As I understand it, this is the reason why the luminosity seems to increase if we compare the two evolutionary tracks at a given effective temperature.

Beaudet: Adding a thin hydrogen shell on top of an homogeneous carbon star does not change the total luminosity but only the effective temperature. It moves the whole track to the right in a HR diagram. Of course, if one adds too much hydrogen, then hydrogen will burn and the above mentioned scheme does not work.

Friedjung: Would determination of the gravitational red shift of the central star be sensitive to asymmetries of the nebula?

O'Dell: I think not, but the problem would be resolved by studying the velocities of different parts of the nebula.

Schwarzschild to O'Dell: Would we commit a very great sin against the observational evidence if we did not consider the 'Harman-Seaton Track' as an evolutionary track but as a patch through which a whole variety of tracks evolve? Could we consider the right hand edge as determined by the condition of the nucleus becoming blue enough to illuminate the nebula and the left hand edge by the condition that the nebula is becoming too dispersed rather than that the nucleus is turning down in luminosity?

O'Dell: I do not think that it is correct to apply such an interpretation. If you look at the average radii of shells about stars of different luminosity, it seems that, even if there are many paths, the general features of the track are clear.

Schwarzschild: If one relaxes the assumption of all nebular stars having the same mass, could you not shift the lower end appreciably?

Tayler (ed.), Late Stages of Stellar Evolution, 220–221. All Rights Reserved.
Copyright © 1974 by the IAU.

O'Dell: I think not as we get the same features by studying both nearby nebulae and others for which we can determine distance and luminosity.

Tayler to O'Dell: The observed density of stars in the diagram which you showed does not agree with the timescale of evolution. Is this due to observational selection?

O'Dell: Yes.

WHITE DWARFS

S. C. VILA

Dept. of Astronomy, University of Pennsylvania, Philadelphia, Pa., U.S.A.

Abstract. The effects of core crystallization and of convection in the envelope on the cooling of white dwarfs are reviewed. The case of a 0.6 $M_\odot$ white dwarf composed of an oxygen core and a helium envelope is taken as example. Also the amount of hydrogen that a white dwarf can accrete before nuclear burning occurs is estimated and possible evolutionary relations between white dwarfs of types DA, DB and DC, as advanced by Baglin and Vauclair, are presented.

1. Introduction

I will review the classical theory of white dwarf cooling and its modifications due to core crystallization and convection in the envelope, the question of how much hydrogen a white dwarf can accrete before nuclear reactions cause a thermal runaway, and finally a possible evolutionary relation between white dwarfs of types DA, DB and DC.

2. Classical Theory

The classical theory of cooling for white dwarfs was proposed by Mestel (1952). It is also found in Schwarzschild (1958) and Weidemann (1968). It considers the star to be composed of an isothermal core and a radiative envelope. The opacity of the envelope is given by Kramer's law $K = K_0 \varrho T^{-3.5}$, and matter obeys the perfect gas law $P = (k/\mu H) \varrho T$. Since the mass and the luminosity are essentially constants in the envelope the stellar structure equations are reduced to the two equations:

$$\frac{dP}{dr} = -\varrho \frac{GM}{r^2} \quad \text{and} \quad \frac{dT}{dr} = -\frac{3}{4ac} \frac{K\varrho}{T^3} \frac{L}{4\pi r^2}.$$

These two equations are integrated with the boundary conditions $P = T = 0$ and the opacity and pressure laws given. One obtains two relations:

$$P = AT^{4.25} \quad \text{and} \quad \varrho = BT^{3.25}, \tag{1}$$

where A and B depend on M, L, and other parameters.

This envelope joins a degenerate core at the point where the pressures of both are equal

$$\frac{k}{\mu_e H} \varrho T_t = K_1 \left(\frac{\varrho}{\mu_e}\right)^{5/3}. \tag{2}$$

Substitution of (2) into (1) gives a condition for T_t of the form

$$L = aT_t^{3.5}. \tag{3}$$

If all the thermal energy is contributed by the ions and those are considered as a perfect gas, the cooling law is

$$L = -\frac{\mathrm{d}}{\mathrm{d}\tau}\left(\frac{3}{2}\frac{kT}{\mu_i H} M\right),$$

(4)

where μ_i is the ion molecular weight, H the unit of atomic weights and τ is time. Substitution of (3) into (4) and integration with the initial condition $\tau = 0$ at $T_t = \infty$ gives a cooling time

$$\tau = b\left(\frac{M}{L}\right)^{5/7}.$$

(5)

Significant deviations from the classical theory are due to crystallization of the ions and convection in the envelope.

3. Ion Crystallization

The theory is found in Salpeter (1961), Van Horn (1968), and Ostriker (1971). The properties of the ion component depend on the parameter Γ defined by

$$\Gamma = \frac{1}{kT}\frac{(Ze)^2}{R} = 2.28\,\frac{Z^2}{A^{1/3}}\frac{\varrho_6^{1/3}}{T_7},$$

(6)

where ϱ_6 is the density in units of 10^6 gm cm^{-3} and T_7 the temperature in units of 10^7 K.

The parameter Γ is the Coulomb energy between one ion and a sphere of uniformly distributed negative charge containing Z electrons divided by kT. If $\Gamma \gg 1$ the ions arrange themselves in a crystalline lattice. If $\Gamma \ll 1$ they form a perfect gas.

The specific heat of a crystalline solid is given by Debye's theory. At high temperatures it is twice the value corresponding to the gaseous case because free particles have three degrees of freedom and bound ones six. At low temperatures the high modes of oscillation of the lattice are not excited, and the specific heat drops in value. The specific heat per ion is given in all cases by

$$C_v = 3kD\,(\Theta/T)$$

(7)

where Θ is the Debye temperature given by

$$\Theta = 1.74 \times 10^3 \left(\frac{2Z}{A}\right)\varrho^{1/2}$$

(8)

The function D tends to unity for $T \gg \Theta$ and varies as $(T/\Theta)^3$ for $T \ll \Theta$.

The decrease in C_v at $T \ll \Theta$ causes a marked drop in the heat capacity of white dwarfs of low luminosity.

At intermediate temperatures, $\Gamma \sim 1$, we have a partially ordered state that may

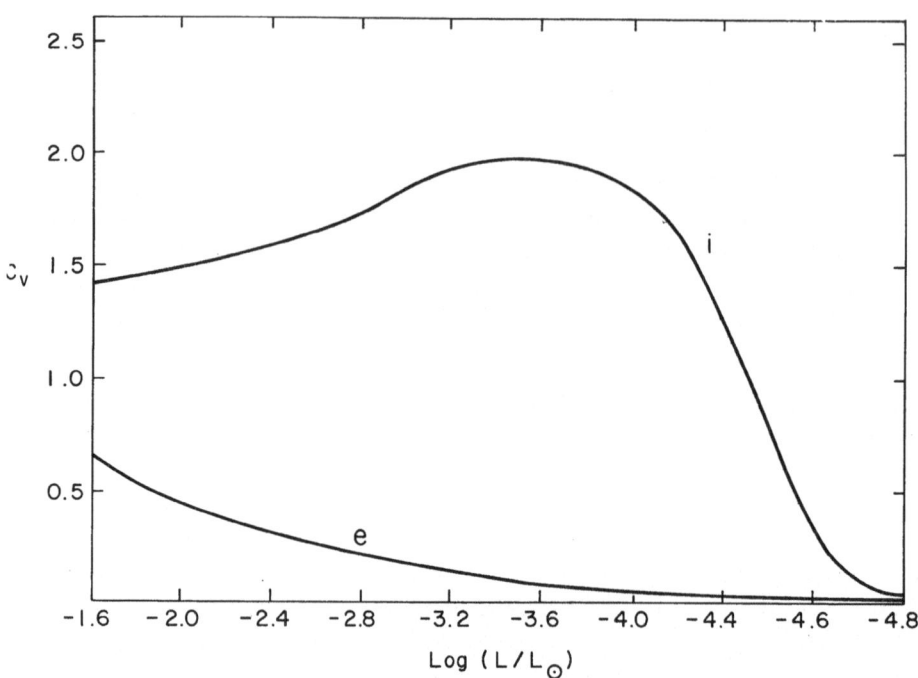

Fig. 1.   Specific heat C_v of an ion (i) and Z electrons (e) for the half mass point of
0.6 $M_\odot$ white dwarf.

be compared to a liquid. Its properties have been studied by Brush *et al.* (1966).
Their values for the specific heat show a gradual transition from the perfect gas
values to the crystal values as Γ increases.

In Figure 1 I present C_v for one ion measured in units of $3k/2$. The values correspond
to the half mass point of a 0.6 $M_\odot$ white dwarf as it evolves. This star, described later
in the text, has an oxygen core and a helium envelope. It is to be noted that there is
an increase to about twice the perfect gas value followed by a drop at low temperatures.
The specific heat of Z electrons is also given.

4. Envelope Convection

The idea of convection in white dwarf envelopes was suggested by Schatzman (1958).
Convection is caused by the decrease in the adiabatic gradient due to partial ionization and the high values of the radiative gradient produced by high opacities. Calculation of this envelope involves the evaluation of the ionization by Saha's equation
including the treatment of pressure ionization and also accurate estimation of the
opacities. Helium envelopes have been calculated for a 0.6 $M_\odot$ star by Böhm (1970)
and Böhm and Casinelli (1971) and hydrogen and helium envelopes for stars of
0.32, 0.57, 0.89 and 1.17 $M_\odot$ by Koester (1972).

In Figure 2 the maximum temperatures in the helium envelopes for a 0.6 $M_\odot$
star by Böhm and a 0.57 $M_\odot$ star by Koester are plotted versus effective temperature.
The agreement is remarkable. It turns out from these results that convection is important for these stars only at temperatures below 15000 K.

5. White Dwarf Evolution

Using the envelopes given by Böhm, I have calculated the evolution of a 0.6 $M_\odot$
white dwarf composed of an oxygen core and a helium envelope. The details of the
model construction are published (Vila, 1971). This evolution includes both the
effects of ion crystallization and core convection and shows the combined influence
of both in shortening the cooling times in advanced stages of white dwarf evolution.

In Figure 2 the upper solid curve gives the central temperature as function of the
effective temperature. The straight line represents the corresponding relation in the
classical treatment. Using Kramer's law for opacity one obtains the proportionality
$T_c \propto T_e^{8/7}$. As can be seen in the figure, this law applies at high temperatures but at
low ones, T_c is lower than the classical value. Hence, for a given central temperature,
convective envelopes require a higher effective temperature than radiative ones and
a correspondingly higher luminosity. The rate of cooling is increased by convection.

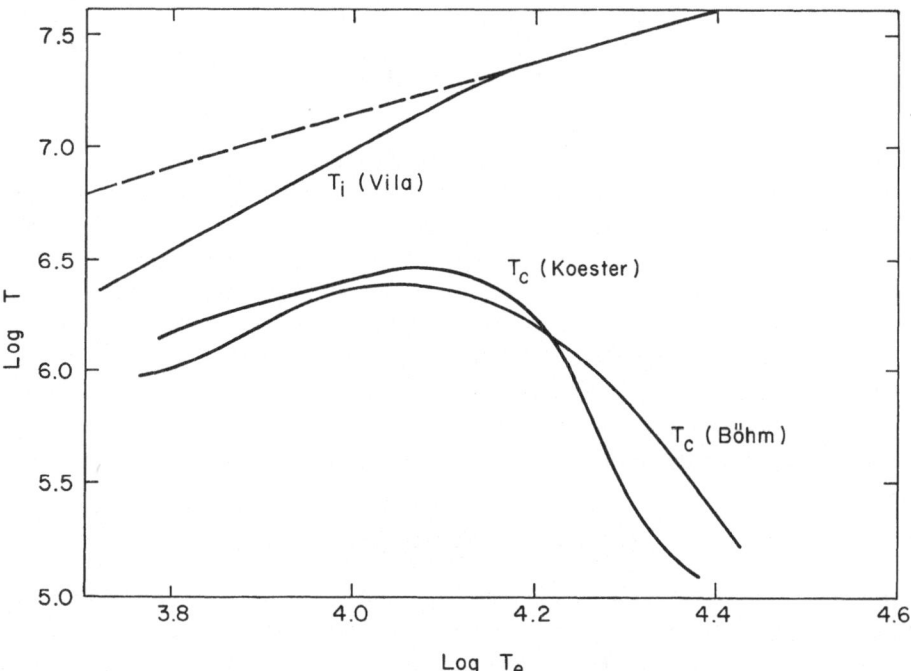

Fig. 2. Maximum temperatures of convective envelopes by Böhm and Koester. At top central
temperatures of models by Vila including convection and of classical models without convection.

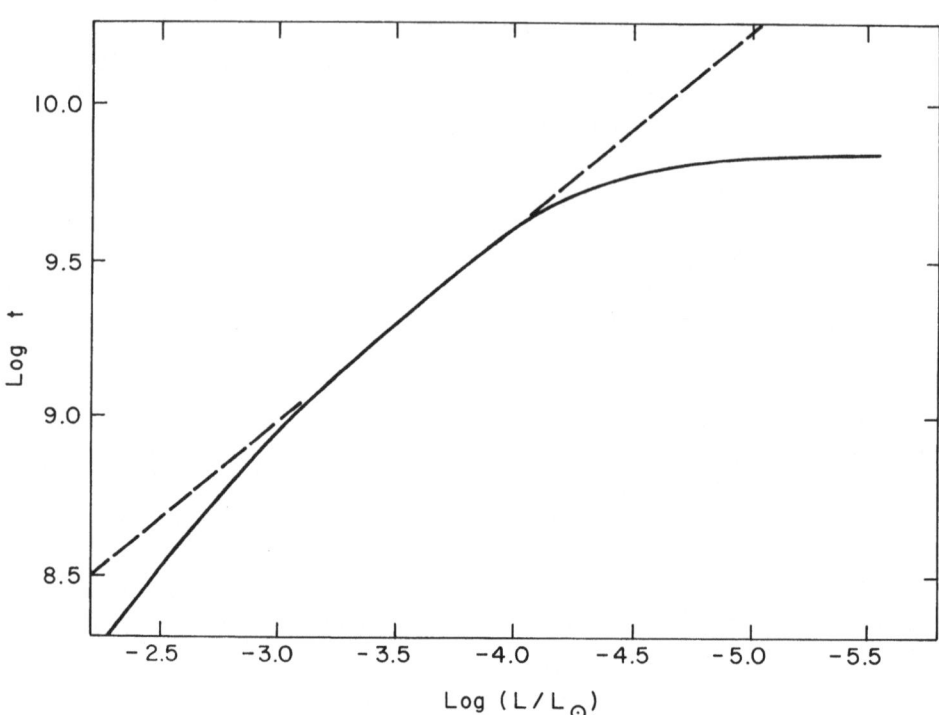

Fig. 3.   Cooling times of a 0.6 $M_\odot$ white dwarf with crystallization and envelope convection compared to the ones in the classical case.

In Figure 3 I give the cooling time in years vs the luminosity in solar units. The
straight line gives the corresponding quantity in the classical case with Kramer's
opacity, that is $\tau \propto L^{-5/7}$. The discrepancy at small times is due to a different time
origin for the two cases. The discrepancy at large times is significant and shows
a rapid cooling in the real case as compared with the idealized one. This has also
been found by Ostriker and Axel (1969). It is due both to the drop in the ion specific
heat and to envelope convection. The mass of the star is the average mass of white
dwarfs. The evolution time was always under 7×10^9 years. This is less than the current
estimates of the age of the universe and supports the suggestion by Weidemann (1968)
that cold degenerate dwarfs may contribute significantly to the local mass density in
the Galaxy.

6. Accreted Hydrogen Envelopes

A problem arises in close binaries in which one star is a white dwarf receiving matter
from its companion: how much matter can be transferred before the ignition of
nuclear reactions causes a thermal runaway? To answer this question I extended the
models by Koester for hydrogen envelopes of white dwarfs of 0.32, 0.57, 0.89 and
1.17 $M_\odot$ and several effective temperatures until nuclear reactions produce energy

equal to the luminosity of the star at the given effective temperature. This is certainly an upper limit to the hydrogen mass.

In this integration the partial degenerate pressure of the electrons has been treated as in Chandrasekhar (1939) and numerically evaluated from the table by Grasberger (1961). The opacities used were from Cox and Stewart (1970) that were extended to higher densities by the tables for conductive opacities by Hubbard and Lampe (1969). The nuclear energy rates for hydrogen burning were taken from Cox and Guili (1968).

In Figure 4 I present the temperatures at the bottom of the hydrogen convective

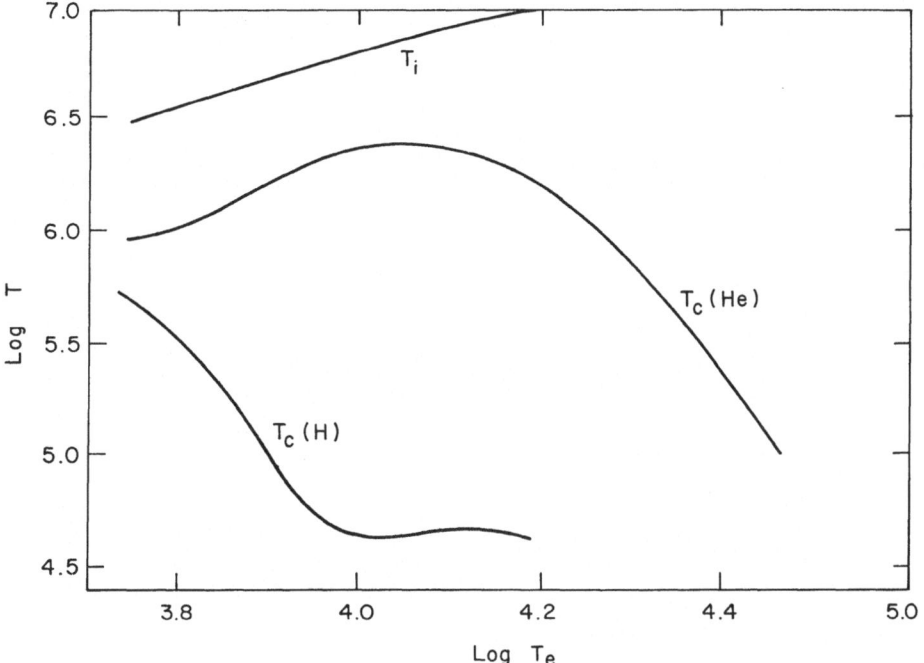

Fig. 4.   Maximum temperature for hydrogen and helium convection zones and maximum envelope temperature for a 0.57 $M_\odot$ white dwarf.

zone and similar quantities for the case of a helium one. The upper curve gives the temperature of the inner boundary of the hydrogen envelope.

In Figure 5 the envelope mass is given as a function of the effective temperature for the 0.57 $M_\odot$ star. The results show that this quantity does not change very much with effective temperature. In Figure 6 the envelope mass is a function of the star's mass. The effective temperature is 10000 K. It shows that the envelope mass strongly increases with decreasing star mass.

What is the future evolution of a white dwarf that has ignited a hydrogen shell and is still accreting more hydrogen? This is an interesting question that will require detailed calculation. Two alternatives seem possible. Either part of the hydrogen will

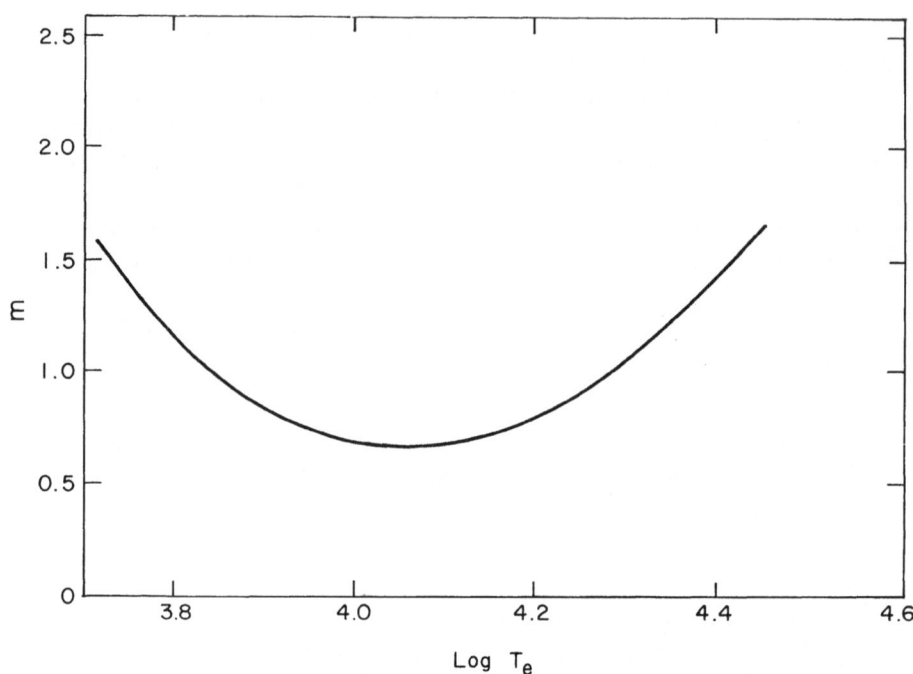

Fig. 5.   Maximum mass of hydrogen envelope for a 0.57 $M_\odot$ white dwarf as function
of effective temperature.

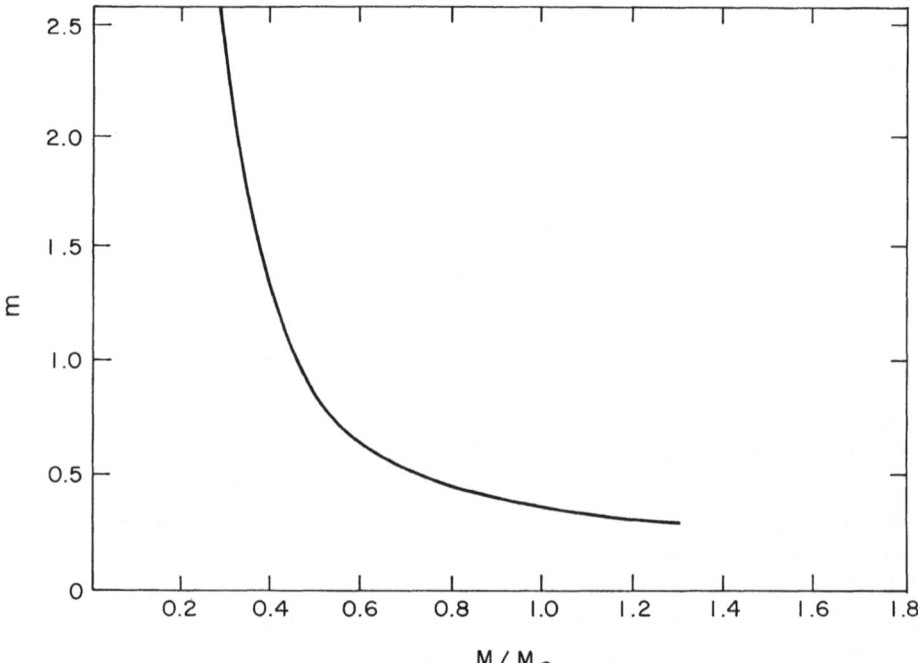

Fig. 6.   Maximum mass of hydrogen envelope for a white dwarf of $T_e = 10000\,K$ as a function
of its mass.

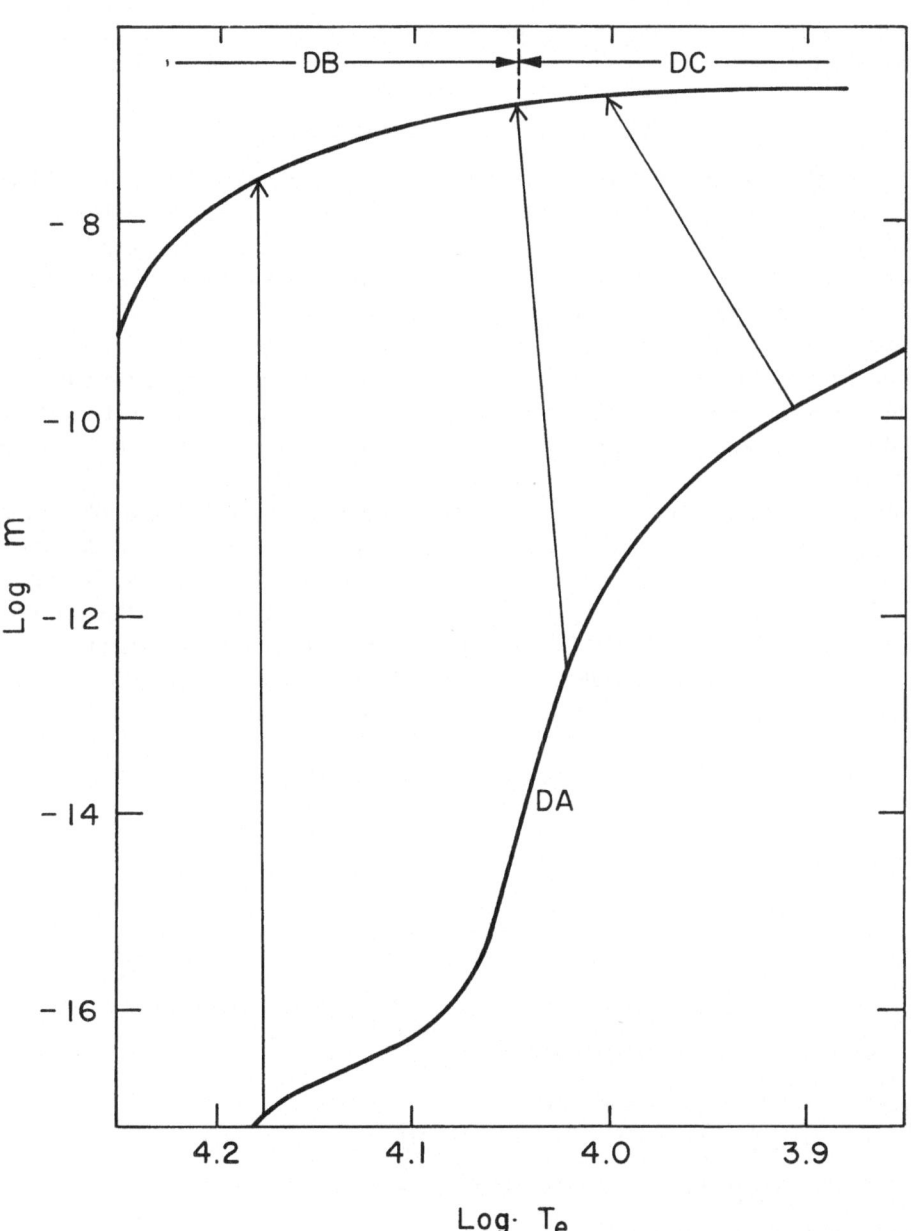

Fig. 7   Schematic evolution through the different spectral types by mixing according to Baglin and Vauclair. See text for explanation.

be ejected in a shock wave and the process will repeat itself or the thermal runaway will take the star to the region of the red giants burning hydrogen in a shell.

7. White Dwarfs of DA, DB and DC Types

The possible interrelations between white dwarfs of types DA, DB, and DC have been considered by Strittmatter and Wickramasinghe (1971), Shipman (1973), Sion (1973), and possibly others. I will only give a brief review of a recent work by Baglin and Vauclair (1973) that was sent to me in preprint form as I was preparing this paper.

Baglin and Vauclair start from the consideration that there exist two classes of white dwarfs; the DA's, that show strong hydrogen features and the rest that show no hydrogen. It is then reasoned that if there is hydrogen left in the star it will show unless it is mixed by convection with the inner layers. They constructed envelopes of different effective temperatures and amounts of hydrogen above a helium envelope. The helium is convective. These envelopes are represented in Figure 7. For a given hydrogen mass a DA star cools to an effective temperature at which the whole mass of hydrogen becomes convective. This temperature is plotted in the lower curve. Mixing of the hydrogen with the helium produces an envelope of almost pure helium. The mass of these helium envelopes is given by the upper curve. The vertical lines show the transition by mixing from a DA to DB or DC according to the initial hydrogen content.

In this work accretion is neglected and it is considered that the gravitational energy liberated by accretion would produce enough energy to cause a coronal outflow preventing accretion from taking place.

References

Baglin, A. and Vauclair, G.: 1973, *Astron. Astrophys.* **27**, 307.
Böhm, K. H.: 1970, *Astrophys. J.* **162**, 919.
Böhm, K. H. and Cassinelli, J.: 1971, *Astron. Astrophys.* **12**, 21.
Brush, S. G., Sahlin, H. L., and Teller, E.: 1966, *J. Chem. Phys.* **45**, 2102.
Chandrasekhar, S.: 1939, *An Introduction to the Study of Stellar Structure*, University of Chicago Press, Chicago.
Cox, J. P. and Giuli, R. T.: 1968, *Principles of Stellar Structure*, Gordon and Breach, New York.
Cox, A. N. and Stewart, J. N.: 1970, *Astrophys. J. Suppl.* **19**, 261.
Grasberger, W. H.: 1961, *U.C.R.L. Report*, No. 6196.
Hubbard, W. B. and Lampe, M.: 1969, *Astrophys. J. Suppl.* **18**, 297.
Koester, D.: 1972, *Astron. Astrophys.* **16**, 459.
Mestel, L.: 1952, *Monthly Notices Roy. Astron. Soc.* **112**, 583.
Ostriker, J. P.: 1971, *Ann. Rev. Astron. Astrophys.* **9**, 353.
Ostriker, J. P. and Axel, L.: 1969, in S. Kumar (ed.), *Low Luminosity Stars*, Gordon and Breach, New York, p. 357.
Salpeter, E. E.: 1961, *Astrophys. J.* **134**, 669.
Schatzman, E.: 1958, *White Dwarfs*, North-Holland Publ. Co., Amsterdam.
Schwarzschild, M.: 1958, *Structure and Evolution of the Stars*, Princeton University Press, Princeton.
Shipman, H. L.: 1972, *Astrophys. J.* **177**, 723.
Sion, E. M.: 1973, *Astrophys. Letters* **14**, 219.
Strittmatter, P. A. and Wickramasinghe, D. T.: 1971, *Monthly Notices Roy. Astron. Soc.* **152**, 47.
Van Horn, H. M.: 1968, *Astrophys. J.* **151**, 227.
Vila, S. C.: 1971, *Astrophys. J.* **170**, 153.
Weidemann: 1968, *Ann. Rev. Astron. Astrophys.* **6**, 351.

THE SOURCES OF ENERGY IN WHITE DWARFS*

(Abstract)

E. V. CHUBARIAN, G. S. SAHAKIAN, and D. M. SEDRAKIAN

Erevan State University, U.S.S.R.

In the theory of the internal structure of white dwarfs it is assumed that there is no other essential source of energy than thermal energy. The thermal energy is about $3 \times 10^{39} T$ erg. When the temperature is 10^7 K, this energy is enough to supply a luminosity of the order of 10^{32} erg s^{-1} for about 10^6 yr.

In our group Sahakian and Avakian (1972) have shown that in white dwarf interiors there must exist some nuclear energy which is connected with the transformation of medium and heavy nuclei into those with more stable nuclear properties defined by the Fermi energy of the degenerate electron gas. This energy can be greater than the thermal energy by one or two orders of magnitude. Perhaps this energy plays an important role in the evolution of these stars.

Now we discuss another source of energy which is connected with the rotation of white dwarfs. Besides the kinetic energy of rotation, the star contains potential energy of deformation. Calculations of the internal structure of rotating white dwarfs show that the maximum angular velocity is of order 1 s^{-1} and that the deformation is of order of 10 to 20% of the radius of the star. It is easy to show that the additional deformation energy is $\Delta E \sim (GM^2/R) (\Delta R/R)$, where M is the mass of the star and R is its radius. If we take $GM^2/R \sim 2 \times 10^{50}$ and $\Delta R/R$ as 20%, then $\Delta E \sim 4 \times 10^{49}$ erg. The exact calculation of the deformation energy gives $W_g = \Delta Mc^2 - W_r = 8.96 \times 10^{-29} N - W_r$, where N is the total number of baryons in the star and $W_r = I\Omega^2/2$ is the kinetic energy of rotation. If we assume that the main source of energy is connected with rotation, then, as through its evolution the angular velocity decreases and the star contracts, gravitational energy of deformation will be released in the volume of the star, while the kinetic energy of rotation is released in the outer region of the star.

Such ideas bring us to the calculations of white dwarf models with internal sources. For solving this problem we need the distribution and temperature of the release of this energy. As a first approximation we assumed that the energy is released at a constant rate and distributed through the star by the law $C\varrho\Omega^2 r^2$, where C is a constant which can be calculated, if we know the rate of release of the energy. Thus the rate of release of energy is one free parameter in the problem. Our calculations show that this store of energy is enough to supply the luminosity of the order of a solar luminosity or more for 10^8 to 10^9 yr.

Reference

Sahakian, G. S. and Avakian, R. M.: 1972, *Astrofizika* **8**, 123.

* This paper was presented by E. V. Chubarian.

Tayler (ed.), Late Stages of Stellar Evolution, 231. *All Rights Reserved.*
Copyright © 1974 *by the IAU*.

DEGENERATE DWARFS WITH LIQUEFYING CORES

(Abstract)

UMBERTO DE ANGELIS

Astronomical Observatory, Naples, Italy

The problem of phase transitions in hot white dwarf matter has been recently investigated and the results applied to follow the cooling of degenerate dwarfs whose core undergoes liquefaction and to build liquefying sequences on the HR diagram.

Gas-liquid phase transitions are found to occur at $\Gamma \simeq 1$ and the transition points are found to be well fitted by the rule

$$T_L = 1.4 \times 10^5 Z^{5/3} \varrho^{1/3}, \tag{1}$$

where T_L is the liquefaction temperature, ϱ the corresponding density and Z the chemical composition.

Equation (1) gives a liquefaction temperature about 100 K higher than the melting temperature (for the same ϱ, Z) as given by Lindemann's melting rule

$$T_m = 3.7 \times 10^3 Z^{5/3} \varrho^{1/3}$$

so that crystallization will set in at $\Gamma \simeq 100$.

Assuming a star model with a degenerate, hot, isothermal core (ϱ, T, Z in the liquefying range) and an envelope acting as a thermal blanket, an opacity was derived in the form:

$$k = k_0 \varrho^n T^{-s}$$

and then the luminosity law for such a model could be written in the form

$$L = L_0 M \varrho_C^{-0.4} T_C^{3.4}, \tag{2}$$

where the subscript C indicates central values and M is the star's mass.

By combining Equations (1) and (2) a luminosity law for stars with liquefying cores results

$$L = L_0' M Z^{1.9} T_C^{1.45}. \tag{3}$$

For such models a central temperature-effective temperature relation was also obtained

$$T_e = \text{constant} \times M^{5/12} (ZT_c)^{1/2}. \tag{4}$$

Equations (3) and (4) were then used to build liquefying sequences on the HR diagram and these came out to be in the region of the diagram dense with observed nuclei of planetary nebulae.

Tayler (ed.), Late Stages of Stellar Evolution, 232–233. All Rights Reserved.
Copyright © 1974 by the IAU.

Further evidence suggesting a possible identification of such models as central stars of planetary nebulae and an overall evolutionary picture from the nebula ejection to a crystalline white dwarf has been considered in (3) and (4).

References

De Cesare, L., Forlani, A., and Platania, G.: 1973, *Astrophys. Space Sci.*, **21**, 461.
De Angelis, U., De Cesare, L., Forlani, A., and Platania, G.: 1973, *Astrophys. Space Sci.* **20**, 875.
De Angelis, U. De Cesare, L., Forlani, A., and Platania, G.: 1973, *Nature Phys. Sci.* **244**, 133.
De Angelis, U., De Cesare, L., Forlani, A., and Platania, G.: 1973, *Astrophys. Space Sci.* **20**, 886.

EVIDENCE FOR g-MODE OSCILLATIONS IN THE
PECULIAR BLUE VARIABLE CD −42°14462*

(Abstract)

J. E. HESSER, B. M. LASKER, and P. S. OSMER

Cerro Tololo Interamerican Observatory**, La Serena, Chile

CD −42°14462 is a star of peculiar spectral properties first noted by Bond and Landolt (1971). The spectroscopic evidence indicates that the star has broad shallow lines of both H and He, features that suggested that it was a binary composed perhaps of two degenerate stars. Photometric variability was discovered by us in 1972, but no coherent variations were discerned. Following Warner's (1973) observation in April 1973 of a 0.003 mag. variation at 29s, a reexamination of our data revealed two variations present with periods of 29.08 s and 30.15 s and mean amplitudes of 0.00028 mag. Variable Hα emission and Hβ and Hγ absorption lines were observed and the continuum colours place it near the 12000 K blackbody line.

The observations are consistent with a model invoking g-mode oscillations on a hot, rotating white dwarf. Whether the oscillating star is a member of a binary system is still somewhat problematical, in view of Greenstein's (1973) observations of apparently single degenerate stars with both H and He lines.

References

Bond, H. E. and Landolt, A. U.: 1971, *Publ. Astron. Soc. Pacific* **83**, 485.
Greenstein, J. L.: 1973, private communication.
Warner, B.: 1973, *Monthly Notices Roy. Astron. Soc.* **163**, 25P.

* This paper was presented by J. E. Hesser.
** Operated by the Association of Universities for Research in Astronomy, Inc., under contract with the National Science Foundation.

CIRCULATION IN DIFFERENTIALLY
ROTATING WHITE DWARFS*

(Abstract)

C. MÖLLENHOFF and R. KIPPENHAHN

Universitäts Sternwarte, Göttingen, F.R.G.

The Eddington-Vogt circulation in differentially rotating white dwarfs has been investigated. Although these stars are nearly isothermal, rather high circulation velocities occur. The reason for this is the appearance of $\delta = (\partial \ln \varrho / \partial \ln T)_P$ in the denominator of the formula for the circulation velocity (c.f. the invited paper by Kippenhahn). With increasing degeneracy δ goes to zero and the circulation velocity increases. Estimates show that the angular momentum of the white dwarf is redistributed within its cooling time. This has to be taken into account if one considers the evolution of rapidly rotating degenerate objects.

* This paper was presented by C. Möllenhoff and will be published in full in *Astrophys. Space Sci.*

DISCUSSION AFTER PAPER BY VILA

Faulkner to Hesser: Can you tell us whether there is any indication of emission features in CD −42°14462?

Hesser: Yes; Hα is in emission and Hβ and Hγ have weak variable absorption.

Ostriker to Vila: Is it certain that accreting white dwarfs cannot burn H stably? Is it not possible if H is accreted sufficiently rapidly? Has this been tried?

Vila: No calculations have been made but I understood that it would not work.

Ostriker: A thermal runaway will occur if burning is in a degenerate region but possibly not if the region is non-degenerate. Such a non-degenerate zone may be produced if the accretion rate is high enough.

Vila: This should be calculated.

Paczyński to Vila: What is the cooling time of a 1 $M_\odot$ white dwarf until it disappears entirely?

Vila: It is shorter than that for 0.6 $M_\odot$.

Ostriker: Cooling times are a maximum somewhere in the middle of the white dwarf mass range. At 0.7 $M_\odot$ the time is about 7×10^9 yr. Above 0.7 $M_\odot$ and up to 1.4 $M_\odot$, Debye effects dominate and the time reduces to 10^9 yr. At 0.3 $M_\odot$ convection dominates and the time is again reduced.

Weidemann to Vila: There is no observational evidence for the existence of either crystallization or liquefaction sequences. John Graham's observations show that the DA white dwarfs form a very well defined and continuous cooling sequence. For the cooler DA's the surface gravity is constan˙, $\log g = 8.3$. For the hotter DA's, the surface gravity might be somewhat lower ($\log g = 8$) as derived by differential comparison for 40 Eri B and from hydrogen line broadening.

Ruben to Hesser: (1) What is the amplitude of the 30 sec variations? (2) Has the star been observed for polarization?

Hesser: The amplitude is very low. No polarization studies have been made.

Tayler (ed.), Late Stages of Stellar Evolution, 236. All Rights Reserved.
Copyright © 1974 by the IAU.

HORIZONTAL BRANCH STARS

At this point in the advertised programme there was expected to be an invited talk by P. Demarque but he was unfortunately unable to be present at the Symposium. At short notice I. Iben gave a very brief introductory survey of the problem on the understanding that he would not have to provide a written version for the published proceedings. As some questions in the General Discussion refer to Iben's survey, an edited version* of Iben's remarks follow.

EDITOR

The properties of horizontal branch stars depend on the amount of carbon, nitrogen and oxygen that they contain but all theoretical models have difficulty in explaining those stars right at the blue end of the horizontal branch. The length of the theoretical evolutionary tracks always appears to be less than what is observed. This might possibly be explained by a variation in mass loss at the top of the giant branch. Another possibility is that allowance for overshooting of convection or semi-convection might lead to an extension of the evolutionary loops in HR diagram.

There has in the past been some difficulty in understanding the relative numbers of RR Lyrae stars of types ab and type c in globular clusters. It now seems possible that it can be explained because there is a region in the instability strip in which stars are unstable in both the fundamental mode and the first overtone. If stars entering this region from the left continue to pulsate in the overtone and those from the right continue in the fundamental, the number of variables of the two types will depend on exactly where, relative to the instability strip, stars enter the horizontal branch.

The giant branch in the globular cluster ω Cen is very wide. This could be so because the observed giant branch contains post-horizontal branch stars as well as pre-horizontal branch stars. However, another possibility is that there is a spread in chemical composition in the stars in the cluster.

---

* The Editor takes full responsibility for this version.

Tayler (ed.), Late Stages of Stellar Evolution, 237. *All Rights Reserved*
Copyright © 1974 by the IAU

FOUR-COLOR OBSERVATIONS OF
BLUE HORIZONTAL-BRANCH STARS*

(Abstract)

A. G. DAVIS PHILIP

State University of New York at Albany and Dudley Observatory, New York, U.S.A.

Effective temperatures and surface gravities can be calculated for early-type stars with $0.5 < \theta_e < 0.7$ and $2.0 < \log g < 4.4$ by means of a grid computed by a graduate student at State University of New York at Albany, (Linda Matlock) which relates the four-color indices $b-y$ and c_1 to θ_e and $\log g$. The four-color indices must be dereddened and deblanketed, then θ_e and $\log g$ can be read off the grid with rms errors of ± 0.015 in θ_e and ± 0.2 in $\log g$. The details concerning these relations will be published soon in another article.

Blue horizontal-branch stars have been measured in the following globular clusters: M4, M5, M13, M55, and NGC 3201. Each star has been observed an average of four times and the rms error of the mean c_1 and $b-y$ indices are approximately ± 0.03 to ± 0.04 mag. About ten blue horizontal-branch stars, well-spaced in $b-y$ color, are selected for measure in each globular cluster. The indices of the measured stars are plotted in Figure 1. The index c_1 is plotted against the index $b-y$ (both indices have been dereddened and deblanketed). Lines of equal θ_e, from 0.5 to 0.7 and lines of equal $\log g$ from 2 to 4 are indicated. A double line indicates an evolutionary blue horizontal-branch model ($M_c = 0.475 M_0$, $M = 0.62 M_0$, $Y = 0.3$, and $Z = 10^{-3}$) calculated by Sweigart and Gross (1973, unpublished). The indices of the blue horizontal-branch stars in the metal-poor globular cluster, M55, scatter about this double line and thus one could expect its parameters to agree with those defined by the blue horizontal branch model above. The indices of the blue horizontal-branch stars in the intermediate metal abundance ($[\text{Fe/H}] \simeq -1.5$) clusters M 13 and NGC 3201 indicate surface gravities approximately 0.2 smaller in $\log g$ while the indices of the blue horizontal-branch stars in the more metal-rich ($[\text{Fe/H}] = -0.6$) clusters M4 and M5 indicate surface gravities approximately 0.2 larger in $\log g$. The bluest stars in the last two clusters scatter about M55 relation. It is not known which parameter, or combination of parameters, is causing this different behavior of blue horizontal-branch stars in the c_1, $b-y$ diagram or in its analogue, the $\log g$, $\log T_{\text{eff}}$ diagram, but it is evident from these data that at least one parameter must vary in some clusters.

Acknowledgement

The partial support of the National Science Foundation is acknowledged.

* Visiting astronomer, Kitt Peak National Observatory and Cerro Tololo Inter-American Observatory which are operated by the Association of Universities for Research in Astronomy, under contract with the National Science foundation.

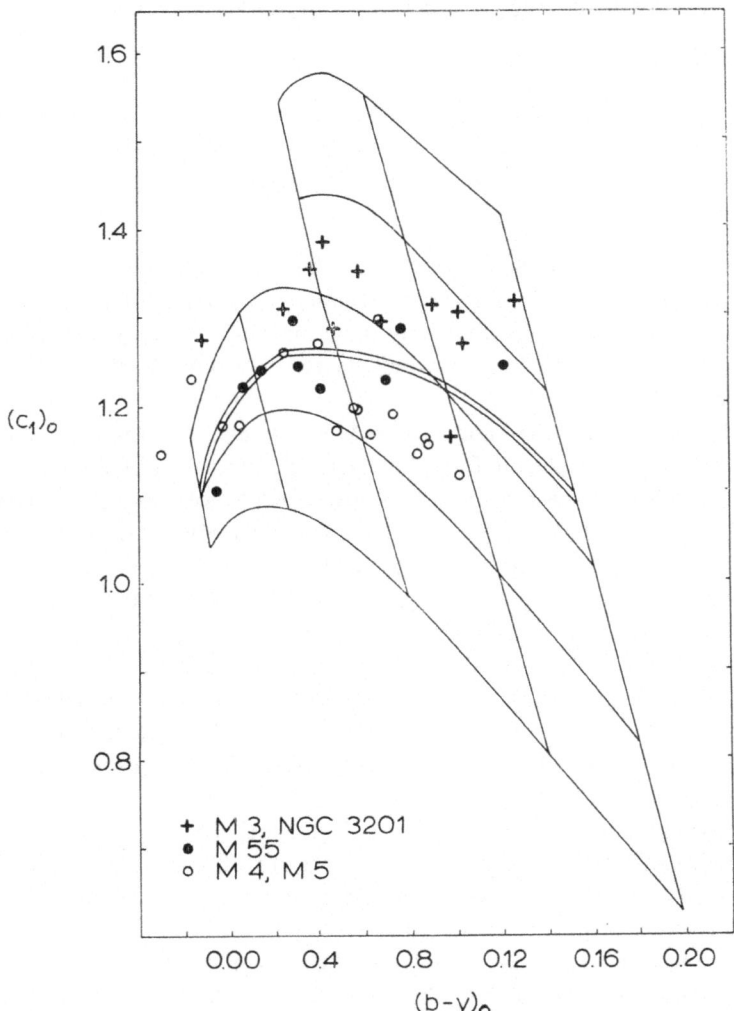

Fig. 1.   Four-color observations of blue horizontal-branch stars.

ON THE BEHAVIOUR OF GLOBULAR CLUSTER RED
GIANTS IN THE COLOR-MAGNITUDE DIAGRAM

(Abstract)

R. M. RUSSEV

Dept. of Astronomy, University of Sofia, Bulgaria

In the last years it has become clear, that a knowledge of the infrared magnitudes of red giant stars in globular clusters is of great importance for the investigation of their behaviour on the color-luminosity diagram. Unfortunately, the data available at present are insufficient. Here we shall try to draw some preliminary conclusions based on available results.

On the basis of a wide-band infrared image-tube photometry ($\lambda_{eff} = 8460$ Å, $\Delta\lambda = 2050$ Å) of selected stars in four globular clusters (Russev, 1972, 1973), we pointed out the possibility of the existence of a special sequence PRG (Peak Red Giants) in the '$M_I - (B-I)_0$' diagram (Figure 1), whose stars are in a stage near to the helium flash. The sequence PRG is in fact the top-end of the giant branch and it begins from the break-point of this branch. This point could be easily determined on all of the diagrams constructed using infrared magnitudes.

Recently, Eggen (1972) published UBVRI electrophotometric observations of red giants in 10 globular clusters, as well as the old disk-population cluster M67 and pointed out the behaviour of stars in the '$M_{bol} - (R-I)$' diagram. In Figure 2, taken from his paper, it is obvious, that the break-points of red giant branches are placed at different values of $(R-I)$ with a dispersion of about $0^{m}.2$.

The diagrams 'color $(B-I)$-magnitude I_J (8800 Å)' for all 11 clusters were constructed (Russev, 1972, 1973) according to Eggen's observations. It seems that the color-index $(B-I)$ is the best effective temperature-index for red giant stars in globular clusters. There are several facts, which lead to this conclusion:

(i) The band of V magnitude includes a spectral region, where the low-temperature stars show a large number of TiO absorption bands. Therefore, as was convincingly proved by Eggen (1969), the color-index $(B-V)$ is not a simple function of the temperature of the red stars.

(ii) The band of R_J (6800 Å) magnitude lies not far from the band I_J, so that even the presence of a small number of absorption lines and bands acts unfavourable on the color $(R-I)$ as a temperature index. On the other hand, the bands R_J and I_J lie near the continuum's maximum of the red stars, which leads to a slow change of the color $(R-I)$ with the star's temperature.

In this way, we can conclude, that color-index $(B-I)$ is a better criterion for temperature classification of red giant stars. In Figure 3 are shown two 'color-magnitude' arrays for stars in 47 Tuc by Eggen's observations (Eggen, 1972). To make the comparison easily observable we have plotted the infrared magnitude I as a function of

Tayler (ed.), Late Stages of Stellar Evolution, 240–243. All Rights Reserved.
Copyright © 1974 by the IAU.

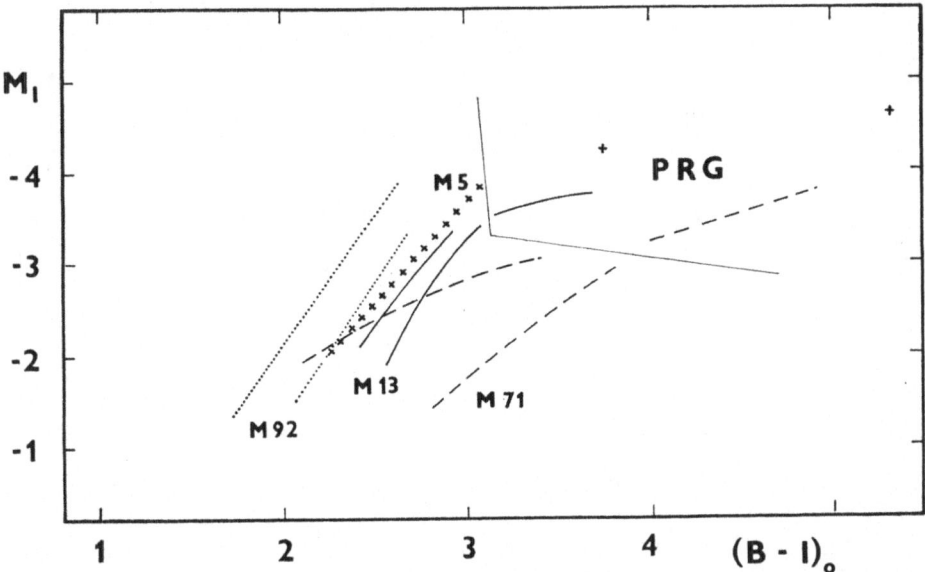

Fig. 1.   Composite '$M_I - (B - I)_0$' diagram for globular clusters M5, M13, M71 and M92 (Russev, 1972, 1973). The region PRG of reddest stars is noted.

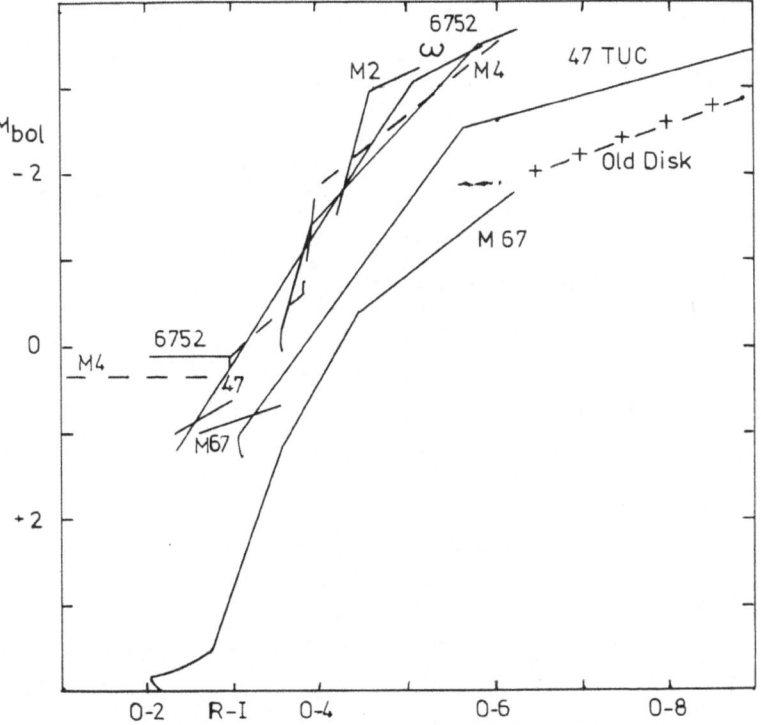

Fig. 2.   Composite '$M_{bol} - (R - I)$'-relation from Eggen's paper (Eggen, 1972).

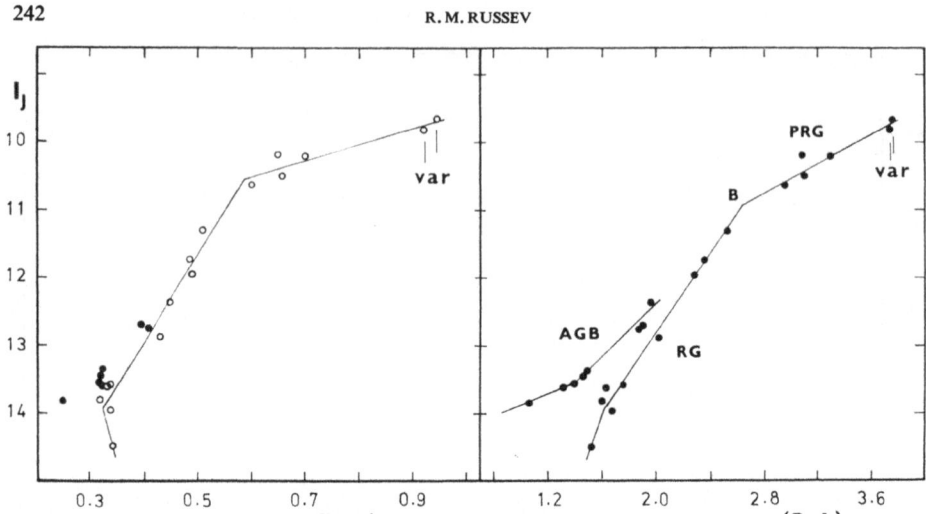

Fig. 3. The '$I-(R-I)$' and '$I-(B-I)$'-relations for red giants in 47 Tuc. According to Eggen (1972), on the left side diagram, the filled circles represent the asymptotic giant stars.

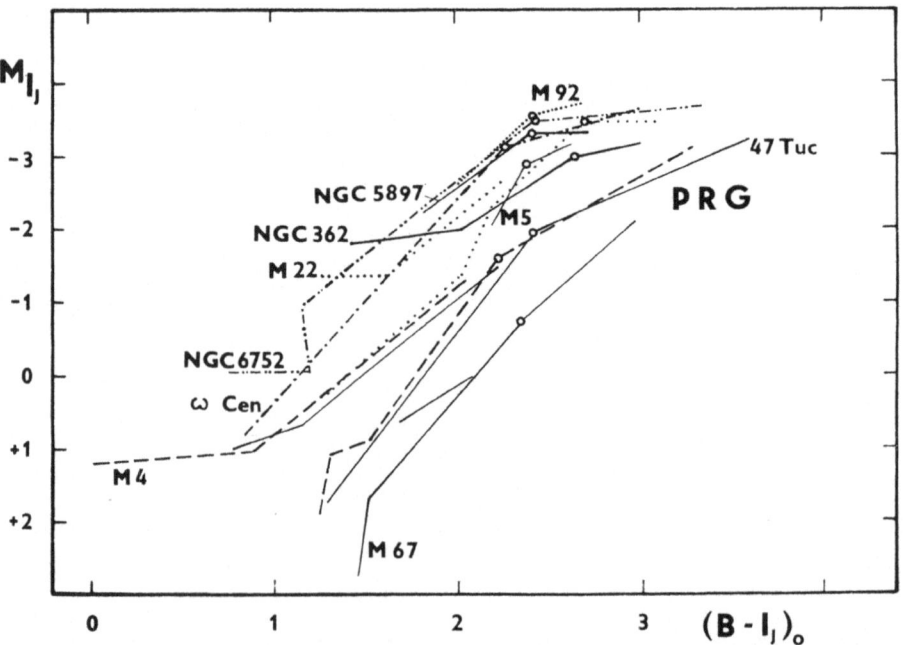

Fig. 4. Composite 'color-luminosity' diagram for ten clusters. The open circles are the break-points of red giant branches.

the color $(R-I)$ (Figure 3 – left) and I as function of the color $(B-I)$ (same figure – right). Obviously it is much easier to draw the red giant branch (RG), the asymptotic branch (AGB) and the sequence PRG on the right-hand side diagram and there the break-point (B) can be defined with better accuracy. This applies to all of Eggen's clusters.

A composite 'M_I-$(B-I)_0$' diagram (Figure 4) was constructed by means of a homogeneous system of distance-moduli and color excesses by Kukarkin and Russev (1972). Only for M67 we adopted $\text{Mod}_{I,0}^{C1}=9\overset{m}{.}29$ and $E_{B-I}=2.5$. $E_{B-V}=0\overset{m}{.}15$ (Russev, 1972, 1973; Eggen and Sandage, 1964). With open circles we denote the break-points of the red giant branches. The dispersion of these points on the $(B-I)_0$ axis is $0\overset{m}{.}46$, which corresponds approximately to half of the dispersion on Figure 2. The average position of the red giant break-points is at $(B-I)_0=2.41\pm0.15$ or at effective temperature about 3800 K. On Figure 4 the line for M2 was not drawn, as no stars situated in the PRG region have been observed.

Finally, we can characterize the red giant sequence PRG by the following:

(1) It has a smaller slope than the red giant branch. In general the brighter the absolute infrared luminosity of the breakpoint the smaller the slope.

(2) It is less populated with stars than the red giant branch, which indicates comparatively short time of evolution on this sequence.

Moreover, the blue boundary of PRG region on the '$M_I-(B-I)_0$' diagram probably disagrees with the red instability region.

References

Eggen, O. J.: 1969, *Astrophys. J.* **158**, 225.
Eggen, O. J.: 1972, *Astrophys. J.* **172**, 639.
Eggen, O. J. and Sandage, A.: 1964, *Astrophys. J.* **140**, 130.
Kukarkin, B. V. and Russev, R. M.: 1972, *Soviet Astron. AJ.* **49**, 121.
Russev, R. M.: 1972, Dissertation, GAISh, Moscow.
Russev, R. M.: 1973, *Soviet Astron. AJ.* **50**, 535.

PHOTOMETRIC BEHAVIOUR OF FIELD RR LYRAE STARS*

(Abstract)

K. STEPIEN

Warsaw University Observatory, Poland

Three colour observations of the rising branches of 25 field RR Lyrae stars are presented. This photometry supplemented with the observations by Sturch (1966) permitted the finding of mean colour indices of a number of single-period variables. Observations of additional single-period variables with well determined light curves were included in the discussion. The unreddened colour indices were found for strong-lined and weak-lined variables and then the differential blanketing effect between the two groups was removed. The results indicate that the width of the instability strip for field RR Lyrae stars is similar to that for cluster variables. Absolute magnitudes and masses of both groups were found. They are: $M_v \approx 0.8$ and 1.3, $M \approx 0.52$ $M_\odot$ and 0.47 $M_\odot$ for weak-lined and strong-lined stars respectively. A difference in mass of 0.05 $M_\odot$ was found from a difference in transition colour index.

All of the stars from the sample discussed which are cooler than $T_e = 6600\,\mathrm{K}$ exhibit bumps on the rising branches of their V curves. $U - B$ excess is associated with the bump. It occurs $0.02P$ earlier than the bump – at the phase of maximum compression and at the phase of the mean V magnitude. The bolometric corrections were found to be very small for RR Lyrae variables.

Reference

Sturch, C.: 1966, *Astrophys. J.* **143**, 774.

* This paper appeared in full in *Acta Astronomica* **22**, 175, 1972.

WHAT SNU? THE CASE OF THE MISSING SOLAR NEUTRINOS

WILLIAM A. FOWLER

California Institute of Technology, Pasadena, Calif., U.S.A.

Abstract. The current attempts to observe solar neutrinos are reviewed. It is concluded that serious consideration must be given to the possibility that the neutrino flux from the Sun is essentially zero even though we are not yet absolutely forced to this conclusion. Transient effects in the Sun are briefly discussed as a possible explanation for the lack of solar neutrinos.

This is a rather late stage in this symposium on the late stages of stellar evolution to call attention to the rather unpleasant fact that there are apparently still severe problems in our understanding of earlier stages and, in particular, of the main-sequence stage. I refer to *the case of the missing solar neutrinos,* which is still with us. It is my intention in this talk to survey the current situation and future prospects without any elaboration of the background of the case because of lack of time. For an excellent presentation of that background I refer you to a recent review by R. W. Kavanagh (1972).

TABLE I

Observations of solar neutrino flux

Shielding	Run no.	Period of exposure	Atoms ^{37}Ar in tank	^{37}Ar production per day
Bare	18	Apr 12–Nov 14, 1970	30 ± 13	0.60 ± 0.26
	19	Nov 14, 1970–Mar 6, 1971	29 ± 14	0.63 ± 0.30
	20	Mar 6–June 17, 1971	8 ± 10	0.19 ± 0.22
Water	21	June 17–Oct 2, 1971	5 ± 16	0.12 ± 0.36
	22	Oct 2–Dec 13, 1971	1 ± 12	0.03 ± 0.31
	23	Dec 13, 1971–Mar 2, 1972	-5 ± 30	-0.13 ± 0.75
	24	Mar 2–May 18, 1972	10 ± 9	0.23 ± 0.23
	27	July 7–Nov 5, 1972	57 ± 23	1.24 ± 0.50
	28	Nov 5, 1972–Jan 26, 1973	16 ± 15	0.39 ± 0.36
Mean	21, 22, 23, 24, 28		5.2 ± 8.0	0.13 ± 0.20
Cosmic-ray background			3.6 ± 1.2	0.09 ± 0.03
Net production rate			1.6 ± 8.1	0.04 ± 0.20

Rate in solar neutrino units	0.2 ± 1.0 SNU [a]
	or
(1 SNU = 10^{-36} events s^{-1} per ^{37}Cl target nucleus)	$\lesssim 1.2$ (1 σ limit) SNU
Recent typical standard solar model calculations (Bahcall *et al.*, 1973)	$5.6^{+1.8}_{-0.8}$ SNU [b]

[a] Two additional runs in 1973 change this to 0.15 ± 0.75 SNU with a 1 σ limit of 0.9 SNU (R. Davis, Jr., private communication).

[b] Errors quoted due to uncertainties in opacity only. Total standard deviation is approximately ± 2.3 SNU (R. Ulrich, private communication).

Tayler (ed.), Late Stages of Stellar Evolution, 245–248. All Rights Reserved.
Copyright © 1974 by the IAU.

Table I presents a summary of recent results obtained by Raymond Davis, Jr. and John C. Evans (1973) at the Brookhaven Neutrino Observatory in the Homestake Gold Mine, Lead, South Dakota, using the Pontecorvo-Alvarez technique of radio-active ^{37}Ar production by neutrino interaction with ^{37}Cl. Results are reported for all runs since Davis introduced the 'double window' identification of ^{37}Ar counts using pulse-rise time as well as energy discrimination. The use of the double window and of 'dry run' background tests has enabled Davis to reduce background counts markedly and, in addition, to estimate the accidental background in individual runs. In the tabulated results accidental backgrounds have been subtracted, leading in one case to a net negative number of counts.

The first three runs were made with the tank containing the target ^{37}Cl (in the form of C_2Cl_4) unshielded from the surrounding rock wall in the mine. There is clearly a background effect – presumably from fast neutrinos produced by (α, n) reactions and spontaneous fission from uranium and thorium contained in the rock. In any case the number of counts decreased when the rock cavity containing the tank was flooded with water producing an effective fast neutron shield by slowing down the neutrons. Run 27 is a marked exception to this statement and Davis and Evans have excluded it from the mean quoted at the bottom of the table "because it probably does not reflect the true neutrino background". They have also excluded the 'bare' runs 18, 19, 20. Davis and Evans have investigated the possibility that run 27 arose from a supernova event and have calculated that 0.03 $M_\odot$ equivalent of neutrino radiation with average energy 10 MeV per neutrino at 10 kpc would be required to produce the 57 atoms of ^{37}Ar recovered from the tank (after correction for decay during exposure, etc.).

Not everyone may concur in the exclusion of run 27 but I do.* In the last analysis we can only await the results of further runs and hope that any further high count runs will be accompanied by observable optical, X-ray, or other type events within the exposure period. It is an exciting prospect to say the least.

From the last entries in Table I it will be noted that the 1 σ upper limit of 1.2 SNU is well below a recent typical standard model calculation by Bahcall et al. (1973) which yielded $5.6^{+1.8}_{-0.8}$ SNU. There have been numerous modifications of the standard solar models which replace the usual virial theorem result for the temperature of the central (c) regions of the Sun

$$kT_c \propto GM_\odot/R_\odot$$

with

kT_c + rotational terms (Demarque et al., 1973)
  + magnetic field terms (Chitre et al., 1973; Bartenwerfer, 1973) $\propto GM_\odot/R_\odot$.

In this way T_c can be lowered and the very temperature sensitive flux of neutrinos from ^{7}Be and ^{8}B decays can be markedly reduced. However, in a recent very general

---

* Including run 27 yields $\sim 1 \pm 1$ SNU or a 1 σ limit of ~ 2 SNU.

analysis Ulrich (1974) has shown that it is very difficult to obtain calculated values below 1 SNU when the nuclear processes are treated correctly in detail.*

It is my personal conviction from the results given in Table I that we must now seriously face the possibility that the neutrino flux from the Sun at the earth is essentially zero even though we are not yet absolutely forced to this conclusion. In this case it seems to me that there are only two viable explanations, either the neutrino decays or the Sun decays! For a discussion of neutrino decay and the significant implications for elementary particle physics I refer to Bahcall *et al.* (1972).

Transient effects in the Sun as an explanation for the lack of solar neutrinos have intrigued me for some years (Fowler, 1969). My most recent discussion (Fowler, 1972, 1973) has been followed by a spate of papers (Dilke and Gough, 1972; Rood, 1972; Ezer and Cameron, 1972; Cameron, 1973) showing that episodic mixing can indeed reduce the solar neutrino flux at the earth to well below 1 SNU and indeed to as low as 0.1 SNU. Mixing of new fuels, ^1H and ^3He, into the solar interior results in heating followed by over expansion and cooling to the point where the nuclear reactions are essentially turned off and the Sun shines for a time on its internal store of thermal and gravitational energy. The solar luminosity does decrease markedly and recovers within the Kelvin-Helmholtz time characteristic of the solar core, namely about 5×10^6 yr. In the model calculated by Rood (1972) the Sun was about 20% more luminous in its steady state about 10^6 yr ago and it will eventually go through a minimum at about 60% of its present luminosity and then return to its steady state luminosity. According to Dilke and Gough (1972) these episodes can occur no more frequently than about once in 250×10^6 yr.

There are a number of important astrophysical and geophysical consequences of this episodic, transient model for the Sun. Terrestrial glaciation and paleoclimatological changes are obvious consequences but there is much controversy about this subject (Öpik, 1952; Emiliani, 1966; Devereux, 1967; Lowenstam and Epstein, 1959). I personally intend to devote considerable study and attention to it in the near future. For the time being one can also point to climatological changes on Mars (Sagan and Young, 1973), to a spread in the main sequence (although this will be small with only one star in 50 deviating at any one time on the episodic model discussed above) and to an increase in the main-sequence lifetime from ~ 12 to $\sim 16 \times 10^9$ yr for $M_\odot$ (Rood, 1973). Most intriguing to me is the fact that standard solar models imply an increase in the solar luminosity of about 40% over the Sun's lifetime. This could imply (Sagan and Mullen, 1972) global mean temperatures below the freezing point of sea water less than about 2×10^9 yr ago, contrary to geologic and paleontological evidence. Rather drastic assumptions about the past composition of the Earth's atmosphere must be made to resolve the problem. The model of Demarque *et al.* (1973) encounters even greater difficulties in this regard. The episodic, transient mixing model resolves this problem quite simply by requiring a greater luminosity

---

* The oblateness of the Sun produced by these terms is an order of magnitude greater than calculated by the quoted authors and similarly greater than observational limits (R. Rood and R. K. Ulrich, private communication).

for the Sun than in the standard model over all its lifetime except during relatively short periods. Thus, mean global temperatures below that of the freezing temperature of water may never have occurred except for short periods which is consistent with, rather than contrary to, geologic and paleontological evidence.

Another suggestion made by me (Fowler, 1972) and independently by Fetisov and Kopysov (1972) was that a resonance in the reaction ^{3}He(^{3}He, 2p)^{4}He might short circuit the production of the neutrino emitters ^{7}Be and ^{8}B. It would have been a shame if this unexpected loophole in our knowledge of nuclear physics had been the solution of the missing neutrino case but recent experiments have shown that such a resonance and its corresponding excited state in ^{6}Be do not exist (Parker et al., 1973; Dwarakanath, 1974; Halbert et al., 1973).

It will be clear that the case of the missing solar neutrinos is still an exciting scientific detective problem for astronomers, physicists, chemists, and geologists. So, what SNU?

References

Bahcall, J. N., Cabibbo, N., and Yahil, A.: 1972, Phys. Rev. Letters 28, 316.

Bahcall, J. N., Huebner, W. R., Magee, N. H., Jr., Merts, A. L., and Ulrich, R. K.: 1973, Astrophys. J. 184, 1.

Bartenwerfer, D.: 1973, Astron. Astrophys. 25, 455.

Cameron, A. G. W.: 1973, Rev. Geophys. Space Phys. 11, 505. This paper contains an excellent account of the history of the development of the episodic, transient mixing model of the Sun.

Chitre, S. M., Ezer, D., and Stothers, R.: 1973, Astrophys. Letters 14, 37.

Davis, R., Jr. and Evans, J. C.: 1973, Proceedings of the 13th International Cosmic Ray Conference, Denver, Colorado, August 17–31.

Demarque, P., Mengel, J. C., and Sweigart, A. V.: 1973, Astrophys. J. 183, 997.

Devereux, I.: 1967, New Zealand J. Sci. 10, 988.

Dilke, F. W. W. and Gough, D. O.: 1972, Nature 240, 262.

Dwarakanath, M. R.: 1974, Phys. Rev. C9, 805.

Emiliani, C.: 1966, Science 154, 851.

Ezer, D. and Cameron, A. G. W.: 1972, Nature Phys. Sci. 240, 180.

Fetisov, V. N. and Kopysov, Yu. S.: 1972, Phys. Letters 40B, 602.

Fowler, W. A.: 1969, Contemporary Physics: Trieste Symposium, 370 (International Atomic Energy Agency, Vienna, 1969); Proceedings of Cosmic Ray Studies in Relation to Recent Developments in Astronomy and Astrophysics, 256 (Tata Institute, Bombay, 1969).

Fowler, W. A.: 1972, Nature 238, 24; 1973, 242, 424.

Halbert, M. L., Hensley, D. C., and Bingham, H. G.: 1973, Phys. Rev. C8, 1226.

Kavanagh, R. W.: 1972, in F. Reines (ed.), Cosmology, Fusion and Other Matters, Colorado Associated University Press, Boulder, Colorado, Chapter 10.

Lowenstam, H. A. and Epstein, S.: 1959, Proceedings of the 20th Congreso Geologico Internacional, p. 65.

Öpik, E. J.: 1952, Irish Astron. J. 2, 71.

Parker, P. D., Pisano, D. J., Cobern, M. E., and Marks, G. H.: 1973, Nature Phys. Sci. 241, 106.

Rood, R. T.: 1972, Nature Phys. Sci. 240, 178.

Rood, R. T.: 1973, private communication.

Sagan, C. and Mullen, G.: 1972, Science 177, 52.

Sagan, C. and Young, A. T.: 1973, Nature 243, 459.

Ulrich, R. K.: 1974, Astrophys. J. 188, 369.

THE INTERIOR STRUCTURE OF THE SUN AS GLEANED FROM THE VARIOUS EXPLANATIONS OF THE NULL SOLAR NEUTRINO EXPERIMENT RESULTS

(Abstract)

CARL A. ROUSE

Institute of Theoretical Science, University of Oregon, Eugene, Ore., U.S.A.
and
Intelcom Rad Tech*, San Diego, Calif., U.S.A.

The experiment of R. Davis, Jr., designed to detect neutrinos from the Sun, was planned as a quantitative test of the conventional ideas of solar evolution, the resulting model of the present Sun deduced therefrom and the corresponding theories of energy generation by nuclear processes. Although a positive counting rate would have been proof of the nuclear processes, the pre-1968 predictions would not have established a unique model for the interior structure of the Sun. Now, however, the null solar neutrino results that have been obtained from the fall of 1967 to the present bring into question all three aspects of the problem. Since several explanations for the null results have been put forth during the past five years, it is of interest to review these explanations and to determine what, if anything, can be gleaned about the interior structure of the Sun. For each published explanation we take this point of view: "if true, what is unambiguously implied about the interior of the Sun?" The various explanations of the null solar neutrino experiment can be divided into five groups, viz., (1) solar neutrinos oscillate or decay; (2) solar neutrinos lose energy through photon or electron scattering before emerging from the Sun; (3) energy generation in the Sun is periodic or is in a transient phase; (4) actual cross sections for the absorption of solar neutrinos by ^{37}Cl are less than the standard cross sections; and (5) measured cross sections for some of the nuclear reactions of the pp-chain are in error when extrapolated to the energies of interest for standard solar models. Any ambiguities regarding the solar interior implicit in any explanation will be discussed with comments on the additional information needed to derive a unique model. After considering the above explanations and considering the methods for performing standard solar evolution and solar model calculations, it is argued that the best place to begin in order to resolve the present dilemma of the null solar neutrino experiment is in the astrophysics, starting with solar models that use real-gas physics throughout and models that match observed boundary conditions without the use of *ad hoc* parameters. Even if a positive solar neutrino counting rate is obtained in the future, it is believed that a realistic model of the present sun will require alternative concepts of solar evolution and solar structure. Clearly, the resolution of the null solar neutrino dilemma will be of fundamental importance for the general understanding of the internal structure and evolution of stars in the late stages of evolution.

* Permanent address: Formerly known as Gulf Radiation Technology.

Tayler (ed.), Late Stages of Stellar Evolution, 249. All Rights Reserved.
Copyright © 1974 by the IAU.

EFFECT OF MAGNETIC FIELD ON SOLAR NEUTRINO FLUX

(Abstract)

K. J. FRICKE

Universitäts Sternwarte, Göttingen, F.R.G.

At Göttingen Observatory Mrs D. Bartenwerfer has considered the effect of a strong magnetic field in the solar core on the solar neutrino flux. The solar neutrino flux is calculated in the same way as in the paper by Bahcall and Ulrich (1971). The unperturbed Sun with composition $X=0.734$, $Y=0.245$ has then a neutrino flux $\Sigma_0 \simeq 10$ SNU. This flux can be reduced when – in average – a magnetic field

$$B^2(M_r) = \tfrac{4}{3} \int\limits_{M_r}^{M} (\omega^2/r)\, \mathrm{d}M_r$$

is assumed in the solar interior. Here ω is an equivalent rotation law

$$\omega = (\omega_c - \omega_s)\exp\left(-b(M_r/M)^2\right) + \omega_s,$$

where for ω_s the observed surface angular velocity is taken and b and ω_c are adjustable parameters; the central field strength, B_c, is then a function of ω_c and b. The choice $B_c = 1.4 \times 10^9$ G, $b = 22.2$ results in the neutrino flux $\Sigma = 0.08\ \Sigma_0 = 0.8$ SNU. Bahcall and Ulrich (1971) did not obtain this strong reduction since they admitted – for stability reasons – only fields which do not peak in the center. Here no stability consideration has been made but it is suggested that the gradient of molecular weight in the Sun prevents the field from bubbling up to the surface.

References

Bahcall, J. N. and Ulrich, R. K.: 1971, *Astrophys. J.* **170**, 593.
Bartenwerfer, D.: 1973, *Astron. Astrophys.* **25**, 455.

PRAESEPE: NO EVIDENCE FOR SOLAR MIXING

(Abstract)

R. C. SMITH

Astronomy Centre, University of Sussex, England

Sagan and Young (1973) have claimed that the spread in the main sequence of Praesepe, which is greater than can be accounted for by observational error, is evidence for episodic mixing of the kind expected on the Dilke-Gough (1972) mechanism for solving the solar neutrino problem. I wish to make two points:

(1) It has been known for some time that the main sequence spread in Praesepe and other clusters can be accounted for by the observed spread in rotational velocities of the stars in the clusters, if the stars are not uniformly rotating. The spread is therefore no evidence for episodic mixing.

(2) As noted by Sagan and Young themselves, the observed distribution of stars in the main sequence band is nearly uniform. This is *not* expected from episodic mixing; if stars mix for only 1/50 of their lifetime, only two per cent of the stars should be displaced from the main sequence which means maybe one star in Praesepe. However, the observed uniform spread is nicely consistent with the observed relatively flat distribution of rotational velocities.

References

Dilke, F. W. W. and Gough, D. O.: 1972, *Nature* **240**, 262.
Sagan, C. and Young, A. T.: 1973, *Nature* **243**, 459.

Tayler (ed.), Late Stages of Stellar Evolution, 251. All Rights Reserved.
Copyright © 1974 by the IAU.

A LISTENER'S DIGEST OF IAU SYMPOSIUM No. 64 ON
GRAVITATIONAL RADIATION AND
GRAVITATIONAL COLLAPSE

WILLIAM A. FOWLER

California Institute of Technology, Pasadena, Calif., U.S.A.

IAU Symposium No. 64 was primarily concerned with observations and theory concerning gravitational waves and black holes. An excellent introduction to the theory was given by Misner (Maryland). It was, however, disappointing to learn from him that the prospects of enhanced gravitational radiation from beaming effects in relativistic orbits were no longer considered to be very great, contrary to original expectations. In the relativistic orbit of a mass about a black hole the gravitational radiation is indeed beamed tangential to the orbit path but since the orbiting mass spirals in toward the black hole the radiation is also beamed toward the hole and is largely captured by it.

Weber (Maryland) reported that he continues to observe pulses of gravitational radiation at the rate of 7 coincidences per day between Argonne and Maryland at 20% detection efficiency so that the overall rate is 10^4 events per year with each event involving 10^{8-9} erg cm^{-2}. This exceeds by many orders of magnitude the maximum radiation possible from the galactic center if reasonable restrictions on mass loss are taken into account. Weber also reported that he had found coincidences between his observations at Maryland and records on tapes sent to him by Douglas (Rochester).

Tyson (Bell Labs) reported that his observations of frequency versus pulse size at 710 Hz were consistent with expected amplifier noise (40 K) and indicated no gravitational wave pulses. Drever (Glasgow) reported 5 ± 11 events in six months of operation. Kafka (Munich-Frascati) reported no evidence for gravitational pulses in his own observations but did report that he found a small effect on tapes sent to him by Weber using the Munich algorithm for what constitutes a gravitational pulse. It was agreed that the exchange of tapes between observers was of the greatest importance in making further progress in this field.

Braginski (USSR) reported only on his future plans which involve the reduction of noise by the use of high Q, single crystals of sapphire. Progress in the production of these crystals was reported. Braginski also discussed the use of Doppler ranging between two drag-free Earth satellites to detect the long wave-length gravitational waves expected from non-relativistic collisions of globular clusters as suggested by Zel'dovich (USSR). Braginski pointed out Doppler ranging techniques can now measure velocity differences of $\sim 10^{-5}$ cm s^{-1} and that this number need only be reduced to 2.5×10^{-7} cm s^{-1} in order to make the technique useful.

Fairbank (Stanford) and Hamilton (LSU) discussed their ambitious plans for low temperature gravitational wave detectors where they hope to gain a reduction in

Tayler (ed.), Late Stages of Stellar Evolution ,252–253. All Rights Reserved.
Copyright © 1974 by the IAU.

noise relative to signal by a factor of 10^5 by operating at 0.003 K! Beautiful slides of the current status of their equipment lent support to their hope that they would be in operation within one year.

Giacconi (Harvard) discussed the continuing observations on binary X-ray sources and the growing evidence for accretion on the compact member of the binary as the source of the X-rays. In terms of implications that the compact member is a black hole he indicated that Cyg X-1 and SMC X-1 were the best candidates in this regard. He was followed by Kraft (Lick) who reported on the recent measurements of his group on Cyg X-1 which is a single line spectroscopic binary. By observations of reddening they conclude that the object is more than 2.5 kpc distant from the Earth. In this way the absolute magnitude of the primary is determined indicating that its mass is $\geqslant 30\ M_\odot$. In turn then the compact secondary must have $\geqslant 6\ M_\odot$ which is rather high for a neutron star on present theories and thus Cyg X-1 may well include a black hole, the first observed.

There was also much ado about gravitational collapse (Penrose), the stability of general relativistic systems (Chandrasekhar), white holes (Zel'dovich, who was elected the most active participant), gravitational induced electromagnetic radiation (Ruffini), observable effects of black holes (Bardeen, Press, et al.) and accretion on black holes (Rees et al.). All in all, it was a very exciting and stimulating symposium and this listener did not get his normal amount of sleep throughout.

GENERAL DISCUSSION

Ostriker to Fowler and others: Is it obvious that the optical star in Cyg X-1 really has the mass 30 $M_\odot$? If you take a 30 $M_\odot$ and let it evolve a bit and remove the envelope, does not the luminosity remain the same?

Paczyński: The mass estimate for Cyg X-1 is based on purely geometrical considerations. If we know the distance, the effective temperature and the apparent magnitude of the B-type star, then we also know the intrinsic luminosity and the radius of the star. Now both the radial velocity amplitude and the orbital period are known. Crudely speaking we may say that the star cannot be larger than the orbit is. Therefore we have an upper limit to the orbital inclination, i, and we have a lower limit to the masses. This limit is independent of the theory of stellar interiors.

Fowler to Fricke: Parker suggests that such strong magnetic fields are very buoyant and will bubble to the surface in a short time.

Kraft: An interesting study of globular cluster giants is in a thesis by L. Cathey at Santa Cruz. From U, B, V, R photometry of M92, M13 and 47 Tuc, Cathey found no Sandage-Walker effect; i.e. at a fixed $(B-V)$, he found essentially no difference in $(U-B)$ between the stars of the subgiant and asymptotic giant branches, in accordance with theoretical expectation.

Tayler: I wish to ask two general questions.

(1) Is it possible that the width of the giant branch in ω Cen and the indication of composition differences between stars in clusters could mean that in massive clusters sufficient gas could be retained from a first generation of stars so that a new generation enriched in metals could be formed?

(2) If the distance to the Hyades is incorrect so that there is indeed no discrepancy between evolutionary and pulsational masses of Cepheids, is Dr Woolf happy that no significant mass loss occurs in the red giant phase?

Woolf: I would like to say, on the first question, that some years ago I tried to estimate which systems would and which would not retain the ejecta from giant stars. It seemed that galaxies like the Draco, Sculptor and Fornax systems would lose the mass but that systems a little more massive like that in Sextans should retain the gas, and indeed we see a few regions of hot gas and young stars in such galaxies.

On the second topic, I can only say that we see matter being lost. If anyone needs stars to evolve without mass loss, they had better find a way of putting it back.

Friedjung: I first have a comment about novae. I think that there may be different physical processes for novae, so-called dwarf novae and recurrent novae. Thus, although I am not certain about the paper on T Coronae Borealis, it is possible that different processes are involved.

Friedjung to Tayler: Is it certain that the initial chemical composition of stars in a cluster is exactly the same? It seems that when grains condense, the chemical composition of the remaining interstellar gas is changed. When stars are formed, there could be composition variations, if there was a grain-gas separation process. Can someone who knows more about the problem than I do express an opinion?

Iben to Tayler: With regard to the problem of the Cepheids, although I agree that mass is lost in the giant region, I do not believe that the timescale is long enough for enough mass to be lost to affect the evolution of the star.

Some of us have been urging observers to look at the chemical composition of more than one star in a cluster to see if there are important variations. If globular cluster stars are first generation, they must have obtained their metals (however few) from some source. If these metals were produced by explosive nucleosynthesis in the first few times 10^8 yr of the history of the Galaxy, might there be a composition gradient across clusters?

With regard to gas retention in clusters, ω Cen does appear to be ten times more massive than any other cluster in the Galaxy.

CONCLUDING REMARKS

M. SCHWARZSCHILD

Princeton University Observatory, U.S.A.

The following remarks cannot possibly be a summary of the wealth of data and ideas which were presented and discussed during this symposium. May I rather indicate a broad-brush picture for late stellar evolution as it seems to me to emerge from the many diverse investigations we have heard about. Before doing so, however, I would like to touch on one subject that is not directly connected with late evolution phases but may turn out to be relevant to it.

A. MISSING SOLAR NEUTRINOS

I consider the negative result of the solar neutrino test a sufficiently serious matter that I would not want simply to ignore it. It is true that our present theory of stellar evolution has so many substantial contacts with observational data that it is hard to believe that any required corrections would entirely alter the picture. Nevertheless, we surely cannot feel safe as long as a test as fundamental as the solar neutrino test is strongly discordant with our predictions. Resolution of this discrepancy may lie in any one of three fields: the solar neutrino detection experiment itself, the nuclear physics built into our stellar structure theory, or the rest of the physics and mathematics that goes into our model stars. For each of us it is easy to persuade himself that the resolution of the discrepancy must lie in one of the two fields other than the one in which our expertise lies. However, if all of us follow this natural reaction and in consequence do nothing about the neutrino discrepancy, its resolution is not likely to come forth soon. Accordingly, it would seem to me more effective if we all accepted the working hypothesis that the actual problem lies in our own field of expertise, kept an active watch for any new ideas relevant to this critical discrepancy, and tried to work them out whenever they lie in our field of specialty.

B. EVOLUTION CLASSES OF STARS

Now to the broad-brush picture of advanced stellar evolution. To sort out the great variety of phenomena we have heard about and have discussed during the past three days it would seem to me useful to consider all stars in terms of four evolution classes. Even though these classes will divide all stars more or less according to their initial mass, it would seem to me more useful to base the definition of the classes on the nuclear processes dominating the entire life of a star, most specifically the late part of its life, rather than on fixed mass limits.

C. FEATHERWEIGHT STARS

Under this name we might understand that class of stars which during their entire life never burn nuclear fuel – except possibly such minor fuels as deuterium and lithium. Such a definition implies masses of less than about 0.07 $M_\odot$. This class of stars has

Tayler (ed.), Late Stages of Stellar Evolution, 256–258. All Rights Reserved.
Copyright © 1974 by the IAU.

not entered the discussions of our symposium, for good reason: their entire life is dominated by nothing other than contraction, cooling and steady decrease in luminosity. However, in the study of the stellar content of our Galaxy featherweight stars may not be ignorable. Indeed, recent new observations have caused a lively discussion about the relative frequency of stars of low luminosity among which featherweight stars might well be the dominant component. Accordingly, it seems to me that the further studies of this class of stars, specifically, the tracks in the Hertzsprung-Russell diagram, and most importantly, their cooling rates, may become an important contribution for the determination of the stellar content of our Galaxy.

D. LIGHTWEIGHT STARS

Let us include in this class all stars which burn hydrogen and helium but never in their life reach carbon burning. This class contains at its bottom end a sub-class of stars burning only hydrogen but never reaching helium burning. We have not considered the evolution of this subclass during the symposium nor do I want to do so now. For the bulk of the stars in lightweight class which burn both hydrogen and helium, there seems to exist ample observational evidence that mass ejection plays a decisive role during the late red-giant phases. A major portion of this mass ejection may appear in the form of planetary nebulae. From the theoretical side, excessive ionization energy, radiation pressure – particularly if ample grain formation occurs in the extended atmospheres – and runaway pulsations appear to be the main causes of this mass ejection. However, our quantitative knowledge of the mass ejection process is clearly still insufficient to determine with satisfactory accuracy the rate and extent of the ejection. One of the consequences of this uncertainty is that we do not yet know with any accuracy the upper limit of the initial mass of the stars belonging to this class. A reasonable working value appears to be $4 M_\odot$. However, we have heard emphatic warnings that this value may require substantial revision. Fortunately one result seems to be little affected by the uncertainties regarding the extent of mass ejection: the end product of a lightweight star is nearly certainly a white dwarf.

E. MIDDLEWEIGHT STARS

It would seem useful to define this evolution class as containing those stars which not only burn hydrogen and helium but reach temperatures required for carbon burning in a degenerate core. This last condition permits one to determine the upper limit for the initial mass of middleweight stars with some accuracy: $8 M_\odot$ appears to be a good working value (barring unexpectedly high mass ejection at relatively early evolution phases). Regarding the final fate of middleweight stars, the discussion during this symposium seems to me to indicate that at this moment we cannot choose between two quite different versions. In version A the carbon burning in a highly degenerate core leads to a complete explosion, with the nuclear reactions going all the way to the iron peak elements and with no dead remnant left. In version B the carbon burning is effectively subdued by URCA neutrino cooling. This postpones the end but not for long since the degenerate core will soon reach the Chandrasekhar limit, whereupon it must catastrophically collapse with the possible consequence of ejection of the envelope and formation of a neutron star. A decision between these two versions from the theoretical side depends, if I understand correctly, mainly on a more precise assessment of the efficiency of URCA cooling, an obviously difficult topic. From the observational side we have listened to discussions suggesting that a plausible decision between the two versions might be achieved either by a comparison of the birth rate of neutron stars (derived from pulsar observations) with the death rate of stars belonging to different evolution classes, or by a comparison of the relatively high rate of iron peak element production under version A with spectroscopic observations. Finally, both versions appear to lead to the conclusion that the death of middleweight stars leads to a type of supernova.

F. HEAVYWEIGHT STARS

This class of stars should comprise the top end of the stellar mass scale and might be defined as containing those heavy stars which are capable, after hydrogen and helium burning, to enter carbon burning in a non-degenerate and hence non-explosive manner. In spite of the success of a heavyweight star in surviving the carbon burning phase, the discussions during the symposium have left me with the impression that there exists no plausible alternative to an eventual gravitational collapse of at least the core if not the entire star. In spite of the remarkable progress that has been made in detailed dynamical calculations, including great grids of nuclear reactions, calculations which surely point the way to the necessary future steps, it still seems hard at this time to speculate regarding the actual outcome of the death of a heavyweight star. It appears to me highly likely, at least for the upper mass end in this class that a substantial fraction of the stellar mass will form a black hole. On the other hand, the amount and, most importantly, the chemical composition of the mass ejected at the death of a heavyweight star still seems fairly uncertain. On one point we seem tacitly to agree: the death of a heavyweight star is not likely to occur with less visibility than that of a supernova event.

G. RESULTS OF STELLAR DEATHS

The simplest, though far from proven, picture that one may draw from our discussions during the last three days seems to me as follows. At their death lightweight stars produce white dwarfs, middleweight stars produce neutron stars and most heavyweight stars produce black holes. Supernovae occur at the death of both middleweight and heavyweight stars. Most of the enrichment of the interstellar matter in heavy elements is due to the ejection of processed material at the death of middleweight and heavyweight stars, though the relative importance of these two evolution classes in this process is far from clear. The role of lightweight stars for the heavy element enrichment of the interstellar matter appears likely to be small and probably restricted to special items such as carbon and the s-elements. Obviously, a broadbrush picture such as this one, with its over-simplifications and its gross uncertainties, should be looked at as nothing but a set of possible targets to be shot down or to be solidified.

SUBJECT INDEX

OBJECT INDEX

AUTHOR INDEX